recent advances in phytochemistry

volume 35

Regulation of Phytochemicals by Molecular Techniques

RECENT ADVANCES IN PHYTOCHEMISTRY

Proceedings of the Phytochemical Society of North America
General Editor: John T. Romeo, *University of South Florida, Tampa, Florida*

Recent Volumes in the Series:

Volume 27 **Phytochemical Potential of Tropical Plants**
Proceedings of the Thirty-second Annual Meeting of the Phytochemical Society of North America, Miami Beach, Florida, June, 1992

Volume 28 **Genetic Engineering of Plant Secondary Metabolism**
Proceedings of the Thirty-third Annual Meeting of the Phytochemical Society of North America, Pacific Grove, California, June-July, 1993

Volume 29 **Phytochemistry of Medicinal Plants**
Proceedings of the Thirty-fourth Annual Meeting of the Phytochemical Society of North America, Mexico City, Mexico, August, 1994

Volume 30 **Phytochemical Diversity and Redundancy in Ecological Interactions**
Proceedings of the Thirty-fifth Annual Meeting of the Phytochemical Society of North America, Sault Ste. Marie, Ontario, Canada, August, 1995

Volume 31 **Functionality of Food Phytochemicals**
Proceedings of the Thirty-sixth Annual Meeting of the Phytochemical Society of North America, New Orleans, Louisiana, August, 1996

Volume 32 **Phytochemical Signals and Plant-Microbe Interactions**
Proceedings of a joint meeting of the Phytochemical Society of North America and the Phytochemical Society of Europe, Noordwijkerhout, The Netherlands, April, 1997

Volume 33 **Phytochemicals in Human Health Protection, Nutrition, and Plant Defense**
Proceedings of the Thirty-eighth Annual Meeting of the Phytochemical Society of North America, Pullman, Washington, July, 1998

Volume 34 **Evolution of Metabolic Pathways**
Proceedings of the Thirty-ninth Annual Meeting of the Phytochemical Society of North America, Montreal, Canada, July, 1999

Volume 35 **Regulation of Phytochemicals by Molecular Techniques**
Proceedings of the Fortieth Annual Meeting of the Phytochemical Society of North America, Beltsville, Maryland, August, 2000

Cover design: Phenotypes of mutants of anthocyanin regulatory genes in *Antirrhinum majus* (see figure 8.2) and disease resistance of transgenic alfalfa over-expressing isoflavone *O*-methyltransferase (see figure 7.2).

recent advances in phytochemistry

volume 35

Regulation of Phytochemicals by Molecular Techniques

Edited by

John T. Romeo
University of South Florida
Tampa, Florida, USA

James A. Saunders
U.S. Department of Agriculture
Beltsville, Maryland, USA

and

Benjamin F. Matthews
U.S. Department of Agriculture
Beltsville, Maryland, USA

2001

PERGAMON
An Imprint of Elsevier Science

Amsterdam - London - New York - Oxford - Paris - Shannon - Tokyo

ELSEVIER SCIENCE Ltd
The Boulevard, Langford Lane
Kidlington, Oxford OX5 1GB, UK

First edition 2001

Library of Congress Cataloging in Publication Data
A catalog record from the Library of Congress has been applied for.

British Library Cataloguing in Publication Data
A catalogue record from the British Library has been applied for.

ISBN: 0 08 043977 2

♾The paper used in this publication meets the requirements of ANSI/NISO Z39.48-1992 (Permanence of Paper).
Transferred to digital printing 2006

PREFACE

The papers assembled in this volume were originally presented at the joint meeting of the Phytochemical Society of North America and the Mid-Atlantic Plant Molecular Biology Society held at the Henry A. Wallace Beltsville Agricultural Research Center of the United States Department of Agriculture in Beltsville, Maryland, August 6-10, 2000. The symposium from which these chapters were prepared was entitled "Regulation of Phytochemicals by Molecular Techniques" and was organized by James Saunders and Ben Matthews. This joint meeting was timely because of recent landmark advances in molecular biology and genomics as well as the renewed interest in phytochemistry as a rich source of nutraceuticals, drugs, and alternatives to synthetic agricultural pesticides. Progress in genome sequencing in plants such as *Arabidopsis* and rice has been remarkable, as have expressed sequence tag (EST) projects in other plants, including maize and soybean. Recently, private and public sector participants of the Human Genome Project announced that a rough draft of the human genome has been constructed. These advances directly influence phytochemical investigations by providing both insight and tools for exploring and manipulating genomes.

The chapters included in this volume cover a wide range of applications from molecular biology to phytochemistry, and from basic studies on promoters and gene expression to pathway regulation and engineering with transformed plants. A number of noteworthy aspects emerge from this volume: applications of molecular biology to phytochemical practical problems are succeeding; newly emerging molecular tools promise to open new doors to discovery; and remarkable progress already has occurred in phytochemical pathway engineering.

The opening chapter by Walbot emphasizes how the genomics era is revolutionizing our ability to gather and analyze data. We continue to depend on detailed observations of individual genes and proteins for maximal insight, but we can now survey expression patterns and protein composition on a massive scale. The first genomics goal of sequencing genes is being followed by the second of pinpointing the exact structure and function of each gene. The annotation phase of gene discovery is a complex and intriguing foray into an understanding of the molecular blueprint of life. Matching phytochemicals to pathways and to the genes encoding the enzymes and regulatory genes of pathways, however, is becoming possible. Her work provides insights into the Maize Gene Discovery Project and the use of transposon insertions to tag genes and disrupt their function. The transposon tagging approach allows disrupted genes to be correlated with distinct phenotypes.

The chapter by Matthews provides an overview of microarrays, a second new tool for the exploration of genomes. This technique enables the expression of thousands of genes at one time so that comparisons can be made. He discusses the application of microarrays to gene expression in the roots of soybeans that are both resistant and susceptible to cyst nematodes. This tool allows researchers to define and predict the nature of genetic resistance to diseases that are important agricultural problems.

Transgenic plants can be viewed as potential biofactories for producing recombinant proteins. Plants have a high capacity for making proteins and have low upstream production costs. Boothe and Markley discuss the design and commercial production of foreign proteins in plants, including industrial proteins, vaccine/antibodies, and therapeutics with specific reference to recombinant proteins in oilseeds. A method for simplifying downstream purification by using subcellular fractionation methods is described. New approaches to producing vaccines and other biomedicals is further discussed by Yusibov *et al.* They describe the expression of foreign DNA sequences in virus-infected plants by using engineered alfalfa mosaic virus.

The production of more or less of a product or the making of a novel product requires engineering the *flux* of a pathway. Transcription factors with similarity to the *myb* gene family have been implicated in the regulation of phenylpropanoid metabolism. This gene family presents a unique opportunity to learn how pathways have evolved and to determine relationships among gene families, biochemical pathways, and evolution. Braun *et al.* have examined more than one hundred distinct members within this family. Redesigning transcriptional factors may also be a strategy to manipulate the production of a pathway's end products. Rahbar *et al.* discuss basic/leucine-zipper (bZIP) transcription factors in detail and suggest that designer bZIP proteins can be made with unique transcriptional activities allowing the manipulation of plant genes. A role in the detoxification of xenobiotics is envisioned.

The phenylpropanoid pathway is a prime target for pathway engineering, mainly because it produces lignin and other important natural products, including flavonoids, isoflavonoids, coumarins, anthocyanins, and some alkaloids. Dixon *et al.* provide specific examples of the genetic engineering of this pathway. Modification of the alfalfa phenylpropanoid pathway *O*-methyltransferases can improve the digestibility of forage crops, while over-expression of isoflavone *O*-methyltransferase can improve disease resistance of transgenic alfalfa. Future pathway modifications may lead to the production of health-promoting nutraceuticals in food crops. Martin *et al.* emphasize the importance of transcriptional control via regulatory genes as the preferred approach to manipulating the phenylpropanoid pathway.

Engineering isoprenoid metabolism is discussed by Schoenbeck and Chappell, who examine the contribution of the ancestral genomes of *Nicotiana* to more recent defense responses. They focus on the 5-epi-aristolochene synthase (EAS) gene family. EAS is a sesquiterpene cyclase catalyzing the first step in the synthesis of the antimicrobial phytoalexin capsidiol. There are 12-15 members of this gene family with different expression patterns and diverse promoter regions.

Direct applications are discussed in detail in the later chapters. Plants used as bioreactors to produce enzymes, such as cellulase, will likely play leading roles in the future use of fermentable sugars to produce alternative transportation fuels. Danna discusses the manufacture of cellulases in transgenic plants that are modified to express maximally thermostable enzymes for biomass conversion. Resveratrol appears to have beneficial affects in humans by decreasing rates of cancer and heart disease. Paiva and Hipskind have genetically engineered alfalfa to accumulate resveratrol glucoside for use in animal feeding studies to assess its nutritional effect. Scheffler *et al.* discuss strategies for enhancing crop allelopathy, which by itself is seldom effective for weed management. Through identification of enzymes and genes important to the production of such compounds as sorgoleone and DIMBOA, they hope to genetically engineer plants for enhanced production of natural herbicidal compounds. Hilliou *et al.* work with terpenoid indole alkaloids (TIA). They have identified regulators of the synthetic TIA pathway by using *Catharanthus roseus*, the source of the pharmaceutical alkaloids, vinblastine and vincristine, and of the natural plant defense deterrents, ajmalicine and vincamine. By using the yeast one-hybrid system and T-DNA activation tagging, transcriptional regulators have been isolated. Lines over-expressing ORCA3, a jasmonate-responsive transcriptional regulator, accumulate higher alkaloid levels than controls.

Finally, the question of how products enhanced by genetic engineering are directed to compartments inside the cell remains. The anthocyanain pathway has been extensively studied, and Mueller and Walbot discuss their studies of vacuolar sequestration in relation to various models. A role of glutathione S transferases in intracellular trafficking and possibly contributing to active and inactive pools of plant hormones that control growth and development seems likely.

Many of the current developments and the knowledge explosion confronting the traditional lab-bench phytochemist are overwhelming. Professor Walbot cautions us, "although genomic approaches promise to organize our knowledge about genes and pathways, future applications of proteomics and metabolic profiling pursue the components of phenotype directly. These newer disciplines... are developing methods for biochemists to perform rapid analysis of cellular proteins and constituent metabolites. Although currently based on extraction and providing a *snapshot* of composition, ... the next frontier will be continuous monitoring of metabolism with nano-machines." Clearly, this is something to contemplate!

Both the PSNA and the MAPMBS thank the local meeting organizers from the USDA and University of Maryland for their hospitality, the meeting facilities and assistance. The Editors enjoyed working with the authors of these chapters and express thanks for their generosity, flexibility, and good will. JTR thanks Darrin T King for his dedication and expertise in the final processing of the volume.

John T. Romeo
University of South Florida

James A. Saunders
United States Department of Agriculture

Benjamin F. Matthews
United States Department of Agriculture

CONTENTS

Chapter One

GENOMICS: NEW TOOLS TO ANALYZE GENETIC AND BIOCHEMICAL DIVERSITY

Virginia Walbot

Department of Biological Sciences
385 Serra Mall
Stanford University
Stanford CA 94305-5020

INTRODUCTION

This essay is based on a presentation at the Phytochemical Society of North America meeting in August 2000. That presentation provided a progress report on gene discovery in maize and illustrated challenges in gene annotation. This essay expands the scope to comment on the more general goals of plant genomics and on the ways in which this new information can be used in phytochemical research.

Genomics projects have two central goals. First, is to sequence all of the genes or the entire genome of a species. Plant biologists will soon benefit from two fully sequenced angiosperm genomes. The *Arabidopsis thaliana* genome should be finished by late summer 2000, and the rice genome should be finished a few years later. Finished means that >95% of the genome is available as high quality DNA sequence and that only defined gaps remain. Obtaining a complete genome DNA sequence is now relatively easy, although it is still quite expensive. Consequently, it is likely that complete genomes will be sequenced only for "model" organisms by 2010. Much less expensive, and hence feasible for many organisms, will be low pass (~80% coverage) genome sequencing and expressed sequence tag (EST) sequencing of 50,000 - 100,000 cDNAs. Both types of projects can define a reasonable fraction of the genes. Substantial EST sequence information is already available for a number of species. We can anticipate rapid growth in the number of plant ESTs and partial genomic sequences for many agronomic species and for species of particular interest in biological research.

Genomics Projects in Plants

Of the 5,161,236 ESTs deposited in GenBank[1] as of July 21, 2000, more than 10% are from plants. There were 580,521 plant ESTs from 13 species with at least 10,000 entries; the vast majority, 540,284, are from flowering plants. With just one pine and one algal representative, coverage is much less complete in non-flowering plants (Table 1.1). Because >60% of all GenBank entries are from human and mouse, the representation of plants is actually very impressive. Four of the "top ten" organisms are flowering plants. Legumes and grasses, reflecting their importance in agriculture, are robustly represented.

Table 1.1: Plant ESTs deposited in GenBank as of July 21, 2000.

Plant Species	Relative Rank	Number of ESTs
Arabidopsis thaliana	5	100,998
Glycine max (soybean)	6	93,817
Lycopersicon esculentum (tomato)	8	74,454
Zea mays (corn)	10	65,389
Oryza sativa (rice)	11	60,074
Medicago truncatula (barrel medic)	13	48,481
Sorghum bicolor (sorghum)	14	34,738
Lotus japonicus (lotus)	17	26,844
Chlamydomonas reinhardtii (alga)	19	23,496
Pinus taeda (loblolly pine)	22	16,761
Gossypium arboreum (cotton relative)	24	13,662
Hordeum vulgare (barley)	29	11,638
Mesembryanthemum crystallinum (ice plant)	31	10,189

As a result of the new plant genomics initiative at the U. S. National Science Foundation, about 3,000 plant ESTs are deposited weekly at GenBank. At this rate, there should be ~650,000 ESTs by the end of 2000. If EST collections in the private sector and some research organizations were deposited, this number would double. Even though cDNA sequencing is not very efficient at identifying rarely expressed genes, it is likely that most angiosperm genes will be represented at least once in the angiosperm EST collection. This prediction could be tested by determining if putative genes in *Arabidopsis* genomic sequences are represented by at least one plant EST.

Estimating the number of gene types in plants is a new kind of molecular biology parlor game. The current running estimate for genes in *Arabidopsis* is about 30,000, however, as will be discussed below, there are a surprisingly large number of duplicate genes as well as large gene families. Analysts of the *Arabidopsis* genome project will provide a more accurate estimate, but a reasonable estimate is 10,000 unique genes and 5,000 genes represented two to many times for a total of about 15,000 types of genes.

Gene Annotation and Biochemical Diversity

The second goal of genomics projects is to pinpoint the exact structure and the function(s) of each gene. This annotation phase is a much more complex task than generating the DNA sequences. In some cases, comparisons to well-studied genes are appropriate for identification of both structure (protein translation start site, location of introns, stop codon) and biological role; histones, cytochrome c, and other genes encoding well-conserved proteins with well-defined functions are appropriate for this method. Annotation becomes much more complicated when a gene can be classified into an enzyme type but the substrate(s) are unknown. For example, alcohol dehydrogenases are a coherent group and can be readily identified by gene alignments across kingdoms. Such identification does not predict, however, when and where a specific ADH enzyme will be expressed in a particular organism, whether ethanol or larger alcohols are the substrates, whether the enzyme functions primarily to produce or eliminate alcohols from cells, etc. Indeed, ADH enzymes perform quite distinct roles in species that utilize fermentation and in obligate aerobes; furthermore, the tuning of ADH expression to the physiological state of a yeast cell is impressive but cannot predict which cell or tissue types in a complex plant or animal will express this enzyme.

Our current inability to infer specific function is compounded when large gene families exist. A number of genes of particular interest to phytochemists, such as cytochrome P450 mono-oxygenases, glutathione *S*-transferases, glycosyl transferases, and membrane transporters exist in high numbers. These families are simply too complex to be annotated successfully by comparison with genes from fungi and animals. Even comparisons among flowering plants can be misleading. A combination of genetic analysis and expression information will be required to determine the contributions of individual genes and to understand the partial or complete redundancy of function.

Our current quandary, in which we have so much data and so little specific knowledge about individual genes, has produced a near schizophrenic approach to gene annotation. Although we appreciate that individual species, by definition, must elaborate characters that distinguish them from all other taxa, the underlying assumption of current annotation is that matches between and among species allow the assignment of (tentative) functions for new genes. The diversity of phytochemicals provides an excellent illustration of the combined power and fallacy of this approach. It may be correct, most of the time, to infer function through a match to another species for which biochemical function is proven, but some fraction of the time this approach will be incorrect. Not only do we not know which assignments are faulty, but we have also missed an opportunity to discover specifies-specific characteristics by assuming that similarity of nucleotides or encoded protein denotes identical function.

Goals of this Essay

In this essay I consider the structure and proliferation of plant genes through duplication, and how gene duplication may be a particularly effective strategy for the emergence of new biochemical pathways. Most of my examples will involve anthocyanin synthesis and its regulation, because this well-understood pathway illustrates many of the principles considered. I will also speculate on the nature of selection that molds new pathways into species-specific characteristics, a key concern for explaining the diversity of phytochemicals. Next, I will discuss how tools emerging from genomics projects may provide new opportunities to understand the gene content of plant biochemical pathways. These new approaches can be combined with mutation and transformation studies to elucidate the full range of *in vivo* phenotypes conferred by a specific gene. Finally, I will offer a few predictions about the future impact that genomics will have on research in plant biochemistry.

GENE FAMILIES IN THE ANGIOSPERMS

Polyploidy

In the mid-20th century the most striking observation about plant genomes was their vast range in size. All mammals have approximately the same genome size, however, the evolutionarily much younger flowering plant genomes are very diverse. Within genera and families, individual species can vary significantly; at least for the grasses, much of the "extra DNA" in large genomes is composed of retrotransposons.[2] The maize genome has been reported to vary by about 30% between populations, and it is likely that this difference reflects the amplification or loss of specific types of highly repetitive DNA.[3] This variability in genome size suggests that plant genomes were formed and molded by different selective pressures from those acting on the much more stable genomes of mammals. As duplications are widespread in extant plants, we can assume that there is some advantage to possessing both the processes that produce duplications and to the resulting duplications.

Early on it was appreciated that many, perhaps all, flowering plants arose after one or more polyploidization events that increased genome size.[4] Autotetraploidy results from the simple doubling of all of the chromosomes in gametes of a species. When crosses occur between species, allopolyploidy, in which the chromosomes of each parental species are doubled in the hybrid, results in a meiosis in which each chromosome has a homologous pairing partner. By either mechanism, polyploidy provides a built-in redundancy of gene function for all, or nearly all, aspects of the phenotype. Current evidence also indicates that polyploidization is recurrent, rather than the result of a single rare combination of genomes.[4] As a consequence, the allelic diversity in polyploids can be very high. It is likely that species with small genomes today and with traces of partial chromosome duplications indicative of ancient polyploidization events, such as *Arabidopsis*, have lost many polyploid segments of the genome. On a chromosome basis, such genomes are effectively diploid, and for a subset of genes there is only a single locus.

Most flowering plants show greater evidence of polyploidization. Based on the analysis of duplicated loci on different chromosomes, it is clear that maize underwent a polyploidization event as recently as 11 MYA.[5] Alignment of the 10 maize chromosomes with each other or in comparison to other grasses demonstrates that large segments of each maize chromosome are duplicated elsewhere in the genome.[6] Despite the conservation of gene order and gene content, pairing in maize is effectively diploid, that is, the chromosomes with highly similar segments are no longer homologs. To the extent that the duplicated genes have retained identical functions, however, maize is effectively tetraploid. This redundancy within the genome likely explains the survival of haploid plants; these plants are sterile, however, as a result of meiotic failure.[7] On the other hand, many single gene mutations in maize produce visible phenotypes, demonstrating that the duplicated genes created by polyploidization have diverged in function or one copy has become non-functional.

Finally there are the obvious polyploids, such as newly generated tetraploids widely used in horticulture to increase flower size. In addition, hexaploid bread wheat is a good example of successive polyploidization events in which there is considerable functional redundancy. The chromosome sets from the three genomes present in bread wheat show preferential homologous pairing, that is pairing between chromosomes derived from the same species. Homoeologous chromosomes, those of similar morphology and gene content but derived from different species, are sufficiently similar, however, that pairing can occur between chromosomes contributed by different parents; the wheat chromosomes are diversifying, however, and eventually homoeologous chromosomes may be as distinctive as the individual chromosome types within a diploid species.[8]

Local Duplications

Historically, among the very first examples of duplicated genes within a locus, many stemmed from analysis of the *R* gene cluster on maize chromosome 10; these genes regulate the production of anthocyanin, a readily scored phenotype.[9] Some chromosomes contain a single copy of *R*. Most commonly used genetic strains have two closely linked functional copies in a span of ~20 - 40 KB. Many of these complex loci also contain additional, non-functional copies of *R*, suggesting that several gene duplication events may have occurred or that unequal crossing over during meiosis resulted in a net increase in gene copy number on some chromosomes. Within a few centiMorgans of *R*, many strains also have an additional locus (*Lc*, leaf color) that contains yet another copy of *R*. These multiple *R* genes are readily recognized because alleles of each gene program anthocyanin pigmentation in a stereotyped suite of plant parts. The components of *R* can be separated by recombination and are separately affected by mutations, allowing assignment of coloration patterns to specific alleles of individual loci.

As might be expected in a tetraploid, there is a duplicate locus on another maize chromosome: *B* on chromosome 2. *B* and *R* encode the same function, a helix-loop-helix transcriptional activator of the anthocyanin pathway structural genes.[10, 11] By particle bombardment, *R*, *Lc* and *B* alleles perform the same function of activating anthocyanin biosynthesis.[12] The various genes and their alleles have distinguishable roles *in planta*, because the promoter divergence means that each directs transcription in unique spatial and temporal patterns in the plant.

The lessons gleaned from the *R* family of genes will be important to apply in interpreting the unexpectedly high frequency (approaching 20%) of loci in *Arabidopsis* that have a nearby duplication. Like polyploidization, local duplications provide the opportunity for divergence of function. One gene member may function primarily in roots and the other in shoots, one member in the sporophyte and one in the gametophyte, etc. Because short motifs in promoters suffice to program transcription, accumulation of even a few base changes in the regulatory region of a gene could create a new competence. When duplicated genes have diverged in function, in any way measurable, geneticists can recover mutations that affect that aspect of the phenotype that is not duplicated.

The duplicated chalcone synthase genes (*C2, Whp*) of maize provide an instructive example of two loci encoding enzymes of identical function that have almost diverged in function. In brief, the *C2* gene is highly expressed in vegetative plant parts and in seeds but is expressed at very low levels in the tapetal cells of the anthers; in contrast, *Whp* is not expressed in the vegetative or seed tissues but is highly expressed in the tapetum. Chalcone synthase is the entry step for flavonoid synthesis, and hence loss of activity is easy to score. Historically the *c2* (*colorless2*) locus was identified as a factor required for aleurone coloration by anthocyanin; *c2* plants have normal, yellow pollen indicating that flavonoids synthesized by the tapetum are deposited onto the pollen. Thus, in a *c2 Whp* line, loss of anthocyanin pigmentation is the only known phenotype. When *C2* is functional, in either a *C2 whp* or *C2 Whp* line, the plants can synthesize anthocyanin pigment and also have yellow pollen. The *c2 whp* double mutant, however, has white pollen. *Whp* is responsible for nearly all of the chalcone synthase synthesized in the tapetum; however, the low amount of *C2* function is just sufficient to confer pollen coloration. We might predict that further specialization of function will eventually result in *C2* being exclusively expressed in vegetative and seed tissues and *Whp* expressed only in tapetal cells; if this scenario occurs, then the two loci would have acquired non-redundant functions in the plant.

Even in large gene families, specialization of function occurs against a backdrop of some overlapping (redundant) gene functions. For example, in the glutathione *S*-transferase (GST) gene family of maize, the GST encoded by the *Bronze2* locus was genetically identified as a single gene required for normal sequestration of anthocyanin into vacuoles.[13] Mutation to *bz2* results in beige kernels, because the anthocyanin pigment remains in the cytoplasm. ABA treatment of developing *bz2* kernels can restore pink pigmentation, suggesting that an ABA-inducible GST can weakly substitute for *Bz2*. In particle bombardment studies, weak complementation of *bz2* aleurone was achieved with maize *Gst3*.[14]

Components of Biochemical Pathways

Phenylpropanoid metabolism is a large branch of secondary metabolism. Components of this pathway are found in primitive plants, but the proliferation of compounds is most astonishing in the angiosperms. These compounds appear to be crucial for plant interaction with microbes,[15] both symbionts and pathogens, and in plant reproduction, both as a visible signal for pollinators and as a pollen growth factor.[16] Flavonoids are also sunscreens, crucial for the maintenance of DNA integrity.[17] In interaction with abiotic factors, the precise structure of a flavonoid may not be crucial, but specificity in biological interactions will depend on the precision in recognition of specific molecules. We can imagine that stringent selection to repel pathogens and attract beneficial organisms has shaped the diversity of flavonoids present today.

We can view the flavonoid pathway as containing three components: the regulators of the pathway, the enzymes required to build the basic structures, and the often species-specific enzymes that modify the structure. As described in the previous section, gene duplication is commonly observed in maize for both the regulatory genes and core enzymes. Further, the regulatory loci often have many allelic variants, if human selection has intervened to favor diversity of color patterns.[18,19] Using the example of anthocyanin synthesis, one could imagine that this pathway was "built" one step at a time, starting with regulatory genes turning on the first step of the pathway, chalcone synthase. This lone enzyme would catalyze the first step, but not until the second step of the pathway, after chalcone isomerase "evolved", would the pathway proceed further. Although logical, this scenario has difficulties. For example, a number of chemical intermediates are unstable or oxidize and crosslink.[20] In this and other pathways, intermediates are toxic or mutagenic.[21] This is an illustration of a central problem in evolution, that is, to understand how favorable outcomes can emerge from presumably unfavorable intermediate states.

An alternative explanation for the assembly of the anthocyanin pathway is that pre-existing cellular enzymes could perform "side reactions" that yielded small amounts of particular flavonoid compounds. Such pre-existing enzymes could be in primary metabolism or more in primitive secondary metabolism already established in the organism's repertoire. It is already clear that the last cytoplasmic step of anthocyanin synthesis involves glutathione *S*-transferases that are highly diverged; furthermore, GSTs closely related to the anthocyanin genes *Bronze2* of maize and *Anthocyaninless9* of *Petunia hybrida* cannot substitute for the anthocyanin pathway function. BZ2 and AN9 likely had other roles before the advent of anthocyanin synthesis; today the promoters of these genes contain motifs that allow activation by the anthocyanin pathway regulatory genes. The outcome of this scenario, in which pre-existing enzymes might synthesize a low amount of a particular compound, is that rare individuals (mutants in a population) might produce more of a particular flavonoid or the plant might come to depend on this flavonoid for interaction with another species. It is the increase in the flavonoid, combined with an advantageous function, that was subject to positive natural selection. Furthermore, because plants lack a germ line, selection can occur within a chimeric individual. For example, if a new compound is a deterrent to herbivory, then vegetative shoots expressing this compound should thrive relative to the rest of the plant and would be expected to produce more progeny. For genetic and epigenetic changes that increase expression of a product affecting survival or reproduction, selection within an individual may have had a profound impact on the transmission of this quantitative trait.

Nature of Selection

One of the most difficult technical problems in evolutionary analysis is determining the precise nature of selective forces that result in speciation events. What variation in a particular trait or traits was of sufficient importance in a limited span of time or during a specific environmental regime to allow natural selection to occur? It is far easier to survey the diversity that we have than to uncover the mechanisms producing it. Given that plants display such diversity in chemical composition and that each species interacts with suites of other organisms, it is my view that the diversification of phytochemicals was a key to speciation in plants. Phytochemical changes could alter mutualistic, symbiotic, and pathogenic outcomes in interactions with microbes. Chemical changes determine fitness vis-à-vis herbivory and pathogenesis. Chemicals underlie the specificity in pollinator behavior, in turn controlling gene flow within the plant population. Consequently, composition contributes to both components of fitness: survivorship and reproductive success. In my view, chemical diversity in plants parallels the importance of behavior in animals in explanatory power for understanding speciation.

USING GENOMICS APPROACHES TO STUDY PATHWAYS

Phytochemists have already catalogued many species-specific chemicals, but in the era of genomics we would like to match these chemicals to pathways and to the genes encoding the enzymes and regulatory genes of a pathway. Furthermore, the allelic variation within pathways may be crucial to understanding the diversity present within species as well as the nature of the changes that occur in pathways as they are elaborated and put to various uses. The advent of vast stores of gene sequence data has sparked innovation in methods in the analysis of gene expression and of genetic variation. The hallmark of genomics approaches is massively parallel analysis, followed by bioinformatics to sort and analyze the data.

Opportunities with Parallel Profiling of Gene Expression

Once large EST collections are available, several types of new methods are available for determining the spatial and temporal patterns of gene expression simultaneously for many genes. Arrays (prints of DNA segments), expression profiles (identifying transcripts by sizing after RT-PCR and restriction digestion) or chips (oligonucleotides synthesized *in situ* in a defined format) allow rapid assessment of the relative levels of expression of thousands of genes. It is axiomatic that genes of a pathway will be co-regulated, and in many cases this regulation is at least partially transcriptional. By searching for genes that are coordinately expressed, particularly those that respond to changes in a regulatory gene, the putative members of a pathway can be identified. A recent application of expression profiling confirmed detection of pathway genes involved in maize anthocyanin and phlobaphene synthesis.[22]

Technologies based on immobilized oligonucleotides, such as microchips and oligonucleotide microarrays, have the ability to detect allelic variation; both single nucleotide polymorphism (SNPs) and insertion-deletion (indel) mutations can be detected. Applied to RNA analysis, if appropriate polymorphisms exist, distinctions can be drawn between the expression of alleles at a locus or among members of a gene family. These approaches are, thus, well-suited to the problem of dissecting complex gene families to assign expression profiles to particular alleles or loci. These methods can also be used to screen for genetic diversity within a species or between species by scoring DNA-based polymorphisms.

Mutational Analysis: Catalogs of Gene Disruptions

The classification of genes showing similar expression patterns into a putative "regulon" will be of little use unless the end products of the pathway are defined. Consequently, high throughput methods to disrupt individual genes and measure the resulting phenotypes are essential for defining the function of genes. T-DNA and transposon insertion mutations have found widespread application in higher plants.[23-27] Most of the mutations are loss-of-function. Originally designed to find mutations in single genes of interest,[26] these gene tagging methods have been adapted for more rapid screening.[24,28] The key design feature is pooling samples so that DNA samples from many individuals can be efficiently screened; when a positive event is detected, subpools and eventually individuals are identified carrying the mutation of interest. Direct sequencing of large collections of insertional mutations has also started, and it will soon be possible to screen databases rather than DNA samples to find mutations of interest.[25]

In addition to insertional mutants, it is also possible to screen for deletions, which are common in transposon-induced mutations and after some mutagen treatments.[29] In organisms with efficient recombination from introduced vectors, site-directed mutagenesis is possible by introducing precise mutations into genes.

In all of these approaches, the key steps are first, generating a large collection of materials that is sufficiently robust to contain mutations in every gene with high probability. Second, DNA samples must be prepared from pools or individuals; finite DNA samples require careful methodology for screening. Alternatively, insertional tags can be cloned and immortalized collections obtained. Third, the plants must be propagated to create a biological collection of the mutations. At all steps, careful sample tracking and documentation are required to insure the integrity of the materials.

Overexpression Analysis in Transgenic Plants

In addition to the loss-of-function phenotypes typically recovered after insertion mutagenesis, much can be learned about specific genes by ectopic activation of their expression. This gain-of-function approach is commonly used in analysis of individual genes in which a constitutive or novel promoter is used to express the gene in novel patterns from a transgene. With vectors designed to provide such promoters only, randomized insertion of the vector will activate genes throughout the genome.[30] Collections of such over-expressing lines can be screened for novel plant phenotypes. Manipulating putative regulatory genes -- components of signal transduction pathways and transcription factors -- is likely to be particularly fruitful. These genes constitute more than 15% of the *Arabidopsis* genome, and in various combinations control the gene expression and protein functionality of the remainder (V. Walbot, unpublished analysis).

FUTURE PROSPECTS FOR MATCHING GENES TO PATHWAYS

I am optimistic about the future of plant genetics and biochemistry. Over the past twenty years many plant genes have been characterized in detail, and for the past fifty years biochemists have defined enzyme properties and products. Both disciplines were working on just a fraction of the total gene and protein content of cells, however, with no idea of the final content. Full genome sequencing provides a limit to the number of genes and gene products to be studied. Furthermore, genomics has provided a suite of new tools in which patterns of expression and variation can be rapidly assessed for thousands of genes simultaneously. The application of this taxonomic approach to describing the diversity of genes and their relationships should yield many specific models of pathways and interlocking patterns of gene expression. The importance of these relationships will be tested in the future through facile loss-of-function and gain-of-function genetic analysis, after retrieval of suites of mutants from already characterized collections.

Although genomics approaches promise to organize our knowledge about genes and pathways, future applications of proteomics and metabolic profiling pursue the components of phenotype directly. These newer disciplines, taking a cue from the high throughput aspect of genomics, are developing methods for biochemists to perform rapid analysis of cellular proteins and constituent metabolites. Although currently based on extraction and providing a snapshot of composition, I am sure that the next frontier will be continuous monitoring of metabolism with nano-machines. Such nano-detectors may also be able to report the presence of particular proteins. It is already possible to measure integrated parameters such as pH and redox state by using dyes, for example, a technology cytologists have used for more than 100 years. More sensitive and more specific detectors already exist for *in vitro* applications, and with time such detection can be done in real time *in situ*.

SUMMARY

The genomics era is revolutionizing our ability to gather and analyze data. We still depend on detailed observations of individual genes and proteins for maximal insight, but we can survey expression patterns and protein composition on a massively parallel scale. Emergent properties of cells and organisms will be more apparent as we learn to integrate between the detail of the fine scale approach and the patterns of expression of many genes, proteins, and metabolites. It will remain an outstanding challenge for phytochemists to match particular genes to pathways, given the extent of duplication of key enzymes with similar or even identical functions *in vitro*. Given the importance of secondary metabolites to plant survival and reproduction, the investment in analyzing pathway composition, regulation, and evolution should be exceptionally useful in understanding the mode and tempo of evolution within the flowering plants.

ACKNOWLEDGMENTS

Research on anthocyanin sequestration in my laboratory is supported by the National Science Foundation, and work on *MuDR/Mu* transposons is funded by the National Institutes of Health. The Maize Gene Discovery, DNA Sequencing and Phenotypic Analysis project is a consortium of ten laboratories funded by the National Science Foundation.

REFERENCES

1. http://www.ncbi.nlm.nih.gov/dbEST/dbEST_summary.html
2. WENDEL, J.F., WESSLER, S.R., Retrotransposon-mediated genome evolution on a local ecological scale, *Proc. Natl. Acad. Sci. USA*, 2000, **97**, 6250-6252.
3. RAYBURN A.L., AUGER, J.A., Genome size variation in *Zea mays* ssp. *mays* adapted to different altitudes, *Theoretical Appl. Gen.*, 1990, **79**, 470-474.
4. SOLTIS, P.S., SOLTIS, D.E., The role of genetic and genomic attributes in the success of polyploids, *Proc. Natl. Acad. Sci. USA*, 2000, **97**, 7051-7057.
5. GAUT, B.S., LE THIERRY D'ENNEQUIN, M., PEEK, A.S., SAWKINS, M.C., Maize as a model for the evolution of plant nuclear genomes, *Proc. Natl. Acad. Sci. USA*, 2000, **97**, 7008-7015.
6. MOORE G., DEVOS, K.M., WANG, Z., GALE, M.D., Cereal genome evolution: grasses, line up and form a circle. Current Biol., 1995, **5**, 737-739.
7. WASSOM, J.J., Development of doubled haploid maize lines by anther culture, *Plant Physiol.*, 1997, **114**, 254-254.
8. FELDMAN, M., LIU, B., SEGAL, G., ABBO, S., LEVY, A.A., VEGA, J.M., Rapid elimination of low-copy DNA sequences in polyploid wheat: A possible mechanism for differentiation of homoeologous chromosomes, *Genetics*, 1997, **147**, 1381-1387.
9. WALKER, E.L., ROBBINS, T.P., BUREAU, T.E., KERMICLE, J., DELLAPORTA, S.L., Transposon-mediated chromosomal rearrangements and gene duplications in the formation of the maize *R-r* complex, *EMBO J.*, 1995, **14**, 2350-2363.
10. LUDWIG, S.R., HABERA, L.F., DELLAPORTA, S.L., WESSLER, S.R., *Lc*, a member of the maize *R*-gene family responsible for tissue-specific anthocyanin production, encodes a protein similar to transcriptional activators and contains the myc-homology region, *Proc. Natl. Acad. Sci. USA*, 1989, **86**, 7092-7096.
11. CHANDLER, V.L., RADICELLA, J.P., ROBBINS, T.P., CHEN, J.C., TURKS, D., Two regulatory genes of the maize anthocyanin pathway are homologous: Isolation of *B* utilizing *R* genomic sequences, *Plant Cell*, 1989, **12**, 1175-1183.
12. LUDWIG, S.R., BOWEN, B., BEACH, L., WESSLER, S.R., A regulatory gene as a novel visible marker for maize transformation, *Science*, 1990, **247**, 449-450.

13. MARRS, K.A., ALFENITO, M.R., LLOYD, A.M., WALBOT, V., A glutathione-S-transferase involved in vacuolar transfer encoded by the maize gene *Bronze-2*, *Nature*, 1995, **375**, 397-400.
14. ALFENITO, M.R., SOUER, E., BUELL, R., KOES, R., MOL, J., WALBOT, V., Functional complementation of anthocyanin sequestration in the vacuole by widely divergent glutathione S-transferases, *Plant Cell*, 1998, **10**, 1135-1149
15. FISHER, R.F., LONG, S.R., Rhizobium:plant signal exchange, *Nature*, 1992, **357**, 655-660.
16. TAYLOR, L.P., HEPLER, P.K., Pollen germination and tube growth, *Annu. Rev. Plant Physiol. Plant Mol. Biol.*, 1997, **48**, 461-491.
17. STAPLETON, A., WALBOT, V., Flavonoids protect maize DNA from UV damage, *Plant Physiol.*, 1994, **105,** 881-889.
18. CLEGG, M.T., DURBIN, M.L., Flower color variation: A model for the experimental study of evolution, *Proc. Natl. Acad. Sci. USA*, 2000, **97**, 7016-7023.
19. SELINGER, D.A., CHANDLER, V.L., Major recent and independent changes in levels and patterns of expression have occurred at the *b* gene, a regulatory locus in maize, *Proc. Natl. Acad. Sci. USA*, 1999, **96**, 15007-15012.
20. STAFFORD, H.A., Flavonoid evolution: An enzymatic approach., *Plant Physiol.*, 1991, **96**, 680-685
21. WALBOT, V., Sources and consequences of phenotypic and genotypic plasticity in flowering plants, *Trends in Plant Science*, 1996, **1**, 27-32.
22. BRUCE, W., FOLKERTS, O., GARNAAT, C., CRASTA, O., ROTH, B., BOWEN, B., Expression profiling of the maize flavonoid pathway genes controlled by estradiol-inducible transcription factors CRC and P, *Plant Cell*, 2000, **12**, 65-79.
23. http://zmdb.iastate.edu
24. WALBOT, V., Saturation mutagenesis using maize transposons. *Current Opinion in Plant Biology*, 2000, **3**, 103-107.
25. WALBOT, V., Genes, genomes, genomics: What can plant biologists expect from the 1998 NSF Plant Genome Research Program?, *Plant Physiol.*, 1999, **119,** 1151-1156.
26. WALBOT, V., Strategies for mutagenesis and gene cloning using transposon tagging and T-DNA insertional mutagenesis, *Annu. Rev. Plant Phys. Plant Mol. Biol.*, 1992, **43**, 49-82.
27. MARTIENSSEN, R.A., Functional genomics: Probing plant gene function and expression with transposons, *Proc. Natl. Acad. Sci. USA*, 1988, **95**, 2021-2026.
28. LEVY, A.A., WALBOT, V., Molecular analysis of the loss of somatic instability in the *bz2::mu1* allele of maize, *Mol. Gen. Genetics*, 1991, **229**, 147-151.
29. BENSEN, R.J., JOHAL, G.S., CRANE, V.C., TOSSBERG, J.T., SCHNABLE , P.S., MEELEY, R.B., BRIGGS, S.P., Cloning and characterization of the maize *An1* gene, *Plant Cell.*, 1995, **7**, 75-84.
30. SUNDARESAN, V., SPRINGER, P., VOLPE, T., HAWARD, S., JONES, J.D.G., DEAN, C., MA, H., MARTIENSSEN, R., Patterns of gene-action in plant development revealed by enhancer trap and gene trap transposable elements, *Genes Dev.*, 1995, **9**, 1797-1810.

Chapter Two

MICROARRAYS: APPLICATION TO SOYBEAN GENE EXPRESSION DURING SOYBEAN CYST NEMATODE INVASION

Benjamin F. Matthews

USDA-ARS
Soybean Genomics and Improvement Laboratory
Bldg 006 Rm 118
10300 Baltimore Avenue
Beltsville, MD 20705-2350

INTRODUCTION

Microarrays and Phytochemistry

Elucidation of the biochemical pathways of secondary natural products in plants has been an arduous task. The work to date has shown that many phytochemical pathways are complex and interactive. Determination of pathways and regulatory control points for the synthesis of some phytochemicals will be difficult and requires broad knowledge of these inter-related pathways, important intermediates, the enzymes, and the chemistry, the biochemistry and molecular biology of the system. A new technique, microarrays, will help researchers decipher phytochemical pathways, elucidate pathway interactions, and determine the function of genes and proteins by providing expression data for hundreds or even thousands of genes in parallel in cell samples. In the future, it may be possible to equate gene expression with the synthesis of phytochemicals by using microarrays. Gene expression data combined with phytochemical profile levels may elucidate pathways, relationships, and control elements, and provide new insight into the synthesis of phytochemicals.

Complete sequencing of genomes of model organisms and high-throughput DNA sequencing of expressed sequence tags (ESTs; partial sequences of cDNA) are providing researchers with information necessary to study whole genomes. Microarrays provide a new tool to study the whole genomes by providing profiles of gene expression in cells.

Microarray Construction

One type of microarray is constructed by printing cDNA inserts on glass slides.[1,2] Clones from a cDNA library are individually placed in wells of a microtiter dish; the cDNAs are amplified by the polymerase chain reaction (PCR), robotically arrayed on coated glass slides in a grid-like fashion, and immobilized (Fig. 1).[1,2] Fluorescent probes are prepared from RNA samples extracted from control and experimental cells. The RNAs are reverse transcribed in the presence of fluorescent compounds, one with Cy3-dUTP and the other with Cy5-dUTP. The labeled cDNA probes are mixed and hybridized to the microarray affixed to the glass slide. The slide is washed at high stringency to remove unhybridized probe, then each cDNA insert printed on the slide is scanned with a laser at two different wavelengths to measure the fluorescent signal of each of the two fluorescent probes. A comparison of the scans can reveal differential expression of genes between the samples labeled with Cy3 and Cy5. This generates large data sets of gene expression that can be used to identify patterns of gene expression related to the physiological state or condition of cells, and offers the possibility of studying numerous genes at once. This high throughput analysis of gene expression in parallel allows rapid comparison of many genes in one experiment, saving time, reducing the amount of RNA required, and providing large amounts of data in a well-organized format.

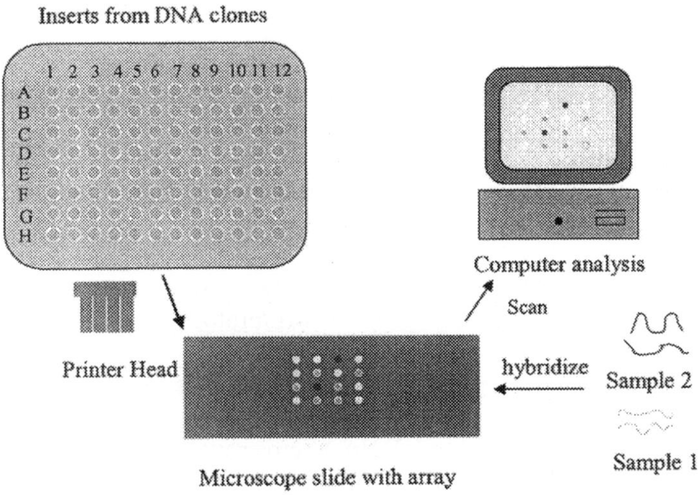

Figure 2.1: Flowchart for microarray analysis. Each archived clone has a unique address in the microtiter plate. The insert of each clone is PCR amplified by using universal primers that match vector sequence flanking the insert. The inserts are printed in a grid on surface-treated microscope slides. Two RNA samples are independently labeled with Cy3 and Cy5, respectively, then mixed and hybridized to the printed cDNAs on the slide. Slides are washed, scanned, and the data are analyzed by computer.

Comparisons using microarrays can be made to monitor changes in cells that are temporal or that are induced by chemicals, pests, pathogens, abiotic stresses, or other environmental conditions. The cells can be from the same tissue under different conditions or different tissues under the same conditions. Microarray analysis reveals the abundance of mRNAs in the cells and how the level of abundance relates to that of other genes, and can be correlated with levels of other compounds or other parameters.

Several resources are available for building microarray equipment, developing microarray techniques, and discussing problems, applications, and results of microarrays. Web sites are available with detailed instructions for building microarray equipment and that provide detailed protocols and other information (Table 2.1). We use the protocol developed by TIGR, and a description of this was recently published by TIGR.[3] It provides in-depth information on preparing microarrays and hybridizing microarrays.

Table 2.1: A sampling of Web Sites providing information and protocols for microarrays.

Building equipment
http://cmgm.stanford.edu/pbrown/ The MGguide: The Complete Guide to MicroArrays Build your own arrayer and scanner.
Protocols
http://arrayit.com/DNA-Microarray-Protocols/ DNA Microarray protocols
http://www.nhgri.nih.gov/DIR/LCG/15K/HTML/protocol.html NHGRI protocols
Resources
http://www.bsi.vt.edu/ralscher/gridit/ Grid IT: Resources for Microarray Technology
http://www.gene-chips.com/ DNA Microarray (Genome Chip)
http://barinth.tripod.com/chips.html DNA Microarrays
http://www.ensam.inra.fr/biochimie/plant_arrays/index.html www virtual library (Botany) Plant-Microarrays

Recent Applications of Microarrays

The first reported use of microarrays was to study gene expression in *Arabidopsis thaliana*.[1] A small array of 48 cDNA inserts from an Arabidopsis cDNA library was printed on a glass microscope slide. In one experiment mRNA from a transgenic Arabidopsis plant containing the transcription factor *HAT4* was compared to that from a wild-type plant. An intense fluorescence signal was obtained at the position containing the transcription factor *HAT4* DNA in the transgenic plant, but not in the wild-type plant. In a second experiment, gene expression in root was compared to that in the leaf of Arabidopsis. Twenty-seven genes differed in expression over five-fold.

Other early examples include monitoring genes in a transgenic line of Arabidopsis,[4] monitoring over 1000 human genes subjected to heat stress,[4] and identifying genes expressed uniquely in human cancer cells.[5-7] The sequencing of the entire yeast genome provided data for designing arrays to survey gene expression of nearly all of the genes in yeast.[8-13]

GENE EXPRESSION IN SOYBEAN ROOTS UPON SCN INVASION

The Soybean Cyst Nematode

My laboratory studies the resistance response of soybean to invasion by the soybean cyst nematode (SCN), *Heterodera glycines* Ichinohe, by monitoring the gene expression of genes in resistant and sensitive soybean genotypes infected and not infected with SCN. Rather than studying only one or a few genes at a time, we developed microarrays for analyzing the expression of many genes at one time.

The soybean cyst nematode is a devastating pest of the soybean worldwide. Estimated yield losses from SCN in the US in 1996 amounted to an estimated 213 million bushels worth $1.5 billion.[14] SCN infection of soybean causes various symptoms that may include chlorosis, root necrosis, loss in seed yield, and suppression of root and shoot growth. In general, nematodes cause an estimated loss of approximately $100 billion per year worldwide.[15]

Chemical control for soybean normally is not used because of economic and environmental costs. Therefore, cultural practices, such as crop rotation and the use of resistant cultivars, are used to decrease the damage of SCN. There are fourteen known races of SCN,[16] which complicates using cultivars resistant to SCN for control and makes rotation schemes more difficult. The limitations of current control practices with soybean dictate investigation of other avenues. The prospect of genetically engineering broad resistance against SCN into soybeans is attractive.

The roles of individual resistance genes in the SCN resistance response are unknown. At the cellular level, several mechanisms have been proposed. One mechanism is formation and thickening of cell wall material to seal off the syncytium.[17] Another is early degeneration of the syncytium,[18] while a third proposal is necrosis of cells that are immediately surrounding the juvenile nematode.[19]

Some of the morphological changes in soybean roots induced by SCN invasion have been documented by light, transmission, and scanning electron microscopy.[17,20-22] Certainly, the area around the site of invasion is metabolically hyperactive.[21] After invasion of a susceptible root, the nematode may feed for up to two months. During this time, a syncytium or giant feeding cell is formed from which the nematode feeds. A number of changes occurs to the cells. The nuclei and nucleoli hypertrophy, cytoplasmic organelles proliferate, the central cell vacuole is reduced or lost, the cell expands, and the cell wall becomes perforated. The perforations enlarge as the cell expands. Surrounding cells merge at the perforations to form a syncytium.

In contrast, in the SCN resistant soybean cv. Forrest, a necrotic layer surrounds the syncytium, and the syncytium becomes necrotic.[23] This is similar to the response noted in the resistant cultivar 'Peking', wherein wall depositions formed and there was an increase in lipid globules before necrosis.[17] Syncytia were degenerating four days after inoculation with SCN.[23] Thus, a major mechanism of resistance may be to wall off the area of nematode invasion with secondary wall in order to either starve the nematode or to allow toxic by-products to build up in the area. Another resistant cultivar 'Bedford' had a somewhat different response, in that the nuclei degraded, then the cytoplasm degraded.[23] Interestingly, syncytia form in both resistant and sensitive cultivars. Because the morphological responses of these cultivars vary, it is likely that their gene expression patterns will vary also.

Four genes conferring resistance to SCN have been reported.[24,25] Three are recessive, rhg_1, rhg_2, rhg_3, and one is dominant Rhg_4. Fourteen different races of SCN have been reported that are defined by their interaction with soybean differentials.[16] The soybean differentials provide cultivars that differ in their response to different races of SCN. The SCN race that is predominant in the U.S. is SCN race 3. Of the genes conferring resistance to SCN, the rhg_1 and Rhg_4 genes found in cv. 'Peking' are most important in conferring resistance to SCN race 3. The function and mechanisms of action of these genes and their products are unknown. Because the responses of the differentials to different races of SCN vary, it is likely that patterns of gene expression will also vary.

There are a few reports concerning the molecular aspects of resistance of plants to nematodes,[26-28] but most discuss the expression of only one or a few genes. However, numerous nematode-induced transcripts from tomato have been sequenced and identified.[29,30] Recently, a gene conferring resistance to beet cyst nematode *Heterodera schachtii* Schmidt was cloned by map position.[27] When this gene was transformed into a sensitive variety, resistance to nematodes was conferred.

Differential Display

Initially, we used differential display to identify a small cadre of approximately forty genes expressed by the resistant soybean cultivar, Peking, during nematode invasion.[31] Peking was inoculated with SCN race 3 and compared to uninoculated and susceptible soybean cultivar Kent. The differential display technique of Liang and Pardee[32] and Liang *et al.*[33] was used with minor modifications. Leaf and root samples from resistant and susceptible plants were examined at 6, 12, 24, 48, 96, and 144 hours after nematode inoculation and nematode invasion was monitored.[34] Genes differentially expressed with these primer sets were visualized by autoradiography, and the differential expression of some was confirmed by using northern blots or by RT-PCR. More than 45 DNA fragments of 300-800 nt were identified by differential display to represent genes expressed in SCN-inoculated resistant plants that were not expressed in controls. These DNA fragments were excised from the differential display gel, subcloned, and DNA sequenced. Several of the DNA bands turned out to encode for the same gene.

A transcription factor cDNA was identified that is induced in the resistant plants as early as 6 hours after the inoculation with the nematodes, and its expression increases through four days after infection.[31] This mRNA was neither detected in the susceptible plants nor in the uninoculated or wounded controls. These results were confirmed by Reverse Transcriptase-based PCR assays (RT-PCR) from mRNA of the resistant and susceptible plants. Another interesting clone has high similarity with a serine/threonine kinase that was induced in resistant plants 24 hours after inoculation, and, therefore, could be involved in signal transduction mechanisms. Two other cDNAs isolated from resistant plants have similarity with cDNAs coding nucleotide-binding proteins and are presently being tested for specificity by northern blot and RT-PCR experiments.

The cDNAs related to general defense responses were isolated from resistant plants 48 to 96 hours after the inoculation with SCN race 3. This group includes genes that encode peroxidases, chitinases, heat shock proteins, cinnamate 4-hydroxylase, and phenylalanine ammonia lyase. Genes related to the nodulation process also were induced in the resistant plant after the inoculation. There were four clones identified by differential display that do not have similarity with any known sequence. We confirmed the specificity of expression of most of these clones by northern blot experiments or by RT-PCR. We are now in the process of evaluating these clones by using northern blots and quantitative RT-PCR.

cDNA Libraries and Database Construction

Recently we switched from using differential display to using microarrays, so that we can determine the expression profiles of many genes in parallel in a high-density, high-throughput format.[1,2] Microarray technology has several advantages over differential display. With an arrayed cDNA library, full length or close to full length clones can be archived and interesting clones can be retrieved after microarray analysis. Differential display provides only short fragments of genes and requires excision and subcloning of fragments from electrophoresis gels. Sometimes two or three different fragments are present in the same excised gel plug, therefore, expression of the clone must be verified. Differential display requires multiple reactions for comparison to avoid false positives. To survey gene expression, numerous primers, reactions, and gels must be run. Quantification of transcript levels is not possible with differential display alone, but requires northern blot analysis or quantitative RT-PCR on each individual fragment.

Microarrays offer (1) a method to monitor the expression of many genes in parallel at higher sensitivity than previously possible; (2) convenient access to clones through archived libraries that contain large inserts or even full length cDNAs; (3) a much more ordered and comprehensive approach to gene analysis. Double spotting of clones within an array, and replication of arrays in a single experiment allow confirmation of gene expression and reduced error variance of transcript expression evaluation. Full-length and nearly full-length clones are present in the archived library and need not be reconstructed through PCR and RT-PCR procedures or obtained by library screening (*i.e.*, colony hybridization) with probes from short DNA fragments from differential display. Genome exploration is more orderly, because each clone has a unique address in a microtiter dish and is archived. A relational database can be constructed for analyzing the large quantities of data and tracking specific genes. New cDNA arrays of known genes can be added as second and third generation tools for further genome exploration.

We constructed a directionally-cloned cDNA library from mRNA isolated from cv. Peking two days post SCN-inoculation.[35] A set of 384 clones was independently isolated. [35] This provided us with a "mini-array" to develop procedures for handling and processing arrays in addition to obtaining and characterizing numerous genes related to the nematode defense response. The clones were one-pass sequenced at the putative 5' end, and some of the clones were sequenced at the 3' end to further characterize the clone and confirm that two different cDNAs were not concatenated within the clone. The amino acid and DNA sequences were compared with those in GenBank for identification, and were compared with each other to determine redundancy within the library. Data representing the redundancy found in 384 clones is provided in Figure 2.2. Four genes were most abundant. Two clones encoded genes with moderate identity to plant β-galactosidase (77%; 34 of 44 amino acids), soybean hydroxyproline rich (HPR) glycoprotein (76%; 54 of 71 amino acids), while the other two either had very low homology or were not represented in GenBank. Approximately 14% of the clones possessed putative methionine start sites.

Although most of the clones analyzed were not highly similar to known genes, some had over 90% identity at the amino acid level with other genes. The genes could represent those induced by biotic or abiotic stress. We added clones to the array from a cDNA library from mRNA isolated from young cv. Century cotyledons grown in the dark to provide genes that probably would be relatively unaffected by invasion of SCN.

A computer database was designed and implemented to organize the clone addresses, DNA sequences, and other data that would be useful for microarray analysis later. A database with our clones is provided for public use and can be visited at our homepage on the web at http://bldg6.arsusda.gov/benlab/. Our web site

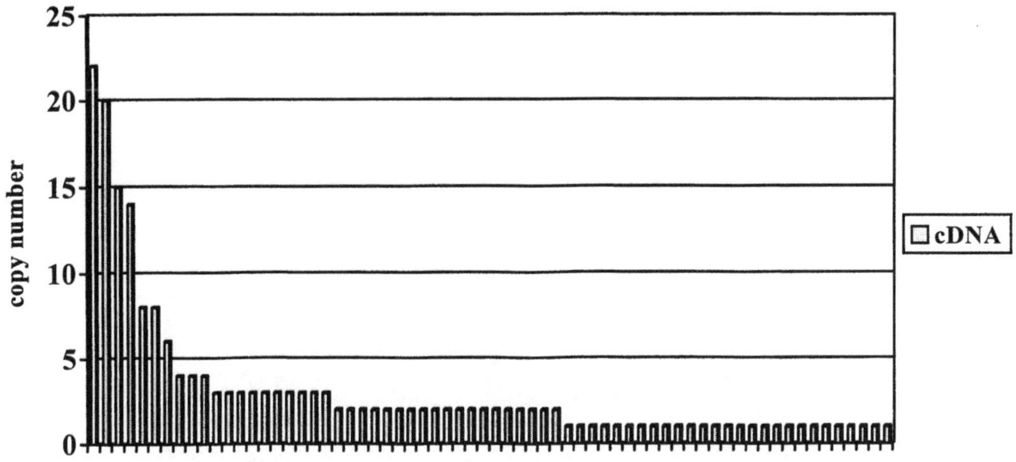

Figure 2.2: Distribution of cDNAs in the cDNA library made from mRNA from Peking soybean roots and leaves two days post inoculation with the soybean cyst nematode race 3. Four clones were found more than ten times in a 384 well plate. The chart shows only a small portion of the clones that were represented only one time in the sampling.

provides addresses, sizes, and BLAST searches for clone identity for over 1000 cDNA clones. A portion of our homepage is password protected and is provided for our collaborators for their personal microarray data. Our collaborators may have provided clones, RNA samples, or other materials and expertise. Behind the password, collaborators find protocols used in our laboratory, can see their microarray scans, review and analyze their data, compare microarray results, check clone identity using BLAST searches, and perform other data mining tasks. The data sets are moved from behind password protection into the public domain as they are published.

Application of Microarrays to Soybean-SCN Interactions

These preliminary findings represent our efforts to profile the expression of genes in roots from SCN-resistant and SCN-susceptible soybean cultivars in response to nematode invasion by using microarrays. RNA was isolated from the SCN-resistant cv. Peking and SCN-sensitive cv. Kent, independently labeled during reverse transcription with the fluorescent tags Cy3 and Cy5, and hybridized to a microarray containing DNA inserts from 468 different cDNA clones (Fig. 2.3). The ratios of fluorescence of Cy3 and Cy5 were compared to determine if each gene was induced or repressed in roots upon nematode invasion in resistant (cv. Peking) and in susceptible (cv. Kent) soybean (Fig. 2.4).

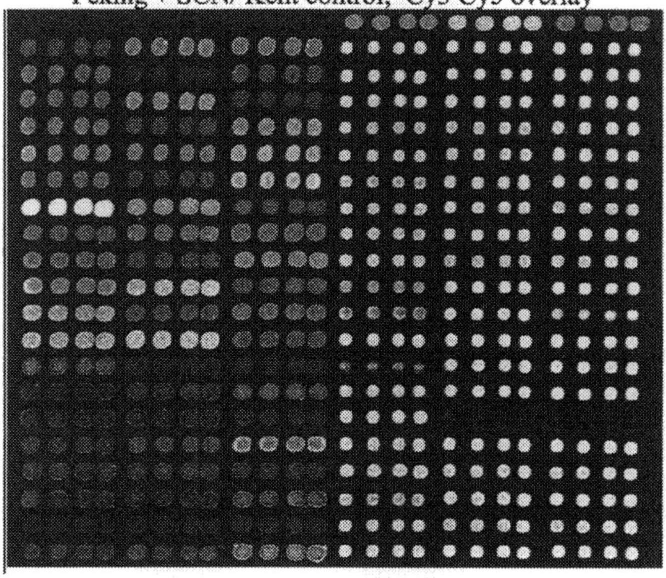

Figure 2.3: A small portion of a two thousand feature microarray after hybridization. The cDNAs are printed in quadruplicate within a column.

The studies indicate that some genes encoding enzymes of the phenylpropanoid pathway may be expressed moderately higher in roots of both resistant and susceptible soybean cultivars two days after SCN invasion as compared to roots not exposed to SCN. This is in agreement with Edens *et al.*,[36] who examined the expression of some of the genes involved in phenylpropanoid synthesis in soybean upon SCN invasion. A portion of the pathway is provided in Fig. 2.5. Transcription of the genes encoding phenylalanine ammonia lyase and 4-coumaryl CoA- ligase increased, as did these enzyme activities in resistant, but not in susceptible cultivars of soybean after infection by SCN. However, transcription of genes encoding enzymes found later in the pathway was enhanced in both resistant and susceptible cultivars after nematode invasion. Interpretation of these data is complicated due to multiple genes encoding isoforms of some of the enzymes.

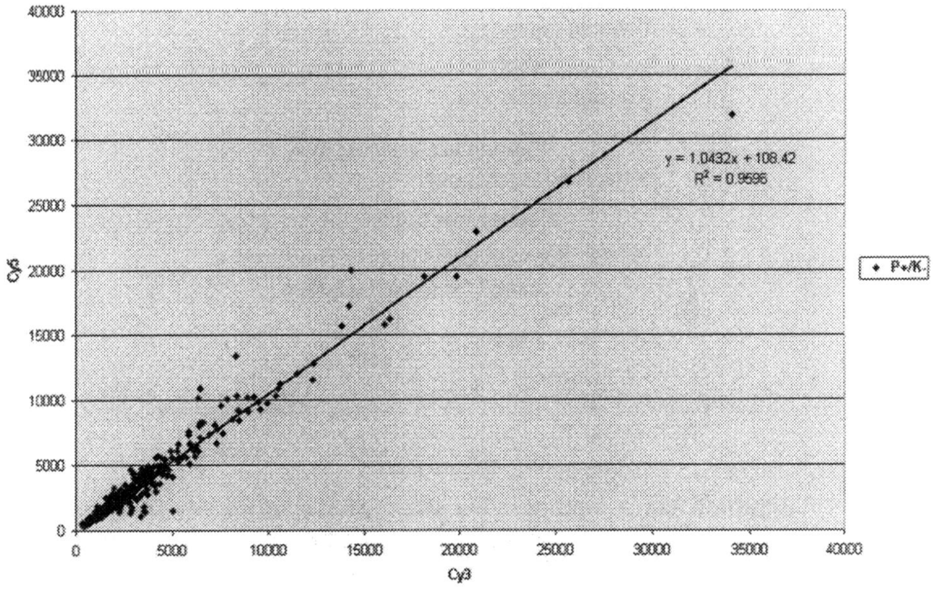

Figure 2.4: A plot of the fluorescence from Cy5 and Cy 3 of each cDNA on the microarray. Data points falling substantially above the line are increased and those substantially below the line are decreased in expression in the resistant cv. Peking in the presence of SCN after two days as compared to the susceptible cv. Kent in the absence of SCN.

The data also suggested that several genes were elevated that have no strong similarity at the amino acid level with those encoded by gene sequences deposited in GenBank (Fig. 2.6). These genes may be important, but have no function assigned to them to explain their role in the resistance response. These preliminary data must be validated with further replications and use of other techniques, such as northern blots and quantitative RT-PCR. Also, several different computer software analysis programs will be used to analyze and compare the data.

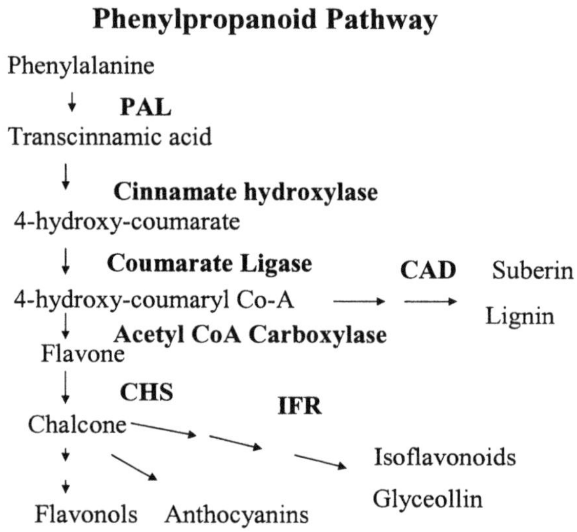

Figure 2.5A: A portion of the phenylpropanoid pathway with some of the enzymes with abbreviations printed in bold; phenylalanine ammonia lyase (PAL), cinnamoyl alcohol dehydrogenase (CAD), chalcone synthase (CHS), isoflavone reductase (IFR).

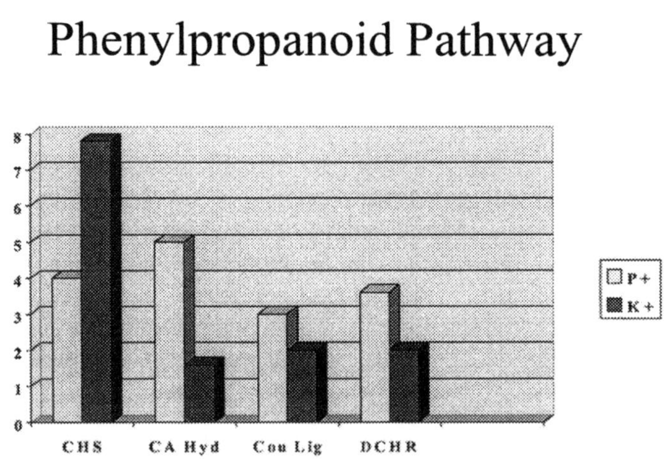

Figure 2.5B: Expression of genes encoding chalcone synthase (CHS), cinnimate hydroxylase (CA Hyd), coumerate CoA ligase (Cou Lig) and desoxychalcone reductase (DCHR) in roots of the resistant cv. Peking (P^+) and the sensitive cv. Kent (K^+) two days post infection with SCN.

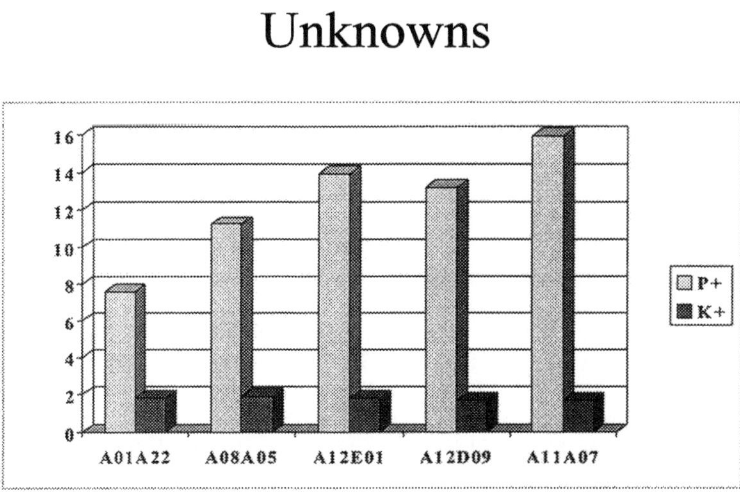

Figure 2.6: Expression of genes with no strong similarity to genes archived in GenBank. Ratio of expression levels are provided for these genes in roots of the resistant cv. Peking (P^+) and the sensitive cv. Kent (K^+) two days post infection with SCN. The genes are identified by their microtiter plate number and well position.

SUMMARY

The microarray is a relatively new and powerful technique that has not reached its full potential. Gene DNA sequences are needed to profile gene expression with microarrays, but EST and genome sequencing projects are in early stages for most organisms. These projects will provide needed information to construct better microarrays that represent whole or targeted portions of genomes for analysis. Equipment prices, software, and other tools for microarray analysis will improve rapidly and become easier to use and more powerful. Chemical profiling of secondary compounds in plant tissues and organs combined with microarrays monitoring gene expression will become powerful allies with other biochemical and molecular techniques for researchers to learn more about and better understand phytochemistry of plants.

ACKNOWLEDGEMENTS

A special thanks to Nadim Alkharouf, Hunter Beard, Rana Khan, Margaret MacDonald, Kris Pilitt, Michael Yang who conducted the research in my laboratory.

Susan Meyer and Sandra Sardanelli provided valuable assistance in growing and maintaining soybean cyst nematodes. USDA-National Research Initiative Grant No. 99-35302-8189 and the United Soybean Board Grant No. 0217 supported this research.

REFERENCES

1. SCHENA, M., SHALON, D., DAVIS, R.W., BROWN, P.O., Quantitative monitoring of gene expression patterns with a complementary DNA microarray, *Science*, 1995, **270**, 467-470.
2. SHENA, M., Microarray Biochip Technology, Eaton Pub. Co., 2000, 205 p.
3. HEGDE, P., QI, R., ABERNATHY, K., GAY, C., DHARAP, S., GASPARD, R., HUGHES, J.E., SNESRUD, E., LEE, N., QUACKENBUSH, J., A concise guide to cDNA microarray analysis, *BioTech.*, 2000, **29**, 548-562.
4. SHENA, M., SHALON, D., HELLER, R., CHAI, A., BROWN, P.O., DAVIS, R.W., Parallel human genome analysis: Microarray-based expression monitoring of 1000 genes, *Proc. Natl. Acad. Sci. USA.*, 1996, **93**, 10614-10619.
5. DERISI, J., PENLAND, L., BROWN, P.O., BITTNER, M.L., MELTZER, P.S., RAY, M., CHEN, Y., SU, Y.A., TRENT, J.M., Use of a cDNA microarray to analyse gene expression patterns in human cancer, *Nature Genet.*, **1996**, **14**, 457-460.
6. WELFORD, S.M., GREGG, J., CHEN, E., GARRISON, D., SORENSEN, P.H., DENNY, C.T., NELSON, S.F., Detection of differentially expressed genes in primary tumor tissues using representational differences analysis coupled to microarray hybridization, *Nucl. Acids Res.*, 1998, **26**, 3059-3065.
7. KHAN, J., SAAL, L.H., BITTNER, M.L., CHEN, Y., TRENT, J.M., MELTZER, P.S., Expression profiling in cancer using cDNA microarrays, *Electrophor.*, 1999, **20**, 223-229.
8. SHALON, D., SMITH, S.J., BROWN, P.O., A DNA microarray system for analyzing complex DNA samples using two-color fluorescent probe hybridization. Genome Research, 1996, **6**, 639-645.
9. LASHKARI, D.A., DERISI, J.L., MCCUSKER, J.H., NAMATH, A.F., GENTILE, C., HWANT, S.Y., BROWN, P.O., DAVIS, R.W., Yeast microarrays for genome wide parallel genetic and gene expression analysis, *Proc. Natl. Acad. Sci. USA*, 1997, **94**, 13057-13062.
10. WODICKA, L., DONG, H., MITTMANN, M., HO, M-H., LOCKHART, D.J., Genome-wide expression monitoring in *Saccharomyces cerevisiae*, *Nature Biotech.*, 1997, **15**, 1359-1367.
11. CHU, S., DERISI, J., EISEN, M., MULHOLLAND, J., BOTSTEIN, D., BROWN, P.O., HERSKOWITZ, I., The transcriptional program of sporulation in budding yeast, *Science*, 1998, **282**, 699-705.

12. EISEN, M.B., SPELLMAN, P.T., BROWN, P.O. BOTSTEIN, D., Cluster analysis and display of genome-wide expression patterns, *Proc. Natl. Acad. Sci. USA*, 1998, **95**, 14863-14868.
13. PELLMAN, P.T., SERLOCK, G., ZHANG, M.Q., IYER, V.R., ANDERS. K., EISEN, M.B., BROWN, P.O., BOTSTEIN, D., FUTCHER, B., Comprehensive identification of cell cycle-regulated genes of the yeast *Saccharomyces cerevisiae* by microarray hybridization., *Molec. Biol. Cell*, 1998, **9**, 3273-3297.
14. WRATHER, J.A., Univ. of Missouri-Delta Center, http://aes.missouri.edu/delta/research/ soyloss.htm
15. SASSER, J.N., FRECKMAN, D.W., A world perspective on nematology: The role of society, *in* Vistas on Nematology (Veech, J.A. and D.W. Dickerson, eds.), Society of Nematologists, 1987, pp. 7-14.
16. RIGGS, R.D., SCHMITT, D.P., Complete characterization of the race scheme for *Heterodera glycines*, *J. Nematol.*, 1988, **20**, 392-395.
17. RIGGS, R.D, KIM, K.S., GIPSON, I., Ultrastructural changes in Peking soybeans infected with *Heterodera glycines*, *Phytopath.*, 1973, **63**, 74-84.
18. ENDO, B.Y., Histopathological responses of resistant and susceptible soybean varieties, and backcross progeny to entry and development of *Heterodera glycines*, *Phytopath.*, 1965, **55**, 375-381.
19. ROSS, J.P., Host-parasite relationship of the soybean cyst nematode in resistant soybean roots, *Phytopath.*, 1958, **48**, 578-579.
20. JONES, M.G.K., DROPKIN, V.H., Scanning electron microscopy of syncytial transfer cells induced in roots by cyst-nematodes, *Physiol. Plant Pathol.*, 1975, **7**, 259-263.
21. ENDO, B.Y., J.A. VEECH, J.A., Morphology and histochemistry of soybean roots infected with *Heterodera glycines*, *Phytopath.*, 1970, **60**, 1493-1498.
22. GIPSON, I., KIM, K.S., RIGGS, R.D., An ultrastructural study of syncytium development in soybean roots infected with *Heterodera glycines*, *Phytophat.*, 1971, **61**, 347-353.
23. KIM, Y.H., RIGGS, R.D., KIM, K.S., Structural changes associated with resistance of soybean to *Heterodera glycines*, *J. Nematol.*, 1987, **19**, 177-187.
24. CALDWELL, B.E., BRIM, C.A., ROSS, J.P., Inheritance of resistance of soybean to the cyst nematode, *Heterodera Glycines*, *Agron. J.*, 1960, **52**, 635-636.
25. MATSON, A.L., WILLIAMS, L.F., Evidence of a fourth gene for resistance to the soybean cyst nematode, *Crop Sci.*, 1965, **5**, 477.
26. GURR, S.J., MCPHERSON, M.J., SCOLLAN, C., ATKINSON, H.J., BOWLES, D.J., Gene expression in nematode-infected plant roots, *Molec. Gen. Genet.*, 1991, **226**, 361-366.
27. CAI, D., KLEINE, M., KIFLE, S., HARLOFF, H.J., SANDAL, N.N., MARCKER, K.A., KLEIN-LANKHORST, R.M., SALENTIJN, E.M.J., LANGE, W., STIEKEMA, W.J., WYSS, U., GRUNDLER, F.M.W., JUNG. C., Positional cloning of a gene for nematode resistance in subar beet, *Science*, 1997, **275**, 832-834.

28. GRUNDLER, F.M.W., SOBCZAK, M., LANGE, S., Defence responses of *Arabidopsis thaliana* during invasion and feeding induction by the plant-parasitic nematode *Heterodera glycines*, *Physiol Molec. Plant Path.*, 1997, **50**, 419-429.
29. BIRD, D.M., WILSON, M.A., DNA sequence and expression analysis of root-knot nematode-elicited giant cell transcripts, *Molec. Plant Microbe Interact.*, 1994, **7**, 419-424.
30. WILSON, M.A., BIRD, D.M., VAN DER KNAAP, E., A comprehensive subtractive cDNA cloning approach to identify nematode-induced transcripts in tomato, *Molec. Plant Path.*, 1994, **84**, 299-303.
31. HEINZ, R.A., MACDONALD, M.H., MATTHEWS, B.F., Differential gene expression in response to soybean cyst nematode, *Heterodera glycines*, infection, *FOCUS*, 1998, **19**, 9-10.
32. LIANG, P., PARDEE, A.B., Differential display of eukaryotic messenger RNA by means of the polymerase chain reaction, *Science*, 1992, **257**, 967-971.
33. LIANG, P., ZHU, W., ZHANG X., GUO, Z., CONNELL, R., AVERBOUKH, L. WANG, F., PARDEE, A.B., Differential display using one-base anchored oligo-dT primers, *Nuc. Acids Res.*, 1994, **22**, 5763-5764.
34. BYRD D.W., Jr., KIRKPATRICK, T., BARKER, K.R., An improved technique for clearing and staining plant tissue for detection of nematodes, *J. Nematol.*, 1983, **15**, 142-143.
35. MATTHEWS, B.F., Using cDNA arrays to examine the response of soybean *Glycine max* to the soybean cyst nematode, *FOCUS*, 1998, **20**, 14-15.
36. EDENS, R.M., ANAND, S.C., BOLLA, R.I., Enzymes of the phenylpropanoid pahtway in soybean infected with Meloidogyne incognita or *Heterodera glycines*, *J. Nematol.*, 1995, **27**, 292-303.

Chapter Three

THE DESIGN AND USE OF TRANSGENIC PLANT EXPRESSION SYSTEMS FOR THE PRODUCTION OF FOREIGN PROTEINS

Joseph G. Boothe and Nancy A. Markley

SemBioSys Genetics Inc.
500, 3605 - 29th Street N.E.
Calgary, AB, T1Y 5W4

INTRODUCTION

With the successful demonstration that foreign proteins can be expressed and stably accumulated in plant cells has come the concept of utilizing transgenic plant systems as bioreactors for commercial protein production. The potential advantages of plant-based systems have been described in numerous reviews[1-5] and include high capacity, low cost of production, post-translational modification capability, and freedom from mammalian pathogens. The feasibility of manufacturing large volumes of recombinant eukaryotic proteins economically has made it possible to contemplate their use in ways previously thought impractical due to high cost or limited availability. As a result, transgenic plants are currently being considered for a wide variety of applications ranging from production of enhanced animal feeds and human nutraceuticals to food and industrial enzymes, vaccines, and therapeutic proteins.

Over the past decade considerable progress has been made in developing the technology for recombinant protein production in plants. As with more established systems in bacteria, yeast, cell culture, and recently, transgenic animals, this development has involved identifying the regulatory elements, targeting sequences, tissues, and host species required to achieve high levels of foreign protein production. Additionally, it has required characterization of the differences between the protein processing and modification reactions in plants and other systems that can potentially have an impact on product stability and efficacy. Commercial production systems are currently under development utilizing a variety of host species and tissues with promoters that permit several options for regulating transgene expression. While differences between plants and animals have been found in some aspects of post-translational modification, results indicate that plants are capable of producing a wide array of active mammalian proteins.[6-10]

The first steps in the commercialization of transgenic plants as recombinant protein production vehicles have been taken. To date, there are two recombinant protein products on the market derived from transgenic plants.[11,12] Several others are reported to be in advanced stages of development, with some of these in clinical trials.[13-16] Studies describing processes for the large-scale recovery of recombinant proteins from plants, together with an economic analysis of production costs, have recently been published.[17,18] From these studies, data regarding the sensitivity of production costs to factors, such as expression levels and the complexity and efficiency of downstream recovery operations, have been obtained.

Here, we consider some of the factors critical in the design of transgenic plant expression systems and describe the features of commercial production systems currently being developed. We also examine the range of applications of the technology and the commercial opportunities that exist. Finally, we review the preliminary data from the large-scale recovery of recombinant proteins from plants and describe strategies for simplifying downstream purification.

EXPRESSION OF FOREIGN PROTEINS IN PLANTS

The primary consideration in selecting a promoter for the commercial production of recombinant proteins is maximizing the transcriptional activation of the transgene. Although properties of the recombinant protein will influence greatly the amount of product obtained, a strong promoter will ensure that the level of transcript is not a limiting factor. A variety of different promoters capable of driving expression in plant cells is available to meet this objective. Depending on the choice of promoter, expression can either be constitutive, developmentally regulated, or inducible. All of these options have been exploited in the design of plant-based production systems.

Constitutive Expression

The most thoroughly characterized of all plant promoters used for heterologous expression is derived from the 5' upstream regulatory region directing transcription of the 35S RNA of cauliflower mosaic virus (CaMV 35S). This promoter is capable of driving high levels of expression in a variety of tissues of dicotyledons[19] and is also active in some monocotyledons.[20] Detailed dissection of the sequence has resulted in the identification of the regions most responsible for strong expression[21] as well as multiple elements that individually confer specificity for different tissues and stages of development.[22] Tandem duplication of a 250 bp fragment of the 35S CaMV promoter has also been shown to increase expression levels at least 10-fold over the native element and to act as a powerful enhancer in chimeric promoter configurations.[23] These properties of the CaMV 35S promoter have led to its extensive use for expression of heterologous genes in plants. Although levels obtained for different proteins and systems vary greatly, expression equal to or greater than 1% of total protein has been achieved.[24] In one exceptional example, a level of 26% of total cellular protein has been reported for the expression of a microbial cellulase in *Arabidopsis*.[25]

Another constitutive promoter capable of driving high levels of expression has been obtained from the ubiquitin-1 gene of maize. For maximum expression in heterologous constructs, a fragment comprising the 5' regulatory region together with the untranslated first exon and first intron was used.[26] This promoter is particularly active in cereals where it gives rise to higher expression levels than those obtained using the CaMV 35S promoter.[27] Conversely, the CaMV 35S promoter appears to be much stronger in dicotyledons.[28] With the maize ubi-1 promoter levels as high as

5.7% of extractable protein have been reported in seeds of transgenic maize plants.[11,29] However, interestingly, a significant difference in the level of expression was observed in the endosperm and embryos of these seeds.[30]

Developmental Regulation of Expression

While constitutive promoters enable high levels of transcription to be achieved, there may be a need for spatial and temporal regulation of expression to avoid problems where the protein produced interferes with plant growth or development. Although for the production of many animal proteins the evolutionary distance between plants and animals makes this less likely, some products will nevertheless require sequestration to avoid detrimental effects on the plant. An example is avidin, a protein found naturally in chicken eggs, which binds biotin with high affinity. When expressed in maize seed using the ubiquitin promoter, production of the protein resulted in male sterility.[11] Although in this case the effect did not prevent the use of the transgenic plants for commercial production, these results indicate the type of problems that may be encountered with unrestricted transgene expression. Apart from physiological effects on the host plant, it may also be desirable from a regulatory perspective to restrict the expression of some proteins to a specific tissue or period of development.

Restricting transgene expression may be achieved through the use of developmentally regulated promoters. Among the strongest members of this class are those responsible for regulating the expression of seed storage proteins. While a number of these promoters have been described, one of the best characterized is responsible for expression of the gene encoding the storage protein phaseolin, in beans of *Phaseolus vulgaris*. As with the 35S promoter from CaMV, the phaseolin promoter has been analyzed in great detail, and the motifs that confer seed specific expression have been identified.[31] By using this promoter we have obtained levels of recombinant protein expression in excess of 2% of total seed protein (unpublished results). Storage protein promoters from a number of monocotyledons also have been described.[32] These promoters may be particularly useful for the production of foreign proteins in the endosperm of cereals.

Oleosins are another type of seed protein that exhibit high levels of expression in oilseed species. Although their function appears primarily to be structural as opposed to serving as a repository for nitrogen, the time of expression of oleosins is similar to that of many storage proteins. Characterization of an oleosin promoter from *Arabidopsis* has revealed that it exhibits maximum expression in the late-cotyledon stage of development and, as with storage protein promoters, is responsive to both ABA and high osmoticums.[33] The strength of an homologous oleosin promoter from *Brassica* has been found to be similar to that of the most active fragment from CaMV 35S promoter over the period of maximum oleosin expression in seeds.[34] We have found that the *Arabidopsis* oleosin promoter also is capable of driving high levels of transgene expression in seeds.

Inducible Expression

Inducible promoters represent another means of restricting transgene expression through exogenous control over the onset of expression. For this strategy to be of use in commercial production systems, it is necessary that induction be simple, cost-effective, and result in sustained expression sufficient to yield high levels of recombinant product. Unfortunately, few inducible promoters meeting these criteria have yet been identified in plants, and hence this approach has not been widely used. An exception, however, is the MeGA™ system currently being developed in tobacco.[35] This system is based on the use of promoters derived from the HMGR family of defense-related genes expressed in response to a variety of stresses, including pathogen attack and wounding or mechanical injury.[36] A key advantage of the system is the post-harvest induction of transgene expression achieved through mechanical shredding of the tissue. This has the potential to reduce dramatically the amount of recombinant protein present in plants growing in the field and restrict the majority of production to such time as the material is received in contained processing facilities.[7] As mentioned, the ability to restrict expression in this manner is attractive from a regulatory perspective. The system also appears capable of achieving high levels of recombinant protein expression with values from 1-10% of total soluble protein reported.[7] Alternative methods for induction of transgene expression in plants, such as through the application of chemical agents, may result in expanded use of this strategy in the future.

SUBCELLULAR TARGETING OF RECOMBINANT PROTEINS

A high level of transcription, while necessary, is not in itself sufficient to guarantee a high level of the corresponding recombinant protein product. To permit stable accumulation, it is also crucial that the protein be targeted to a location in the cell where it will be properly processed and folded. Within the cell, the first step in the process of protein sorting involves the discrimination between cytosolic proteins and those destined for transport through the secretory membrane system. This is accomplished through the presence of a signal peptide that specifies translocation of the protein to the ER membrane. Studies examining heterologous expression in yeast, plant, and animal systems have established the underlying conservation among eukaryotes of this aspect of the protein processing machinery. It has been shown that mammalian secretory proteins with their native signal sequences are recognized and correctly processed when expressed in plant cells.[24,37] Furthermore, it has been demonstrated that hybrid proteins comprising plant signal sequences translationally fused with mammalian proteins also undergo correct targeting and processing.[38] Studies examining the targeting of proteins comprised of plant signal peptides fused to intracellular prokaryotic proteins have revealed that the information for ER translocation is carried within the signal sequence itself and is not a property of the attached protein.[39,40]

Successful targeting to the ER is an essential first step in the expression of most proteins that require post-translational modification for stability. In addition to a number of protein chaperones that assist in the folding process, the lumen of the ER provides an oxidizing environment necessary for disulfide formation and enzymes responsible for the initial steps in protein glycosylation[41] and for disulfide isomerization.[42] As with signal peptide processing, these elements of the protein processing machinery appear to be reasonably well-conserved. Consistent with this role in protein folding, significant differences in the levels of recombinant proteins have been observed depending on whether or not they are targeted for translocation to the ER. This perhaps has been best illustrated with the production in plants of mammalian antibodies and antibody fragments. In pioneering experiments by Hiatt et al.,[24] co-expression of immunoglobulin heavy and light chains with their native signal sequences resulted in levels of antibody more than 200-fold higher than were obtained when these sequences were not included. It was shown subsequently that antibodies assembled within the ER are secreted from the cell and accumulate within the apolplast.[43,44] Secretion to the apoplast in fact represents the default pathway in the plant secretory membrane system that operates in the absence of additional sequences specifying targeting or retention in other subcellular compartments.[40,41] A number of studies also have examined the accumulation of single chain variable region antibody fragments in plant cells (scFvs).[45-48] Attempts to synthesize these proteins in the more reducing environment of the cytosol have generally resulted in low levels of accumulation, most probably due to misfolding and elimination via the cellular protein degradation pathway. However, addition of ER targeting and retention signals to scFvs has enabled levels as high as 6.8% and 4.0% of total soluble protein to be achieved in leaves and seeds, respectively.[46] In one case, where high levels of a scFv were found to accumulate within the cytosol, disulfide bridge formation was not required for correct folding and stability.[49]

For cytoplasmic proteins that do not contain disulfide bridges, a wider array of targeting options may be available. In some cases, subcellular compartmentation of these proteins may even improve levels of accumulation. Targeting to protein bodies in transgenic barley grains increases levels of accumulation of a recombinant wheat thioredoxin when compared to the same protein expressed in the cytoplasm.[50] Finding alternative sites for the accumulation of animal proteins expressed in plants also may be necessary when the native targeting mechanisms or the organelles themselves do not exist in plant cells. Because the optimal site for accumulation of a foreign protein is somewhat unpredictable, ultimately it may be desirable to explore a variety of targeting options when designing a heterologous expression system.

GLYCOSYLATION AND PROTEIN AUTHENTICITY

Although the basic mechanisms of signal peptide processing and protein folding are similar in both plants and animals, all post-translational processing events are not as well-conserved. Of particular note are the reactions involved in protein glycosylation. Glycosylation is a natural modification of many eukaryotic proteins that occurs during trafficking through the secretory membrane system. The core reactions that take place in the ER and result in the addition of high mannose glycans to specific asparagine residues (N-linked glycosylation) are identical in both plants and mammals.[51] However, subsequent trimming and addition reactions that occur downstream in the Golgi and other organelles to produce complex glycans differ markedly among host systems. Whereas sialic acid is added to glycoproteins in mammalian systems, this does not occur in plants.[51] Conversely, plant glycans contain some unique xylosyl residues and fucosyl linkages that are not found on mammalian proteins.[51] These differences present a potential problem in the use of plant-based systems for the production of therapeutic glycoproteins in that protein glycans are known to affect a number of important pharmacokinetic properties.[52] Of even greater concern is the fact that unique oligosaccharides associated with plant complex glycans may be immunogenic.[53] However, depending on the specific product and application, the presence of complex plant glycans need not seriously limit therapeutic use. An example is the expression of a therapeutic mammalian antibody produced in plants for use in the passive mucosal immunization against herpes.[54] Although typical plant glycosylation patterns were obtained, tests performed with a mouse model revealed no significant differences in the stability or efficacy of plant-derived antibodies compared to those from mammalian cell culture. Furthermore, because these antibodies are intended for external application, they would be unlikely to cause adverse immunogenic reaction since the mucosal immune system is exposed regularly to plant glycans. Several approaches are available currently for removing or preventing the addition of complex plant glycans, including chemical deglycosylation and removal of asparagine residues by genetic engineering of the target protein. Although non-glycosylated antibodies produced in maize have reduced effector functions of complement-mediated cytotoxicity and cell-mediated, antibody-dependant cytotoxicity, they were otherwise identical to the mammalian form in pharmacokinetic properties.[55] In the future, it may also be possible to engineer plants to prevent addition of immunogenic glycans. Mutants of *Arabidopsis* have been created that are deficient in N-acetylglucosaminyltransferase and do not perform the complex glycosylation reactions resulting in proteins possessing only the core high-mannose glycans.[56] While these proteins might not be identical to the authentic forms produced in mammalian systems, they may be sufficiently similar in pharmacokinetic properties to be effective therapeutically and not promote an immunogenic response. Further refinements may allow production of proteins with authentic mammalian glycan structure, through introduction of the appropriate genes, such as those required for synthesis and transfer of sialic acid, into plants.

TRANSGENIC CROPS AS PRODUCTION VEHICLES

One of the advantages most often associated with plant-based production systems is ease of scale-up. Once transgenic plants have been obtained, standard agricultural practices may be employed in their propagation. Within a few generations, sufficient biomass can be available to fulfill the market requirements of even the largest volume protein products. This contrasts sharply with fermentation technology, where scale-up is more complicated and requires a significant investment in sophisticated infrastructure. The development of plants as production vehicles has benefited enormously from our extensive knowledge of the biology, agronomics, and processing of the major crop species. Breeding programs, undertaken over thousands of years aimed specifically at increasing yield in a variety of crops have optimized production capacity in these species. However, these advantages do not necessarily extend to the use of non-traditional crop species. Therefore, while the number of species amenable to transformation and regeneration in tissue culture continues to increase, in most cases, the common crop plants remain the vehicles of choice for molecular farming.

Some of the factors to be considered in selecting a particular crop as a production vehicle include ease of transformation and regeneration, the time required to obtain a stable transgenic line and scale up production, and the organ in which the recombinant protein is to be expressed. This final consideration, figures prominently in the selection process since different species have been bred to concentrate biomass in different organs. The choice of organ will, in turn, influence the storage stability of the recombinant protein and determine the initial steps used to process the transgenic plant material. Essentially, three options exist for production of recombinant proteins in plants, vegetative tissue, seeds, and edible storage organs such as tubers and fruits.

Vegetative Tissue

Among the plant species currently being considered for use in the commercial production of recombinant proteins in vegetative tissue are tobacco and alfalfa. Both of these crops are capable of producing large amounts of biomass per acre of land planted. For example, tobacco yields on average approximately fifty tons of leaves per acre based on multiple cuttings, and even higher levels may be achieved by using intensive agriculture practices. At fresh weight protein contents as high as 4.5%[57] and a recombinant protein expression level of 1% of total cellular protein, this translates into greater than 20 kg of recombinant protein per acre. An additional advantage of tobacco as a host species is the rapid scale-up made possible by the large number of seeds obtained per plant. Tobacco plants may produce in excess of ten thousand seeds enabling production to be quickly increased from a single high-expressing line. As with tobacco, alfalfa also offers the advantage of multiple harvests from the same planting. However, a disadvantage of production in leaves and other vegetative organs is the relatively high level of proteases present in comparison to plant storage tissues. This can give rise to increased levels of protein degradation, particularly following harvest when the tissue begins to undergo senescence. As a result, production in these systems may require immediate processing following harvest to limit the amount of degradation. However, reports have indicated that protein stability may be increased somewhat by drying leaf material after harvest.[46]

Seeds

As a natural protein storage organ, seeds may overcome some of the stability problems associated with production in leaves. Typically, seeds accumulate protein to levels of 15 – 30% of their dry weight and at maturity represent a low hydrolytic environment with moisture contents often below 10%. Much of the work in the development of seed-based production systems has been conducted in maize[5,11,55,58] and oilseed crops such as canola.[59-61] Typical seed yields for maize inbreds are in the range of 0.5-2 tons per acre[62] and those in canola are approximately 0.5-1 ton per acre. Studies examining the stability of two recombinant proteins, β-glucuronidase (GUS) and avidin, produced in maize have determined that the transgenic seed may be stored for periods of at least three months at 10°C, and up to two weeks at 25°C, without significant loss of activity.[11,29,30] Recent results have shown that GUS is stable for more than one year at 10°C.[29]

We have obtained similar results for recombinant proteins produced in seeds of canola. By using immunoblotting to examine the stability of an oleosin-hirudin fusion protein expressed in these seeds, we have detected no evidence of protein degradation over a period of at least two and one-half years.[63] The extended stability of proteins in dry seed represents a significant advantage in that it enables the spatial and temporal uncoupling of production and processing of the recombinant protein. As a result, the management of both inventory and supply is simplified, and centralization of processing facilities is made possible. Together, these factors help to reduce overall production costs.

Edible Storage Organs

Expression of foreign proteins has been reported in a variety of tubers and fruits including potato,[38,64] tomato,[65] and banana.[66] In terms of post-harvest stability, these organs are somewhat intermediate to that of fresh vegetative tissue and mature seeds. However, with levels in the range of 2% of fresh weight, the protein content of these tissues is relatively low. Apart from use in the production of purified proteins, edible storage organs have received considerable attention for applications where the product can be consumed without purification. One such application is in the production of proteins for use in oral vaccines.[16,67]

APPLICATIONS AND COMMERCIAL OPPORTUNITIES FOR PLANT SYSTEMS

The plant molecular farming industry has achieved proof-of-concept for the production of a wide array of proteins by using a variety of plant hosts and expression systems (reviewed in Cramer *et al.*[35]). Today, the repertoire of proteins being produced with plant-based expression systems includes biopharmaceuticals (human therapeutics, antibodies, vaccines), industrial and food enzymes, and animal health products (Table 1). At present, two industrial proteins are commercially available, and several biopharmaceuticals are being evaluated in clinical trials. The following provides a brief overview of several applications of plant production systems now in development.

Industrial Proteins

The first commercial recombinant proteins derived from transgenic plants belong to a category of products that can be broadly described as industrial proteins. Avidin[11] and β-glucuronidase[12] were expressed to commercially viable levels in maize using the ubiquitin promoter, and then recovered using standard seed processing and protein purification methods. Currently, both are marketed by Sigma Chemical Co. for use as research and diagnostic reagents. In addition to these relatively small volume research products, this category of proteins includes an array of enzymes used in large quantities in the food, chemical, and other industries. From a commercial perspective, the chief differences among these products and biopharmaceuticals is the level of purity required and the degree of regulatory oversight involved in their testing, approval, and manufacture. Industrial proteins often can be used as crude extracts or partially purified fractions, and the regulatory requirements for their sale and use generally are much less stringent than for biopharmaceuticals. Other industrial enzymes that have been produced in plants

Table 3.1: Commercial developments in plant molecular farming

Plant Company	Product (Development Partner)	Application	Host (Tissue)	Ref.
Industrial Proteins and Enzymes				
Prodigene	Avidin, ß-Glucuronidase (Sigma)	Reagents and diagnostics	Maize (seed)	11,12, a
	Brazzein (NeKtar Worldwide)	Natural dietetic sweetener	Maize (seed)	b
	Undisclosed industrial enzymes (Genencor)	Not reported	Maize (seed)	b
SemBioSys Genetics	Food Enzyme (undisclosed partner)	Food Application	Safflower (seed)	c
Vaccines / Antibodies				
Planet Biotech	*Guy's-13 Mab	Passive immunotherapy for S. mutans (dental caries)	Tobacco (vegetative)	13
Axis Genetics	* *E. coli* LT-B protein (Boyce Thompson Institute)	Travelers diarrhea vaccine	Tobacco (vegetative), Potato (tuber) Maize (seed)	15,70
	*Hepatitis B surface antigen (Boyce Thompson Inst.)	Hepatitis vaccine	Potato (tuber) Tobacco (vegetative)	14,77
	Norwalk virus capsid protein	Gastroenteritis vaccine	Potato (tuber)	16
Epicyte	HSV-2 Mab (ReProtect)	Passive immunotherapy for genital herpes	Soybean (vegetative)	54,74
	Anti-spermatozoa antibodies (ReProtect/Prodigene)	Contraception	Not reported	74
Large Scale Biology Corp	idiotype-specific scFv	Non-Hodgkin's lymphoma vaccine	Tobacco (vegetative)	80
Prodigene	TGEV spike protein	Porcine diarrhea vaccine	Maize (seed)	5
Integrated Protein Technologies	NR-LU-10 Mab (NeoRx) BR-96 Mab (Bristol-Meyers Squibb)	Pancarcinoma immunotherapy / immunodetection	Maize (seed)	55
Therapeutics				
Medicago	Beta-interferon (BioInFiNix Ltd)	Respiratory syncytial virus infections (pneumonia)	Alfalfa (vegetative)	c
Agriculture and Agri-Food Canada	Interleukin-10	Inflammatory bowel syndrome/Crohn's disease	Tobacco (vegetative)	c
SemBioSys Genetics Inc.	Hirudin	Anticoagulant	Canola (seed)	60
CropTech Development Corp	Glucocerebrosidase	Gaucher's disease therapy	Tobacco (vegetative)	93

*Products have entered clinical trials; (a) avidin and GUS are currently marketed by Sigma; (b) company disclosure www.prodigene.com; (c) company disclosure.

include a fungal phytase in tobacco,[68] a fungal xylanase targeted to seed oil bodies in *B. napus*,[59] and a bacterial cellulase (endo-1,4-ß-D-glucanase) in the leaves of *Arabidopsis thaliana*.[25] A number of industrial proteins from transgenic plants are now in either early or advanced stages of commercial development (Table 3.1).

Vaccines

The use of plant systems for the production of human and animal vaccines is an area of intense commercial interest, with several products currently under development (Table 3.1). Plants are particularly attractive hosts for oral vaccine candidates since the product can be produced in edible plant tissue and consumed without purification, significantly reducing production costs while at the same time making distribution and administration easier. Most of the human vaccines being considered for production in plants target bacterial or viral pathogens that initiate infection at mucosal surfaces. These include vaccines for diarrheal diseases (Norwalk virus,[69] Enterotoxigenic *E.coli*,[5,70] *Vibrio cholera*[71,72]), respiratory diseases (*P. aeruginosa*[73]), sexually transmitted diseases (Herpes simplex-2 virus,[54,74] HIV[75]), tooth decay (*S. mutans*[13]), and opportunistic infections (*S. aureus*[76]). Other human disease targets for plant-derived vaccines include hepatitis,[77] malaria,[78] and a variety of cancers (pancarcinoma antigens,[55] colon,[79] non-Hodgkin's lymphoma[80]). The animal vaccine candidates under development target primarily viral pathogens (rabies,[81] mink virus,[82] porcine diarrhea[5]). With animal models, it has now been demonstrated in a number of cases that plant-produced antigenic proteins/epitopes administered either orally[64,83] or parenterally[84,85] can elicit an immune response that provides protection against a subsequent exposure to the pathogen. The field has advanced to the point where several protein antigens are being evaluated in human clinical trials. In 1998, Ma *et al.*[13] described the first human clinical trial for a plant-derived protein in the form of a secretory monoclonal antibody (Guy's-13). The antibody, directed against the bacterium *Streptococcus mutans*, was shown to be an effective passive immunization agent in the prevention of dental caries. The first plant-derived oral vaccines against enterotoxigenic *E. coli*[15] (completed Phase I), Norwalk virus[16] (Phase I), and the Hepatitis B virus[14] (Phase I) also are now in human clinical trials.

Antibodies

Production of mammalian immunoglobulins in plants was first reported in 1989 by Hiatt *et al.* who described the assembly of a fully functional murine monoclonal IgG antibody in tobacco leaves.[24] It was later demonstrated by Ma *et al.* that plants, somewhat surprisingly, were also capable of producing the structurally more complex forms associated with secretory antibodies.[86] Potential applications include passive immunotherapy treatment and the production of therapeutic and diagnostic agents. In addition to full-size recombinant antibodies, plants have been used to produce Fab fragments[87] and scFvs.[47,88-90] One of these antibodies, being developed for passive immunotherapy against *S. mutans* as described above, has been evaluated in clinical trials.[13] There are several antibodies in development for various other applications including treatment of genital herpes[54] and as therapeutic agents in the treatment of various forms of cancer.[79,80] Antibodies also are being developed for the targeted delivery of therapeutic drugs, specifically to the body's epithelial tissues, with the potential to treat a wide range of diseases such as cystic fibrosis, pneumonia, and asthma.[74]

Human Therapeutics

In addition to vaccine antigens and antibodies, there is a growing list of therapeutic biopharmaceutical proteins that have been expressed in plants (reviewed in Cramer *et al.*[35]). These include serum proteins (haemoglobin,[10] human serum albumin,[38] protein C,[7] cytokines, growth regulators, and interferons,[6,9] granulocyte-macrophage colony stimulating factor,[8] epidermal growth factor,[91] erythropoietin,[92] antibiotics (NP1 defensin[9]), anticoagulants (hirudin[60]), and lysosomal enzymes (glucocerebrosidase,[93] iduronidase,[35] ∝-galactosidase[9]). Feasibility of expression in plants has now been firmly established, and, in many cases, the plant-derived protein has been shown to be structurally similar to the human protein and to be enzymatically active.[7,8,60,93] However, many of these therapeutic proteins are glycosylated, and require authentic mammalian glycan moieties for therapeutic function.[7,35] For this reason, use of plant-based systems for commercial production of such proteins may await further engineering of plant pathways to produce humanized glycosylation patterns as described earlier.

Commercial Opportunities

Capitalizing on the technological developments in plant genetic engineering, the plant molecular farming industry is now well-positioned to take advantage of the growing interest in protein-based biopharmaceuticals. The pharmaceutical biotechnology market is expected to experience substantial growth over the next several years, fueled by development of a large number of new protein products identified through advances in areas of genomics, proteomics, and immunology. The market is projected to expand from $20 billion (1998) to $34 billion by 2002, with strong growth in vaccines ($7 billion) and monoclonal antibodies ($9.7 billion). Other markets that are increasingly being impacted by biotechnology and, therefore, may be addressable by molecular farming technology include animal health ($15.1 billion, 1997), nutraceuticals ($7.5 billion, 1996), cosmeceuticals (>$3 billion, 2000), and food and industrial enzymes ($2 billion, 1997).

A key determinant of success in these markets will be the ability of molecular farming companies to prove that their system offers cost advantages and/or value-added benefits over conventional production systems. Selection of the appropriate target markets and product viability within each market will be influenced by parameters such as product volume, demand, margin, life-cycles, purity, and cost of production. In the case of industrial enzymes, where product lifecycles are short and profit margins are low, there may be little opportunity to capitalize on the low cost of production position offered by plants. Nevertheless, there are exceptions, as illustrated by the production of avidin in corn where plant-based production offers considerable cost savings. It takes 1 ton of eggs costing $1000 to produce 20 grams of avidin, whereas the same amount can be obtained from just four bushels of transgenic corn costing approximately $20.[58] Another example is the use of recombinant xylanse in the conversion of plant biomass into simple sugars for ethanol production.[25] Here, the huge volumes of enzyme that would be required, and the fact that a crude plant extract could be used in the process, make plant-based production attractive.

In today's cost conscious health care environment, pharmacoeconomics is a key determinant influencing viability of pharmaceutical drugs. In some cases, plant production systems may permit development of new protein therapeutics that are otherwise economically marginal or impractical. Examples include new biopharmaceuticals for treatment of chronic conditions, such as cystic fibrosis and multiple sclerosis where development is not pursued due to large capital investment requirements, high production costs, and long-term treatment costs. Plant systems also may offer the potential to expand the range of applications of some existing therapeutic proteins, such as antibodies where use is now limited by availability and cost of production. Currently, the high cost of production ($300-1000/gram) and huge volume of antibodies required (100 kg/year) make use of conventional cell fermentation systems unfeasible for some applications. Plants also offer opportunities for reducing the cost of production for drugs, such as insulin where patents restricting competitive commercial manufacture are about to expire.

Beyond low cost, plants offer value-added benefits that may be leveraged in one or more areas. This is exemplified by human vaccines, where the high cost of production and large volume requirements limits availability in many developing countries.[66] Edible plant vaccines offer a convenient and economical alternative that may allow broader access to vaccination in these countries. Such applications of plant-based production systems provide not only for cost-effective production but also for novel methods of drug delivery.

RECOVERY OF RECOMBINANT PROTEINS FROM PLANTS

Lower production costs in comparison to competing technologies is another of the potential advantages often cited when advocating the use of plant-based production systems. Quantifying this advantage, however, requires a determination not only of the costs associated with transgenic crop production but also of the costs for all downstream operations necessary to recover a purified product. Although in some applications, such as the production of oral vaccines, purification of the recombinant product may not be necessary, for many others, cost-effective recovery of a pure protein will be an essential element of the production process. Initial steps of protein recovery can be adapted from existing methods used in the food and chemical industries for the processing and extraction of components from plant tissue. Subsequent purification of recombinant proteins may then be accomplished through conventional separation methods such as chromatography and filtration. However, to realize an advantage in cost of production over competing systems, it is important that savings in upstream production in plants not be offset by higher costs for downstream recovery of recombinant proteins.

The cost of large-scale recombinant protein production in a plant-based system recently has been determined in two independent studies.[17,18] In both studies, costs associated with production of transgenic crops, processing, purification, and capital expenses were evaluated. Evangelista and colleagues evaluated the process economics for production of recombinant β-glucuronidase (rGUS) in maize seed.[17] The process was modeled on a laboratory-scale extraction and purification procedure using whole seed. The model was based on a rGUS expression level of 0.015% of seed weight and an annual product volume of 137 kg. Transgenic seed cost was estimated at \$0.20 / kg. Based on the process simulation, a total production cost of approximately \$43 / g of purified enzyme was projected. Mison and Curling examined production of a generic protein also expressed in maize seed.[18] The process simulation was based on pilot-scale experience obtained with 500-kg batches of transgenic seed. An expression level of 0.15% of seed dry weight was employed in the model, a value 10-fold higher than that used in the study by Evangelista et al. This was offset to some extent by a higher cost used for transgenic seed, which in this model was estimated at \$0.40 / kg. Production costs of \$49.4 / g – \$4.5 / g were calculated for product volumes ranging from 0.1 – 100 tons annually.

While the data presented do not permit a rigorous comparison of the two analyses, the results do show some interesting similarities in the cost breakdowns. In both cases, crop production represents a relatively low proportion of the total production cost when product volumes are less than or equal to 1 ton of protein per year. As product volumes increase from 0.1 – 100 tons per year, the proportion of total cost attributable to crop production increases from 1.4% to 13.3%. This change is due to the economy of scale obtained in processing, while crop production costs remain relatively fixed. The analyses are also similar in indicating that post-extraction chromatography steps account for a significant proportion of the total production costs. Both processes envisage three such operations that in the 0.1-ton output range combine to represent up to 59% of the cost of production. Another common feature to emerge from these analyses is sensitivity to expression level. It was determined that the impact of expression is greatest in the earliest steps in the process, extraction and capture. During these steps the concentration of the recombinant protein significantly influences the amount of material that must be processed to obtain the desired volume of product.

In addition to presenting a cost model for protein production in seed, Mison and Curling also compared the costs of plant-based production with that of competing technologies in yeast, mammalian cell culture, and transgenic animals by using published data for these systems.[18] The comparison for a 1-ton product volume range revealed a clear cost advantage of plants over the other systems. As product volumes increase to 10 tons per year or more, production costs in transgenic plants and animals are projected to be similar. Thus, this study seems to support the early predictions of low production costs associated with plant-based systems, particularly for large volumes of recombinant protein. However, it is necessary to exercise caution when comparing the cost figures cited from different systems and studies, since the assumptions on which the process models are based may differ markedly. Further refinement of initial cost projections for plant systems should be possible in the near future, as an increasing number of plant-derived products find their way to market.

STRATEGIES FOR SIMPLIFYING DOWNSTREAM PURIFICATION

From the above examples, it is apparent that any operation performed early in processing that simplifies downstream purification has the potential to reduce significantly the final cost of production. One way in which simplification might be achieved is through the use of subcellular fractionation. Here, low cost methods such as decantation, centrifugation, or filtration may be used to replace more expensive chromatographic separations. With this approach, we have developed a strategy for simplifying downstream purification based on the specific targeting of recombinant proteins to seed oil bodies. Similar approaches also are being developed based on other organelles and cellular components.

Seed Oil Bodies as Purification Tools

Oil bodies are natural storage organelles for triacylglycerides (TAG) found in a variety of plant tissues including seeds, pollen, and some fruits.[94] They are especially abundant in oilseed crops such as canola, sunflower, and safflower in which triacylglyceride forms the primary storage reserve, constituting up to 48% of the seed dry weight in some species. Oil bodies are comprised of triacylglyceride surrounded by a half-unit phospholipid membrane. In tissues such as seed that undergo desiccation, they also contain a unique protein known as oleosin. The oleosin protein is comprised of three discrete domains, a central hydrophobic domain that anchors the protein within the triacylglyceride core, and hydrophilic N- and C-terminal domains that reside on the cytosolic side of the organelle.[95-97] Oleosins are expressed and targeted to the surface of oil bodies during their biogenesis on the ER. At maturity, the oleosins form a hydrophilic shell covering the oil body surface and preventing coalescence. They also have been proposed to serve as docking sites for lipolytic enzymes involved in the mobilization of the triacylglyceride during seedling germination.[98]

The ability to target proteins specifically to the oil body surface enables these organelles to be used as carriers to facilitate the purification of recombinant proteins.[61] We have demonstrated that genetically engineered fusions between foreign proteins ranging in size from 7 – 68 kDa and oleosins retain their oil body targeting capability. Fusions can be made to either the N- or C-terminus of the oleosin such that following targeting the recombinant protein is situated on the outer, cytosolic side of the oil body. Subcellular fractionation is accomplished following aqueous extraction through separation based on differential density. Due to the high proportion of TAG in oil bodies (ca. 85 – 95% of the dry weight), they rapidly rise to the surface with centrifugation and are separated from the bulk of the aqueous extract containing the soluble seed impurities. The oil body fraction can then be recovered, resuspended in fresh buffer, and the separation repeated until the majority of soluble impurities, including endogenous seed proteins, have been removed. At this point in the process, a semi-purified product immobilized on the oil body surface is obtained. We have shown that many proteins retain their activity while associated with the oil body, and we are currently developing a number of applications for the use of these novel immobilized products. These include the production of enzyme bioreactors,[59] feed additives,[99] and vaccines.[100]

If a purified protein is desired, a specific cleavage recognition sequence is engineered into the oleosin-recombinant protein junction. Addition of a cleavage agent, such as a site-specific protease to washed oil bodies, followed by a final round of centrifugation enables recovery of the recombinant protein in the aqueous fraction. It is important that this cleavage step be economical so that it does not offset the lower costs achieved through subcellular fractionation. To meet this requirement we are currently developing a repertoire of cost-effective cleavage methods for use with oleosin targeting technology. One such method is based on the autocatalytic maturation of common proteases, such as those found in the digestive tract of animals.[101] These proteases generally are economical and in some cases may themselves be produced in plants, such that costs incurred in cleavage are minimized. The utility of the oil body fractionation approach has been demonstrated on a laboratory-scale with the production and purification of an anticoagulant protein, hirudin.[60] An oleosin-hirudin fusion protein containing a factor Xa protease recognition site was expressed in seeds of *Brassica*. Following proteolytic cleavage of the fusion protein from washed oil bodies and centrifugation, active hirudin was recovered in the aqueous fraction. As a result of the oil body fractionation process, 90-95% of the endogenous seed protein was removed without the need for chromatography.

Enrichment of recombinant proteins through organellar targeting followed by subcellular fractionation can reduce the costs associated with downstream chromatographic purification steps in a number of different ways. Implemented prior to the first chromatographic separation, typically a capture step, enrichment reduces the amount of media required by ensuring that more of the total protein binding capacity is available for the protein of interest. Fewer protein species in the extract following fractionation may also reduce the total number of chromatographic steps necessary to obtain a purified product. Finally, steps that are required may be performed more simply, for example, through batch as opposed to column separation or by using a step gradient in place of a continuous gradient elution.

Additional Methods of Subcellular Fractionation

Similar subcellular targeting and fractionation strategies currently are under development in other laboratories. In one example, granule bound starch synthase fusions are employed to target proteins to starch granules found in cereal grains such as maize.[102] Starch granules, like oil bodies, are discrete organelles delimited by a membrane with associated proteins. Following aqueous extraction, these organelles with their complement of proteins can be recovered by centrifugation. Cellulose binding domains, motifs associated with a number of cellulolytic enzymes, also have been used as affinity tags for the selective binding and recovery of recombinant protein fusions.[103] In this case, the fusion proteins can be adsorbed to the cellulose present in plant cell walls and sedimented with this fraction following extraction and centrifugation. In both cases, the fractionation step has the potential to remove substantial amounts of endogenous plant protein, simplifying subsequent downstream purification. As with oleosin technology, these strategies also require cleavage of the fusion protein in order to obtain a purified product.

SUMMARY

In a relatively short period of time, the production of foreign proteins in plants has progressed from a theoretical concept to the early stages of commercialization. Several plant species currently are being developed for use as production vehicles, and different strategies for regulating transgene expression in a variety of tissues are being explored. Each system under development has its particular advantages, and it is impossible to say at this time that any one is superior for all applications. It is more likely that a small number of successful strategies will be developed, each tailored to meet the requirements for a specific range of products.

Early work on heterologous protein expression in plants, and basic research in areas of subcellular targeting and protein modification have established the underlying conservation of protein-processing machinery upon which the development of recombinant plant production systems depends. Although differences exist between plants and animals with respect to some specific post-translational modification reactions, for many mammalian proteins, properties such as activity and stability that are most important for therapeutic applications are not affected. As demonstrated with the production of mammalian antibodies, the addition of novel plant glycans, often perceived as the most serious limitation to the use of plant-derived products, may have no effect for some applications and may be prevented when necessary. Thus, the selection of a host production system for a particular recombinant protein may depend ultimately less on product authenticity than on system production capacity and process economics.

With the current state of technology, expression of greater than 1% of total cellular protein is obtained routinely for many recombinant proteins in plant systems. If technological improvements mirror those observed previously for more tradition production systems, then significantly higher levels of expression may yet be achieved. Economic analysis of preliminary data from recombinant protein production in high-yielding crop species and process simulations of downstream recovery operations have provided support for the predicted cost advantage of plant systems. Continued development of strategies to simplify downstream purification, such as through the use of subcellular fractionation methods, will result in still lower production costs in the future. Therefore, it seems likely that plants will be increasingly considered as the primary production vehicle, particularly for recombinant proteins required in large volumes or where high production costs now limit their application.

REFERENCES

1. WHITELAM, G.C., COCKBURN, B., GANDECHA, A.R., OWEN, M.R., Heterologous protein production in transgenic plants, *Biotechnol. Genet. Eng. Rev.*, 1993, **11**, 1-29.

2. GODDIJN, O.J.M., PEN, J., Plants as bioreactors, *Trends Biotechnol.*, 1995, **13**, 379-387.
3. PEN, J., Comparison of host systems for the production of recombinant proteins, *in* Transgenic Plants: A Production System for Industrial and Pharmaceutical Proteins (M. R. L. Owen and J. Pen eds.), John Wiley & Sons, Chichester, 1996, pp. 149-167.
4. FISCHER, R., DROSSARD, J., COMMANDEUR, U., SCHILLBERG, S., EMANS, N., Towards molecular farming in the future: moving from diagnostic protein and antibody production in microbes to plants, *Biotechnol. Appl. Biochem.*, 1999, **30**, 101-108.
5. HOOD, E.E., JILKA, J.M., Plant-based production of xenogenic proteins, *Curr. Opin. Biotechnol.*, 1999, **10**, 382-386.
6. ZHU, Z., HUGHES, K.W., HUANG, L., SUN, B., LUI, C., LI, Y., HOU, Y., LI, X., Expression of human a-interferon cDNA in transgenic rice plants, *Plant Cell Tissue Org. Cult.*, 1994, **36**, 197-204.
7. CRAMER, C.L., WEISSENBORN, D.L., OISHI, K.K., GRABAU, E.A., BENNETT, S., PONCE, E., GRABOWSKI, G.A., RADIN, D.N., Bioproduction of human enzymes in transgenic tobacco, *Ann. N.Y. Acad. Sci.*, 1996, **792**, 62-71.
8. GANZ, P.R., DUDANI, A.K., TACKABERRY, E.S., SARDANA, R., SAUDER, C., CHENG, X., ALTOSAR, I., Expression of human blood proteins in transgenic plants: the cytokine GM-CSF as a model protein, *in* Transgenic Plants: A Production System for Industrial and Pharmaceutical Proteins (M. R. L. Owen and J. Pen eds.), John Wiley & Sons, Chichester, 1996, pp.
9. GRILL, L.K., Viral-vectored, large scale production of drugs and pharmaceuticals in plants, Paper read at IBC 3rd Annual International Symposium on Producing the Next Generation of Therapeutics: Exploiting Transgenic Technologies, Feb. 5-6, 1997, at West Palm Beach.
10. DIERYCK, W., PAGNIER, J., POYART, C., MARDEN, M.C., GRUBER, V., BOURNAT, P., BAUDINO, S., MEROT, B., Human haemoglobin from transgenic tobacco [letter], *Nature*, 1997, **386**, 29-30.
11. HOOD, E.E., WITCHER, D.R., MADDOCK, S., MEYER, T., BASZCZYNSKI, C., BAILEY, M., FLYNN, P., REGISTER, J., MARSHALL, L., BOND, D., KULISEK, E., KUSNADI, A., EVANGELISTA, R.L., NIKOLOV, Z., WOOGE, C., MEHIGH, R., HERNAN, R., KAPPEL, W., RITLAND, D., LI, C.P., HOWARD, J.A., Commercial production of avidin from transgenic maize: characterization of transformant, production, processing, extration and purification, *Mol. Breeding*, 1997, **3**, 291-306.
12. WITCHER, D.R., HOOD, E.E., PETERSON, D., BAILEY, M., BOND, D., KUSNADI, A., EVANGELISTA, R.L., NIKOLOV, Z., WOOGE, C., MEHIGH, R., KAPPEL, W., REGISTER, J., HOWARD, J.A., Commercial production of β-glucuronidase (GUS): a model system for the production of proteins in plants, *Mol. Breeding*, 1998, **4**, 301-312.
13. MA, J.K., HIKMAT, B.Y., WYCOFF, K., VINE, N.D., CHARGELEGUE, D., YU, L., HEIN, M.B., LEHNER, T., Characterization of a recombinant plant monoclonal secretory antibody and preventive immunotherapy in humans, *Nat. Med.*, 1998, **4**, 601-606.

14. KAPUSTA, J., MODELSKA, A., FIGLEROWICZ, M., PNIEWSKI, T., LETELLIER, M., LISOWA, O., YUSIBOV, V., KOPROWSKI, H., PLUCIENNICZAK, A., LEGOCKI, A.B., A plant-derived edible vaccine against hepatitis B virus [published erratum appears in *FASEB J.*, 1999 Dec, **13(15)**, 2339], *Faseb J.*, 1999, **13**, 1796-1799.
15. TACKET, C.O., MASON, H.S., LOSONSKY, G., CLEMENTS, J.D., LEVINE, M.M., ARNTZEN, C.J., Immunogenicity in humans of a recombinant bacterial antigen delivered in a transgenic potato, *Nat. Med.*, 1998, **4**, 607-609.
16. TACKET, C.O., MASON, H.S., LOSONSKY, G., ESTES, M.K., LEVINE, M.M., ARNTZEN, C.J., Human immune responses to a novel norwalk virus vaccine delivered in transgenic potatoes, *J. Infect. Dis.*, 2000, **182**, 302-305.
17. EVANGELISTA, R.L., KUSNADI, A.R., HOWARD, J.A., NIKOLOV, Z.L., Process and economic evaluation of the extraction and purification of recombinant beta-glucuronidase from transgenic corn, *Biotechnol. Prog.*, 1998, **14**, 607-614.
18. MISON, D., CURLING, J., The industrial production costs of recombinant therapeutic proteins expressed in transgenic corn, *Biopharm*, 2000, **14**, 48-54.
19. ODELL, J.T., NAGY, F., CHUA, N.H., Identification of DNA sequences required for activity of the cauliflower mosaic virus 35S promoter, *Nature*, 1985, **313**, 810-812.
20. BATTRAW, M.J., HALL, T.C., Histochemical analysis of CaMV 35S promoter-beta-glucuronidase gene expression in transgenic rice plants, *Plant Mol. Biol.*, 1990, **15**, 527-538.
21. FANG, R.X., NAGY, F., SIVASUBRAMANIAM, S., CHUA, N.H., Multiple cis regulatory elements for maximal expression of the cauliflower mosaic virus 35S promoter in transgenic plants, *Plant Cell*, 1989, **1**, 141-150.
22. BENFEY, P.N., REN, L., CHUA, N.H., Tissue-specific expression from CaMV 35S enhancer subdomains in early stages of plant development, *Embo J.*, 1990, **9**, 1677-1684.
23. KAY, R., CHAN, A., DALY, M., MCPHERSON, J., Duplication of CaMv 35S promoter sequences creates a strong enhancer for plant genes, *Science*, 1987, **236**, 1299-1302.
24. HIATT, A., CAFFERKEY, R., BOWDISH, K., Production of antibodies in transgenic plants, *Nature*, 1989, **342**, 76-78.
25. ZIEGLER, M.T., THOMAS, S.R., DANNA, K.J., Accumulation of thermostable endo-1,4-ß-D-glucanase in the apoplast of *Arabidopsis thaliana* leaves, *Mol. Breeding*, 2000, **6**, 37-46.
26. CHRISTENSEN, A.H., QUAIL, P.H., Ubiquitin promoter-based vectors for high-level expression of selectable and/or screenable marker genes in monocotyledonous plants, *Transgenic Res.*, 1996, **5**, 213-218.
27. BASSIE, L., NOURY, M., LEPRI, O., LAHAYE, T., CHRISTOU, P., CAPELL, T., Promoter strength influences polyamine metabolism and morphogenic capacity in transgenic rice tissues expressing the oat adc cDNA constitutively [In Process Citation], *Transgenic Res.*, 2000, **9**, 33-42.
28. HOLTORF, S., APEL, K., BOHLMANN, H., Comparison of different constitutive and inducible promoters for the overexpression of transgenes in *Arabidopsis thaliana*, *Plant Mol. Biol.*, 1995, **29**, 637-646.

29. KUSNADI, A.R., EVANGELISTA, R.L., HOOD, E.E., HOWARD, J.A., NIKOLOV, Z.L., Processing of transgenic corn seed and its effect on the recovery of recombinant beta-glucuronidase, *Biotechnol. Bioeng.*, 1998, **60**, 44-52.
30. KUSNADI, A.R., HOOD, E.E., WITCHER, D.R., HOWARD, J.A., NIKOLOV, Z.L., Production and purification of two recombinant proteins from transgenic corn, *Biotechnol. Prog.*, 1998, **14**, 149-155.
31. SENGUPTA-GOPALAN, S., REICHERT, M.A., BARKER, R.F., HALL, T.C., Developmentally regulated expression of the bean ß-phaseolin gene in tobacco seed, *Proc. Natl. Acad. Sci. U.S.A.*, 1985, **82**, 3320-3324.
32. RUSSELL, D.A., FROMM, M.E., Tissue-specific expression in transgenic maize of four endosperm promoters from maize and rice, *Transgenic Res.*, 1997, **6**, 157-168.
33. PLANT, A.L., VAN ROOIJEN, G.J., ANDERSON, C.P., MOLONEY, M.M., Regulation of an *Arabidopsis* oleosin gene promoter in transgenic *Brassica napus*, *Plant Mol. Biol.*, 1994, **25**, 193-205.
34. KEDDIE, J.S., TSIANTIS, M., PIFFANELLI, P., CELLA, R., HATZOPOULOS, P., MURPHY, D.J., A seed-specific *Brassica napus* oleosin promoter interacts with a G-box- specific protein and may be bi-directional, *Plant Mol. Biol.*, 1994, **24**, 327-340.
35. CRAMER, C.L., BOOTHE, J.G., OISHI, K.K., Transgenic plants for therapeutic proteins: linking upstream and downstream strategies, *Curr. Top. Microbiol. Immunol.*, 1999, **240**, 95-118.
36. YANG, Z., PARK, H., LACY, G.H., CRAMER, C.L., Differential activation of potato 3-hydroxy-3-methylglutaryl coenzyme A reductase genes by wounding and pathogen challenge, *Plant Cell*, 1991, **3**, 397-405.
37. DURING, K., HIPPE, S., KREUZALER, F., SCHELL, J., Synthesis and self-assembly of a functional monoclonal antibody in transgenic *Nicotiana tabacum*, *Plant Mol. Biol.*, 1990, **15**, 281-293.
38. SIJMONS, P.C., DEKKER, B.M., SCHRAMMEIJER, B., VERWOERD, T.C., VAN DEN ELZEN, P.J., HOEKEMA, A., Production of correctly processed human serum albumin in transgenic plants, *Biotechnology (*N.Y.*)*, 1990, **8**, 217-221.
39. ITURRIAGA, G., JEFFERSON, R.A., BEVAN, M.W., Endoplasmic reticulum targeting and glycosylation of hybrid proteins in transgenic tobacco, *Plant Cell*, 1989, **1**, 381-390.
40. DENECKE, J., BOTTERMAN, J., DEBLAERE, R., Protein secretion in plant cells can occur via a default pathway, *Plant Cell*, 1990, **2**, 51-59.
41. CHRISPEELS, M.J., Sorting of proteins in the secretory system, *Annu. Rev. Plant Physiol. Mol. Biol.*, 1991, **42**, 21-53.
42. SHORROSH, B.S., DIXON, R.A., Molecular cloning of a putative plant endomembrane protein resembling vertebrate protein disulfide-isomerase and a phosphatidylinositol-specific phospholipase C, *Proc. Natl. Acad. Sci. U.S.A.*, 1991, **88**, 10941-10945.
43. HEIN, M.B., TANG, Y., MCLEOD, D.A., JANDA, K.D., HIATT, A., Evaluation of immunoglobulins from plant cells, *Biotechnol. Prog.*, 1991, **7**, 455-461.
44. MA, J.K., HEIN, M.B., Immunotherapeutic potential of antibodies produced in plants, *Trends Biotechnol.*, 1995, **13**, 522-527.

45. ARTSAENKO, O., PEISKER, M., ZUR NIEDEN, U., FIEDLER, U., WEILER, E.W., MUNTZ, K., CONRAD, U., Expression of a single-chain Fv antibody against abscisic acid creates a wilty phenotype in transgenic tobacco, *Plant J.*, 1995, **8**, 745-750.
46. FIEDLER, U., PHILLIPS, J., ARTSAENKO, O., CONRAD, U., Optimization of scFv antibody production in transgenic plants, *Immunotechnology*, 1995, **3**, 205-216.
47. FIREK, S., DRAPER, J., OWEN, M.R., GANDECHA, A., COCKBURN, B., WHITELAM, G.C., Secretion of a functional single-chain Fv protein in transgenic tobacco plants and cell suspension cultures [published erratum appears in Plant Mol Biol 1994 24 :833], *Plant Mol. Biol.*, 1993, **23**, 861-870.
48. OWEN, M., GANDECHA, A., COCKBURN, B., WHITELAM, G., Synthesis of a functional anti-phytochrome single-chain Fv protein in transgenic tobacco, *Biotechnology* (N.Y.), 1992, **10**, 790-794.
49. DE JAEGER, G., BUYS, E., EECKHOUT, D., DE WILDE, C., JACOBS, A., KAPILA, J., ANGENON, G., VAN MONTAGU, M., GERATS, T., DEPICKER, A., High level accumulation of single-chain variable fragments in the cytosol of transgenic *Petunia hybrida*, *Eur. J. Biochem.*, 1999, **259**, 426-434.
50. CHO, M.J., WONG, J.H., MARX, C., JIANG, W., LEMAUX, P.G., BUCHANAN, B.B., Overexpression of thioredoxin h leads to enhanced activity of starch debranching enzyme (pullulanase) in barley grain, *Proc. Natl. Acad. Sci. U.S.A.*, 1999, **96**, 14641-14646.
51. CHRISPEELS, M.J., FAYE, L., The production of recombinant glycoproteins with defined non-immunogenic glycans, *in* Transgenic Plants: A Production System for Industrial and Pharmaceutical Proteins (M. R. L. Owen and J. Pen eds.), John Wiley & Sons, Chichester, 1996, pp. 99-114.
52. CUMMING, D.A., Physiological relevance of protein glycosylation, *Dev. Biol. Stand*, 1992, **76**, 83-94.
53. FAYE, L., GOMORD, V., FITCHETTE-LAINE, A.C., CHRISPEELS, M.J., Affinity purification of antibodies specific for Asn-linked glycans containing alpha 1-->3 fucose or beta 1-->2 xylose, *Anal. Biochem.*, 1993, **209**, 104-108.
54. ZEITLIN, L., OLMSTED, S.S., MOENCH, T.R., CO, M.S., MARTINELL, B.J., PARADKAR, V.M., RUSSELL, D.R., QUEEN, C., CONE, R.A., WHALEY, K.J., A humanized monoclonal antibody produced in transgenic plants for immunoprotection of the vagina against genital herpes, *Nat. Biotechnol.*, 1998, **16**, 1361-1364.
55. BAEZ, J., RUSSELL, D., CRAIG, J., Corn seed production of therapeutic proteins moves forward: one company's experience, *Biopharm*, 2000, **13**, 50-54.
56. VON SCHAEWEN, A., STURM, A., O'NEILL, J., CHRISPEELS, M.J., Isolation of a mutant *Arabidopsis* plant that lacks N-acetyl glucosaminyl transferase I and is unable to synthesize Golgi-modified complex N-linked glycans, *Plant Physiol.*, 1993, **102**, 1109-1118.
57. DEJONG, D.W., SAUNDERS, J.A., Fluctuations in protein levels of tobacco leaves and consequences for extractability, *Beitrage zur Tabakforschung International*, 1986, **13**, 139-149.

58. HOOD, E.E., KUSNADI, A., NIKOLOV, Z., HOWARD, J.A., Molecular farming of industrial proteins from transgenic maize, *Adv. Exp. Med. Biol.*, 1999, **464**, 127-147.
59. LIU, J.H., SELINGER, B.L., CHENG, K.J., BEAUCHEMIN, K.A., MOLONEY, M.M., Plant seed oil-bodies as an immobilization matrix for a recombinant xylanse from the rumen fungus *Neocallimastix patriciarum.*, *Mol. Breeding*, 1999, **3**, 463-470.
60. PARMENTER, D.L., BOOTHE, J.G., VAN ROOIJEN, G.J., YEUNG, E.C., MOLONEY, M.M., Production of biologically active hirudin in plant seeds using oleosin partitioning, *Plant Mol. Biol.*, 1995, **29**, 1167-1180.
61. VAN ROOIJEN, G.J., MOLONEY, M.M., Plant seed oil-bodies as carriers for foreign proteins, *Biotechnology (N.Y.)*, 1995, **13**, 72-77.
62. RUSSELL, D.A., Feasibility of antibody production in plants for human therapeutic use, *Curr. Top. Microbiol. Immunol.*, 1999, **240**, 119-138.
63. PARMENTER, D.L., Production of Hirudin in Plant Seeds, Ph.D. Thesis, 1998, University of Calgary, Calgary.
64. HAQ, T.A., MASON, H.S., CLEMENTS, J.D., ARNTZEN, C.J., Oral immunization with a recombinant bacterial antigen produced in transgenic plants, *Science*, 1995, **268**, 714-716.
65. MCGARVEY, P.B., HAMMOND, J., DIENELT, M.M., HOOPER, D.C., FU, Z.F., DIETZSCHOLD, B., KOPROWSKI, H., MICHAELS, F.H., Expression of the rabies virus glycoprotein in transgenic tomatoes, *Biotechnology (N.Y.),* 1995, **13**, 1484-1487.
66. RICHTER, L., KIPP, P.B., Transgenic plants as edible vaccines. *Curr. Top. Microbiol. Immunol.*, 1999, **240**, 159-176.
67. WALMSLEY, A.M., ARNTZEN, C.J., Plants for delivery of edible vaccines, *Curr. Opin. Biotechnol.*, 2000, **11**, 126-129.
68. PEN, J., VERWOERD, T.C., PARIDON, P.A., BEUDEKER, R.F., VAN DEN ELZEN, P.J.M., GEERSE, K., VAN DER KLIS, J.D., VERSTEEGH, H.A.J., VANOOYEN, A.J.J., HOEKEMA, A., Phytase-containing transgenic seeds as a novel feed additive for improved phosphorous utilization., *Bio/Technol.*, 1993, **11**, 811-814.
69. MASON, H.S., BALL, J.M., SHI, J.J., JIANG, X., ESTES, M.K., ARNTZEN, C.J., Expression of Norwalk virus capsid protein in transgenic tobacco and potato and its oral immunogenicity in mice, *Proc. Natl. Acad. Sci. U.S.A.,* 1996, **93**, 5335-5340.
70. MASON, H.S., HAQ, T.A., CLEMENTS, J.D., ARNTZEN, C.J., Edible vaccine protects mice against *Escherichia coli* heat-labile enterotoxin (LT): potatoes expressing a synthetic LT-B gene, *Vaccine*, 1998, **16**, 1336-1343.
71. ARAKAWA, T., CHONG, D.K., MERRITT, J.L., LANGRIDGE, W.H., Expression of cholera toxin B subunit oligomers in transgenic potato plants, *Transgenic Res.*, 1997, **6**, 403-413.
72. ARAKAWA, T., YU, J., CHONG, D.K., HOUGH, J., ENGEN, P.C., LANGRIDGE, W.H., A plant-based cholera toxin B subunit-insulin fusion protein protects against the development of autoimmune diabetes, *Nat. Biotechnol.*, 1998, **16**, 934-938.
73. BRENNAN, F.R., JONES, T.D., GILLELAND, L.B., BELLABY, T., XU, F., NORTH, P.C., THOMPSON, A., STACZEK, J., LIN, T., JOHNSON, J.E., HAMILTON, W.D., GILLELAND, H.E., Jr., *Pseudomonas aeruginosa* outer-membrane protein F epitopes are highly immunogenic in mice when expressed on a plant virus, *Microbiology*, 1999, **145**, 211-220.

74. POTERA, C., Epicyte produces antibodies in plants, *Gen. Eng. News*, 1999, **19**, 22.
75. PORTA, C., SPALL, V.E., LOVELAND, J., JOHNSON, J.E., BARKER, P.J., LOMONOSSOFF, G.P., Development of cowpea mosaic virus as a high-yielding system for the presentation of foreign peptides, *Virology*, 1994, **202**, 949-955.
76. BRENNAN, F.R., BELLABY, T., HELLIWELL, S.M., JONES, T.D., KAMSTRUP, S., DALSGAARD, K., FLOCK, J.I., HAMILTON, W.D., Chimeric plant virus particles administered nasally or orally induce systemic and mucosal immune responses in mice, *J. Virol.*, 1999, **73**, 930-938.
77. THANAVALA, Y., YANG, Y.F., LYONS, P., MASON, H.S., ARNTZEN, C., Immunogenicity of transgenic plant-derived hepatitis B surface antigen, *Proc. Natl. Acad. Sci. U.S.A.*, 1995, **92**, 3358-3361.
78. TURPEN, T.H., REINL, S.J., CHAROENVIT, Y., HOFFMAN, S.L., FALLARME, V., GRILL, L.K., Malarial epitopes expressed on the surface of recombinant tobacco mosaic virus, *Biotechnology (N.Y.)*, 1995, **13**, 53-57.
79. VERCH, T., YUSIBOV, V., KOPROWSKI, H., Expression and assembly of a full-length monoclonal antibody in plants using a plant virus vector, *J. Immunol. Methods*, 1998, **220**, 69-75.
80. MCCORMICK, A.A., KUMAGAI, M.H., HANLEY, K., TURPEN, T.H., HAKIM, I., GRILL, L.K., TUSE, D., LEVY, S., LEVY, R., Rapid production of specific vaccines for lymphoma by expression of the tumor-derived single-chain Fv epitopes in tobacco plants, *Proc. Natl. Acad. Sci. U.S.A.*, 1999, **96**, 703-708.
81. MODELSKA, A., DIETZSCHOLD, B., SLEYSH, N., FU, Z.F., STEPLEWSKI, K., HOOPER, D.C., KOPROWSKI, H., YUSIBOV, V., Immunization against rabies with plant-derived antigen, *Proc. Natl. Acad. Sci. U.S.A.*, 1998, **95**, 2481-2485.
82. DALSGAARD, K., UTTENTHAL, A., JONES, T.D., XU, F., MERRYWEATHER, A., HAMILTON, W.D., LANGEVELD, J.P., BOSHUIZEN, R.S., KAMSTRUP, S., LOMONOSSOFF, G.P., PORTA, C., VELA, C., CASAL, J.I., MELOEN, R.H., RODGERS, P.B., Plant-derived vaccine protects target animals against a viral disease, *Nat. Biotechnol.*, 1997, **15**, 248-252.
83. RAKAWA, T., CHONG, D.K., LANGRIDGE, W.H., Efficacy of a food plant-based oral cholera toxin B subunit vaccine [published erratum appears in *Nat Biotechnol*, 1998, **16**, 478], *Nat. Biotechnol.*, 1998, **16**, 292-297.
84. CARRILLO, C., WIGDOROVITZ, A., OLIVEROS, J.C., ZAMORANO, P.I., SADIR, A.M., GOMEZ, N., SALINAS, J., ESCRIBANO, J.M., BORCA, M.V., Protective immune response to foot-and-mouth disease virus with VP1 expressed in transgenic plants, *J. Virol.*, 1998, **72**, 1688-1690.
85. YUSIBOV, V., MODELSKA, A., STEPLEWSKI, K., AGADJANYAN, M., WEINER, D., HOOPER, D.C., KOPROWSKI, H., Antigens produced in plants by infection with chimeric plant viruses immunize against rabies virus and HIV-1, *Proc. Natl. Acad. Sci. U.S.A.*, 1997, **94**, 5784-5788.
86. MA, J.K., HIATT, A., HEIN, M., VINE, N.D., WANG, F., STABILA, P., VAN DOLLEWEERD, C., MOSTOV, K., LEHNER, T., Generation and assembly of secretory antibodies in plants, *Science*, 1995, **268**, 716-719.

87. DE NEVE, M., DE LOOSE, M., JACOBS, A., VAN HOUDT, H., KALUZA, B., WEIDLE, U., VAN MONTAGU, M., DEPICKER, A., Assembly of an antibody and its derived antibody fragment in *Nicotiana* and *Arabidopsis*, *Transgenic Res.*, 1993, **2**, 227-237.
88. TAVLADORAKI, P., BENVENUTO, E., TRINCA, S., DE MARTINIS, D., CATTANEO, A., GALEFFI, P., Transgenic plants expressing a functional single-chain Fv antibody are specifically protected from virus attack, *Nature*, 1993, **366**, 469-472.
89. FECKER, L.F., KAUFMANN, A., COMMANDEUR, U., COMMANDEUR, J., KOENIG, R., BURGERMEISTER, W., Expression of single-chain antibody fragments (scFv) specific for beet necrotic yellow vein virus coat protein or 25 kDa protein in *Escherichia coli* and *Nicotiana benthamiana*, *Plant Mol. Biol.*, 1996, **32**, 979-986.
90. FIEDLER, U., CONRAD, U., High-level production and long-term storage of engineered antibodies in transgenic tobacco seeds, *Biotechnology (N.Y.)*, 1995, **13**, 1090-1093.
91. HIGO, K., SAITO, Y., HIGO, H., Expression of a chemically synthesized gene for human epidermal growth factor under the control of cauliflower mosaic virus 35S promoter in transgenic tobacco, *Biosci. Biotechnol. Biochem.*, 1993, **57**, 1477-1481.
92. MATSUMOTO, S., IKURA, K., UEDA, M., SASAKI, R., Characterization of a human glycoprotein (erythropoietin) produced in cultured tobacco cells, *Plant Mol. Biol.*, 1995, **27**, 1163-1172.
93. CRAMER, C.L., WEISSENBORN, D.L., OISHI, K.K., RADIN, D.N., High-level production of enzymatically activehuman lysosomal proteins in transgenic plants, *in* Transgenic Plants: A Production System for Industrial and Pharmaceutical Proteins (M. R. L. Owen and J. Pen eds.), John Wiley & Sons, Chichester, 1996, pp. 299-310.
94. HUANG, A.H., Oleosins and oil bodies in seeds and other organs, *Plant Physiol.*, 1996, **110**, 1055-1061.
95. TZEN, J.T., LIE, G.C., HUANG, A.H., Characterization of the charged components and their topology on the surface of plant seed oil bodies, *J. Biol. Chem.*, 1992, **267**, 15626-15634.
96. HILLS, M.J., WATSON, M.D., MURPHY, D.J., Targeting of oleosins to the oil bodies of oilseed rape (*Brassica napus* L.), *Planta*, 1993, **189**, 24-29.
97. ABELL, B.M., HOLBROOK, L.A., ABENES, M., MURPHY, D.J., HILLS, M.J., MOLONEY, M.M., Role of the proline knot motif in oleosin endoplasmic reticulum topology and oil body targeting, *Plant Cell*, 1997, **9**, 1481-1493.
98. MURPHY, D.J., Storage lipid bodies in plants and other organisms, *Prog. Lipid. Res.*, 1990, **29**, 299-324.
99. MAHMOUD, S.S., Production of rcGH in *B. napus* Seed, Ph.D. Thesis, 1999, University of Calgary.
100. ALCANTARA, J., HUTCHINS, W., YU, R.-H., SCHRYVERS, A.B., MOLONEY, M.M., Evaluation of plant oilbodies as a vaccine delivery system, Paper read at 6th International Congress of Plant Molecular Biology, June 18-24, 2000, at Quebec City.
101. KUHNEL, B., ALCANTARA, J., VAN ROOIJEN, G.J., BOOTHE, J., MOLONEY, M., A novel method for cleavage of recombinant fusion proteins. Paper read at 6th International Congress of Plant Molecular Biology, June 18-24, 2000, at Quebec City.

102.GUAN, H., MCKEAN, A., KEELING, P., Starch encapsulation technology as a vehicle for producing recombinant proteins in plants, Paper read at 6th International Congress of Plant Molecular Biology, June 18-24, 2000, at Quebec City.
103.SHPIGEL, E., Recombinant protein purification using CBDS, Paper read at 219th ACS National Meeting: Division of Biochemical Technology, March 26-30, 2000, at San Francisco.

Chapter Four

EXPRESSION OF FOREIGN SEQUENCES IN VIRUS-INFECTED PLANTS USING ALFALFA MOSAIC VIRUS COAT PROTEIN

Vidadi Yusibov*, Nina Fleysh, Sergei Spitsin, Tatiana Mikheeva, and Hilary Koprowski

Biotechnology Foundation Laboratories
Thomas Jefferson University
Philadelphia, PA 19107, USA

**Author to whom correspondence should be addressed.*

INTRODUCTION

Progress in molecular biology has greatly enhanced the ability to characterize the molecular bases of biological processes, particularly in identifying proteins associated with various diseases. Many of these proteins have been produced in heterologous expression systems, such as *Esherichia coli*, yeast, animal cell cultures, and transgenic animals. Bacterial expression systems are relatively easy to manipulate, and the yield of the product is high. However, many human or animal proteins require extensive posttranslational modification for functional activity, which is a limiting factor in bacterial expression systems. Cell cultures (mammalian, human, and insect) are convenient for the production of complex proteins. However, long lead time, low recovery of the product, possible pathogen transfer, high capital and production costs present serious concerns. A relatively new expression system, transgenic animals, may provide an unlimited supply of complex proteins. The limitation of this system is the long cycle time for new and improved products in addition to pathogen transfer. In light of safety concerns, high production costs, and long lead time associated with existing protein production systems, new approaches are needed for the safe and inexpensive production of biologically active proteins.

Plant systems expressing foreign proteins represent a particularly promising avenue since they have several advantages over recombinant microorganisms, human fluids, animal cell lines, or transgenic animals as a source of bioactive products. For example, plants are free of contamination with animal pathogens, relatively easy to manipulate genetically, inexpensive, and use eukaryotic protein modification machinery. Edible plant tissue is a potential vehicle for oral delivery of vaccine antigens. Thus, genetically engineered plants and plant viruses can serve as "green factories" to produce specific proteins of interest, including vaccine antigens and antibodies.

A variety of proteins of mammalian origin have been expressed in transgenic plants, including vaccine antigens,[1-6] a diabetes-associated autoantigen[7] (glutamic acid decarboxylase), the anticoagulant hirudin,[8] glucocerebrosidase,[9] and epidermal growth factor.[10] Each of these transgenic plant-produced proteins was shown to have functional activity, and the plant-produced vaccine antigens induced serum IgG specific for the original viral antigen in immunized mice. Oral administration of Norwalk virus capsid protein to human volunteers through ingestion of transgenic potato tubers induced production of mucosal IgA.[11] More importantly, feeding of unprocessed transgenic lettuce plants expressing hepatitis B surface antigen to human volunteers resulted in significant levels of antigen-specific antibody in sera.[2] Clearly, antigens expressed in plants and delivered orally can elicit an antibody response, suggesting the promise of vaccines based on antigens produced in plants.

In recent years, plant RNA viruses have attracted increasing interest from the scientific community as potential tools for producing foreign proteins in plants. Plant viruses have certain advantages that make them attractive as vectors for foreign protein expression. The molecular biology and pathology of a number of plant RNA viruses are well-characterized, and there is considerable knowledge of virus biology, genetics, and regulatory sequences. Most plant RNA viruses have small genomes, and infectious cDNA clones are available to facilitate genetic manipulation. Once the infectious virus material enters the susceptible host cell, it replicates to high levels and spreads rapidly (10-20 days post inoculation) throughout the entire plant. Virus particles are easily and, therefore, economically recovered from infected plant tissue. Viruses have a wide host range, enabling use of a single construct for infection of several susceptible plant species. Finally, the available information on regulatory elements enables engineering of virus vectors for minimal effect on infected plants.

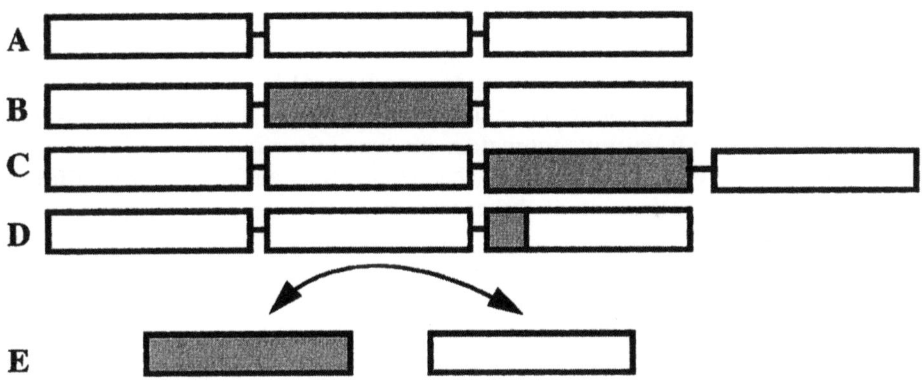

Figure 4.1: Schematic representation of different strategies for foreign gene expression using plant virus-based vectors. (A) Simplified schematic representation of plant virus genome. (B) Expression of foreign gene by replacing one of the viral genes (shaded area). (C) Expression by insertion of foreign gene into the virus genome. (D) Expression of foreign sequences as fusions with structural proteins of virus and (E) functional complementation.

Figure 4.1 schematizes different strategies to express foreign genes by using plant viruses. Foreign sequences can be expressed by replacing one of the viral genes with the desired sequence, by inserting foreign sequences into virus genomes at an appropriate position, or by fusing foreign peptides to the structural proteins of viruses. Moreover, any of these approaches can be combined to express foreign sequences by complementing vital functions of viruses from another construct. Here, we focus on different strategies that use tobacco mosaic virus (TMV) and alfalfa mosaic virus (AlMV), particularly the AlMV coat protein (CP), as tools to express foreign sequences in virus-infected plants.

ALFALFA MOSAIC VIRUS AND TOBACCO MOSAIC VIRUS

AlMV was first described by Weimer[12] in alfalfa plants. The virus belongs to the genus *Ilarvirus* and is a member of the *Bromoviridae* family. The genome of AlMV consists of three genomic RNAs (RNAs 1-3) and a subgenomic RNA (RNA4) (Fig. 4.2). Genomic RNAs 1 and 2 encode virus replicase proteins P1 and P2, respectively. Genomic RNA3 encodes the cell-to-cell movement protein P3 and the CP. The CP is translated from subgenomic RNA4, which is synthesized from genomic RNA3 and is required for the start of infection. Studies have demonstrated the involvement of the CP in multiple functions, including genome activation, replication, RNA stability, symptom formation, and RNA encapsidation.[13-24] Encapsidation is essential for the long distance movement of viruses from inoculated to uninoculated leaves and for systemic infection. Virions of AlMV are encapsidated by a unique CP (24 kD), forming more than one type of particle. The size (30- to 60-nm in length and 18 nm in diameter) and shape (spherical, ellipsoidal, or bacilliform) of particles depend on the size of the encapsidated RNA. Upon assembly, the N-terminus of the AlMV CP is thought to be localized on the surface of the virus particles and does not appear to interfere with virus assembly.[25] We previously showed that the AlMV CP with an additional 38-amino acid peptide at its N-terminus forms particles *in vitro* and retains biological activity.[26]

AlMV has a wide host range, which includes a number of agriculturally valuable crop plants. Together, these characteristics make the AlMV CP an excellent candidate as a carrier molecule and AlMV an attractive candidate vector for the expression of foreign sequences. Moreover, upon expression from heterologous vectors such as TMV, the AlMV CP is capable of forming virus particles without interfering with virus infectivity.[27] This allows the use of TMV as a carrier virus for AlMV CP fused to foreign sequences. TMV, a member of the *Tobamovirus* group, has a genome consisting of a single plus-sense RNA (Fig. 4.2) encapsidated with a 17.0 kD CP, which results in rod-shaped particles (300 nm in length). The CP is the only structural protein of TMV and is required for encapsidation and long distance movement of the virus in an infected host.[28] The 183 and 126 kD proteins are translated from genomic RNA and are required for virus replication.[29] The 30 kD is a cell-to-cell movement protein of virus.[30] Movement and coat proteins are translated from subgenomic mRNAs.[31-33]

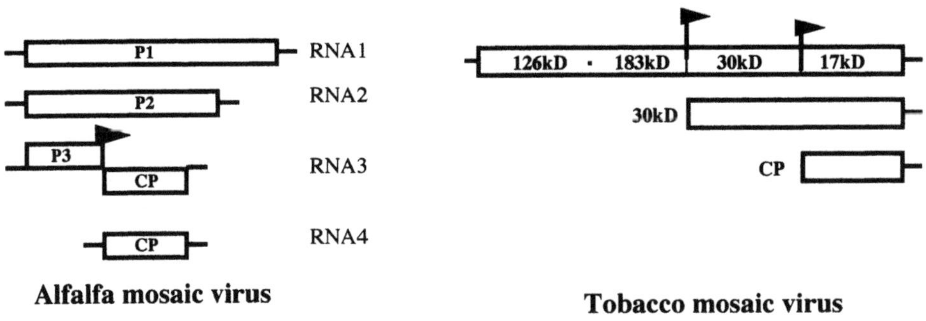

Figure 4.2: Schematic representation of AlMV and TMV genomes. RNAs1 and 2 of AlMV encode replicase proteins P1 and P2, respectively; genomic RNA3 encodes cell-to-cell movement protein P3 and the viral coat protein (CP). The CP is translated from subgenomic RNA4 synthesized from genomic RNA3. The 126 kD and 183 kD proteins of TMV are required for replication; the 30 kD protein is the viral cell-to-cell movement protein; the 17 kD is the CP of virus. The CP and the 30 kD protein are translated from subgenomic RNA. Arrows indicate position of subgenomic promoters.

EXPRESSION OF FOREIGN SEQUENCES AS TRANSLATIONAL FUSION PRODUCTS WITH ALMV CP

Studies of human and animal diseases and their causative agents have led to the identification of gene sequences that can be used for prevention or treatment of disease. Frequently, these peptides or proteins are expressed in heterologous systems as translational fusions by using different tags such as poly-His or carrier molecules to facilitate purification. Generally used carrier molecules are the capsid proteins of bacterial,[34-36] mammalian,[37-39] or plant viruses.[27, 40-50] Upon expression, these carrier molecules, along with the fused peptides, self-assemble into virus-like particles that are easily purified. Coat proteins of several plant viruses, including TMV,[40-43] AlMV,[27, 48, 49] cowpea mosaic virus,[44-47] and tomato bushy stunt virus[50] have been used successfully as carrier molecules. Some of the antigens expressed with plant virus vectors exhibited high immunogenicity and induced pathogen-specific antibody production in immunized mice.[41-42] By using cowpea mosaic virus, Dalsgaard *et al.*,[47] induced protective immunity in mink against mink enteritis virus.

The CP of many plant viruses is involved in encapsidation, which is critical for generating transmissible infectious virus. When foreign sequences are incorporated into the virus genome as translational fusions with the CP, faulty particle assembly and a resulting inhibition of virus spread throughout the plant are commonly observed. Peptides less than 25 amino acids in length have been expressed successfully by using the CP of TMV or tomato bushy stunt virus, whereas expression of larger peptides as translational fusions without interfering with viral infectivity has been problematic. This is a significant limitation for the production of various molecules of biomedical importance. For example, a chimeric peptide antigen that successfully protects mice against rabies virus and is, therefore, suitable for use in plant-based vaccine development[51] is 38 amino acids in length, far in excess of the capacity of the TMV, cowpea mosaic virus, or tomato bushy stunt virus CP. Our study of *in vitro* assembly of *Escherichia coli*-expressed AlMV CP containing a 38-amino acid tag at the N-terminus[26] suggested that AlMV CP can accommodate larger peptides as compared to the CP of TMV or tomato bushy stunt virus. Indeed, by using AlMV CP as a carrier molecule, we were able to express a variety of peptides from different sources as translational fusions with CP in virus-infected plants (Table 4.1). The AlMV CP can assemble into particles independent of carrier virus upon expression from a heterologous virus vector such as TMV, and this might avoid potential problems with replication, movement, or assembly of recombinant virus resulting from incorporation of foreign sequences. Thus, we used TMV-based 30B[52] and Av[53] vectors in addition to the AlMV-based expression system to produce recombinant AlMV CP fused with foreign peptides.

30B

Because the CP of many plant viruses is essential for long distance movement, assembly, and viral spread, foreign sequences incorporated into the CP may result in virus defective in these functions. Therefore, we used a hybrid virus expressing two CPs from two different subgenomic promoters. Vector 30B (a gift from Dr. William Dawson, University of Florida) is based on the U1 strain TMV genome. In this vector, the open reading frame of U1 strain TMV CP is replaced with that of U5 strain TMV including the upstream subgenomic promoter sequences, thus creating a vector with 2 subgenomic promoters. Multiple cloning sites are introduced for subcloning genes under the control of the U1 strain TMV CP subgenomic promoter, in addition to a ribozyme placed at the 3' end of the virus for self-cleavage of in vitro RNA transcripts. Thus, foreign genes can be incorporated by using multiple cloning sites and translated from subgenomic RNA (Fig. 4.3). The ability of heterologously expressed AlMV CP to assemble into particles makes it more attractive for expressing larger peptides compared to other plant virus systems where foreign peptides are fused with the only CP that is essential for assembly and long distance movement. Use of a second CP to serve as a carrier molecule may prevent interference of larger (>30 amino acid) foreign peptides with both virus assembly

Table 4.1: Peptides expressed as fusions with AlMV CP using plant virus vectors 30B, Av or AlMV.

PEPTIDE	SOURCE	SIZE (amino acids)	EXPRESSION AND ASSEMBLY		
			30B	Av	AlMV
V3 loop	HIV-1	35	Yes	Yes	Yes
Vpu	HIV-1	83	Yes	No	No
Vpr	HIV-1	93	Yes	N.D.**	N.D.
RSV* G	RSV G protein	24	N.D.	N.D	Yes
Sand	Sandostatin	10	N.D.	N.D.	Yes
Rg24	Rabies G protein	40	Yes	Yes	Yes
NV10c	Rabies N protein	40	Yes	Yes	Yes
GMC-SF***	Human	145	Yes	No	No

*RSV- respiratory syncytial virus
**N.D.- not done
***GMC-SF- human granulocyte macrophage colony-stimulating factor

and systemic infection. Indeed, several peptides, such as a 40-amino acid chimeric epitope from rabies virus,[27] the 35-amino acid V3 loop of human immunodeficiency virus[27] (HIV), and the 83- and 93-amino acid Vpu and Vpr proteins of HIV, respectively (Yusibov, unpublished data) were successfully expressed by using recombinant AlMV CP fused with foreign peptides and incorporated into the 30B vector under the control of the U1-TMV CP subgenomic promoter (Fig. 4.3). To date, the largest protein we have been able to express and recover as a fusion product with AlMV CP by using 30B vector is the 145-amino acid human granulocyte macrophage colony-stimulating factor (GMC-SF) (Fig. 4.4). Note that expression of human GMC-SF as a fusion with AlMV CP in systemically infected plant tissue with either a hybrid Av vector or an AlMV complementation system was unsuccessful (see *Av* and *AlMV* below). Peptides expressed as fusions with CP, by using 30B were recovered by simple virus purification. However, only 10-20% of purified viral protein (100-200 ug/mg of total virus) consists of recombinant AlMV CP, which may not be a sufficient quantity for manufacturing purposes. To increase the yield of recombinant protein, we adapted another vector, the TMV-based Av vector (a gift from Dr. William Dawson, University of Florida), which expresses

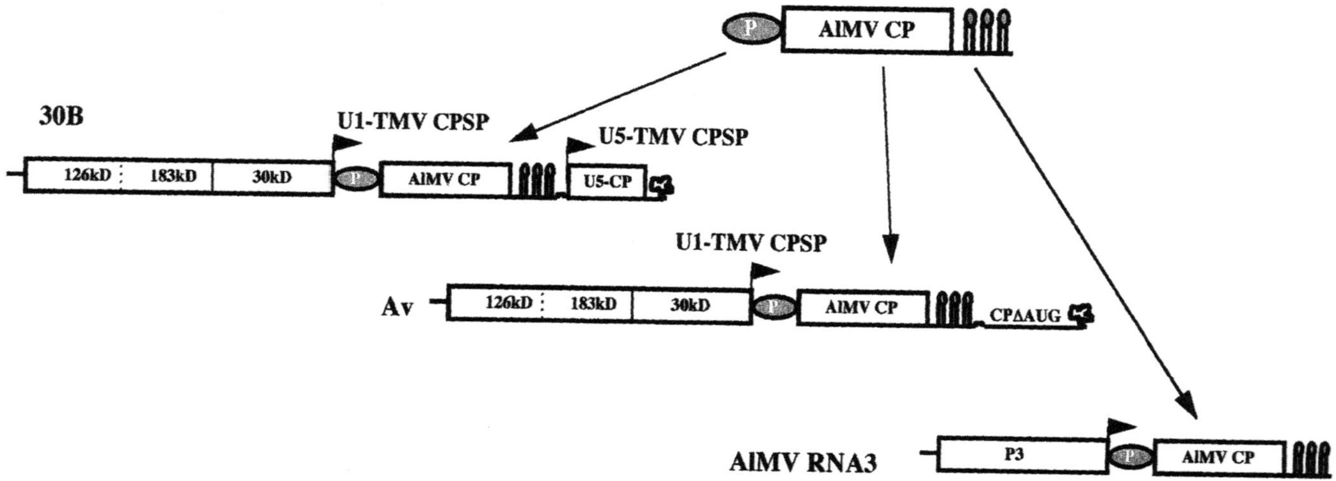

Figure 4.3: Schematic representation of strategies for cloning and expression of AlMV CP fused with foreign sequences. 30B- recombinant protein expression is controlled by the U1-TMV CP subgenomic promoter (U1-TMV CP SP). The systemic spread of 30B in infected plants is supported by expression of the second CP (U5-CP) which is controlled by U5-TMV CP subgenomic promoter (U5-TMV CP SP). Av- recombinant protein expression is controlled by the U1-TMV CP subgenomic promoter (U1-TMV CP SP). Systemic spread of this virus is supported by the recombinant AlMV CP. AlMV-recombinant CP is translated from subgenomic RNA4 which is synthesized from genomic RNA3. The P3 is cell-to-cell movement protein of AlMV.

AlMV CP as the unique coat protein. This was based on our earlier demonstration that AlMV CP can encapsidate TMV genomic RNA, establish systemic infection, and generate infectious virus particles.[53]

Av

The TMV-derived vector was constructed such that the translation start codon ATG of TMV CP was replaced with AGA, and multiple cloning sites were introduced 42 nucleotides downstream of the mutated ATG codon. Av (Fig. 4.3) contains the full-length TMV genome but is defective in CP production, limiting infection to locally inoculated leaves. However, a chimeric Av expressing the AlMV CP (Av/A4) was highly infectious and resulted in high-level accumulation of AlMV CP (0.8-1.0 mg/g fresh tissue) in *Nicotiana benthamiana* plants, suggesting potential for producing high levels of recombinant AlMV CP fused to a foreign peptide. In fact, inoculation of plants with Av engineered to contain recombinant AlMV CP fused with a chimeric epitope from rabies virus (Av/A4g24) or the V3 loop (Av/A4V3) of HIV resulted in systemic infection and accumulation of easily recoverable virus particles in infected plants (Fig. 4.5). Compared to Av/A4, which establishes systemic infection in *N. benthamiana* plants 6-7 days after inoculation, Av/A4g24 and Av/A4V3 moved into the upper uninoculated leaves noticeably late (15-18 days after inoculation), and plants displayed characteristic symptoms (curling and yellowing of leaves with occasional vain necrosis). The delay in virus movement, however, did not have a substantial effect on virus accumulation. Infection of plants with hybrid Av (Av/A4g24 and Av/A4V3) resulted in a higher accumulation of recombinant AlMV CP (0.5-0.7 mg/g fresh tissue) compared to that with 30B (0.1-0.2 mg/g fresh tissue). Unlike Av/A4g24 and Av/A4V3, an Av construct containing AlMV CP fused with the HIV Vpu protein (Av/A4Vpu) did not establish systemic infection; movement of this virus from locally inoculated leaves resulted in severe necrosis and death of leaf petioles and stem tissue. On the other hand, the 30Bvpu construct expressing AlMV CP fused with HIV Vpu moved from locally inoculated into upper uninoculated leaves and resulted in systemic infection and recombinant virus accumulation (Steplewski K., *et al.*, unpublished data), suggesting the advantage of a double CP system for expressing large peptides. Despite the size limitation, our data demonstrate that hybrid Av expressing recombinant AlMV CP fused with foreign peptides is an efficient vector for producing a variety of peptides. An additional advantage is the presence of unique CP expressed from this hybrid vector, which provides 100% of recombinant protein carrying the foreign peptide. Because of the characteristics of the AlMV CP, this hybrid vector can accommodate larger peptides (34-40 amino acids) compared to TMV-based vectors, where foreign sequences are fused with TMV CP. Although we were unsuccessful in expressing recombinant AlMV CP fused with HIV Vpu (83 amino acid protein) or the 145-amino acid human GMC-SF in Av, this vector is excellent for the production of foreign peptides. Since the recombinant AlMV CP fused with HIV Vpu was detectable in locally inoculated leaves and protoplasts (our unpublished data), the impairment of viral systemic infection might be related to the active interference of Vpu with cell membranes, a known function of Vpu in HIV-infected mammalian cells. Thus, Av might still be useful in expressing peptides larger than 40 amino acids. This hybrid virus is also useful for the production of full-length proteins by functional complementation.

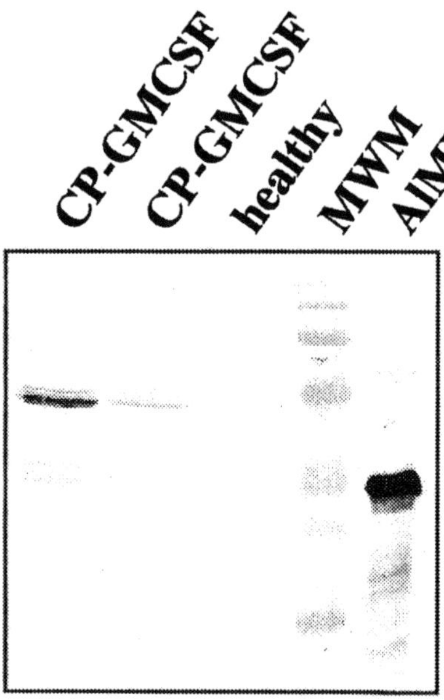

Figure 4.4: Western blot showing accumulation of recombinant AlMV CP fused to human GM-CSF in tissue from systemically infected leaves of *Nicotiana benthamiana* plants. Leaves were inoculated with transcripts of recombinant 30B-CP-GMCSF. Proteins were separated by electrophoresis on a 12% SDS-polyacrylamide gel and detected with monoclonal antibodies specific for the AlMV CP. Antibodies reacted with the expected size recombinant protein (41.0 kD) and with wild-type AlMV CP, used as control (24 kDa). MWM, molecular weight markers.

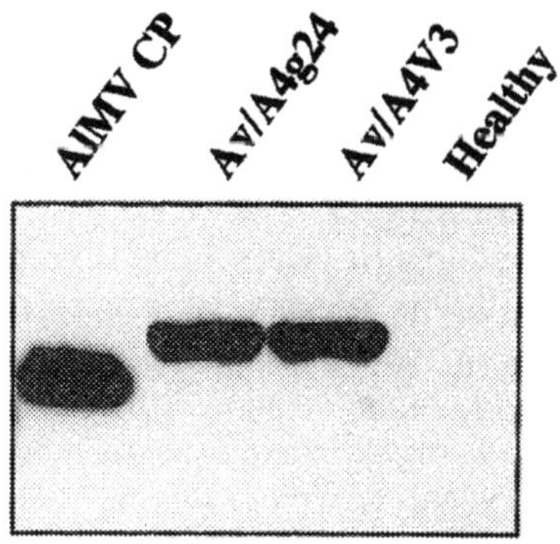

Figure 4.5: Western analysis of recombinant Av/A4 particles containing epitopes from rabies virus glycoprotein (Av/A4g24) and HIV gp120 protein (Av/A4V3). Recombinant Av/A4g24 and Av/A4V3 particles were recovered from systemically infected leaves of *Nicotiana benthamiana* plants. Proteins were separated by electrophoresis on a 12% SDS-polyacrylamide gel and detected with monoclonal antibodies specific for the AlMV CP. Antibodies reacted with expected size recombinant proteins (28.9 kD). Wild type AlMV coat protein (24 kDa) was used as control.

AlMV

A third strategy for the expression of foreign peptides is the system based on AlMV in which peptides of interest are translationally fused to the N-terminus of AlMV CP and placed into full-length infectious AlMV RNA3 (Fig. 4.3). Full-length infectious RNA3 containing foreign sequences can then be mixed with viral RNAs 1, 2, and 4 for inoculation of susceptible host plants such as tobacco, lettuce, spinach, alfalfa, and soybean. Like the constructs with 30B and Av, the resulting virus is infectious and has the potential for transmission to other cultivars, raising safety concerns about genetically engineered organisms. One approach to overcoming these concerns is to engineer the system such that virus infection is limited to a particular host and transmission of virus to other plant species is impaired by introduced mutations. Such a system is available for AlMV.[54] The viral replicase genes P1 and P2 were incorporated into *Nicotiana tabacum* cv. Samsun NN plants to obtain transgenic plants (P12) expressing the AlMV replicase complex.[54] Before incorporation into transgenic plants, the virus replicase genes were modified so that the resulting RNA is not recognized by AlMV replicase. Thus, transgenically generated virus RNAs 1 and 2 are not replicated by the products of these genes, whereas the transgenically expressed replicase efficiently replicates AlMV RNA3, resulting in infection of the entire plant. Infection of P12 plants with AlMV RNA3 yields virus impaired in transmission to other plant species susceptible to AlMV, creating an excellent biocontainment system. Thus, we used a functional complementation system with transgenic P12 plants (the source of virus replicase) and recombinant RNA3 (the source of cell-to-cell movement protein P3 and CP) to express foreign sequences. With this strategy, we successfully expressed a 40-amino acid peptide from rabies virus glycoprotein (our unpublished results), different domains of HIV gp120 (30-60 amino acids, Spitsin *et al.*, unpublished results), sandostatin (8 amino acids, our unpublished results), and a 24-amino acid peptide from the respiratory syncytial virus (RSV) G protein.[49] Mice immunized parenterally with recombinant AlMV particles containing RSV G peptide were protected against challenge infection with RSV Long strain. Immunization of mice with recombinant AlMV particles containing peptide from rabies virus G protein induced neutralizing antibody production (unpublished results). In addition to providing a valuable biocontainment system, P12 plants support high levels of recombinant RNA3 replication, resulting in the accumulation of significant quantities of chimeric CP (0.5-1.0 mg/g fresh tissue) that is assembled into virus particles and easily recovered. More importantly, immunization of mice with recombinant virus particles generated from these constructs induced pathogen-specific antibody production. Thus, this system has great potential for manufacturing large quantities of foreign peptides in a cost-effective and environmentally safe manner.

FULL-LENGTH PROTEIN EXPRESSION

Expression and inexpensive recovery of peptides with adequate biological activity is important for different applications, including development of subunit vaccines. Some applications, however, require full-length, biologically active proteins. Like conventional expression systems, plant virus vectors also can be used to produce full-length proteins, including enzymes, blood component substitutes, and antibodies. By using tobamoviral vectors, McCormick *et al.*[55] and Kumagai *et al.*[56] expressed high levels of idiotype-specific single-chain Fv fragment of the immunoglobulin from the 38C13 mouse B cell lymphoma and rice alpha-amylase, respectively. Verch *et al.*[57] expressed a full-length monoclonal antibody specific for colorectal-cancer associated antigen in *N. benthamiana* plants co-inoculated with two 30B (TMV-based vector) constructs containing heavy and light chains, respectively. Thus, plant virus vectors have a demonstrated ability to express short peptides as well as large complex proteins. To express full-length proteins in virus-infected plants, we used a functional complementation approach.

Functional Complementation System

We used a system where foreign proteins are produced by co-inoculating plants with 2 complementing virus constructs (virus-virus complementation). In virus-virus complementation, the vector consists of a component that is functional in CP production and systemic movement (Av/A4), and a component that is functional in foreign protein expression and depends on the first component for systemic movement (Fig. 4.6). This was achieved by using the TMV-based Av vector.[53] Av lacks functional TMV CP, so that infection is limited to locally inoculated leaves. However, Av/A4, containing the cloned AlMV CP expressed as the unique CP, is highly effective in long distance movement of virus and systemic infection. We hypothesized that co-inoculation with a mixture of two hybrid constructs, AvA4 (source of AlMV CP) and Av/YFG (source of YFG-your favorite gene), might be used for expression of foreign proteins in systemically infected leaves. While infection of plants with Av/YFG alone would be limited to locally inoculated leaves, due to the lack of CP, co-inoculation with Av/A4 would provide CP to encapsidate and move not only Av/A4 but also Av/YFG into the upper uninoculated leaves, thereby supporting YFG expression in the whole plant. In fact, *N. benthamiana* plants co-inoculated with a mixture of *in vitro* synthesized transcripts of Av/A4 and Av/GFP (Fig. 4.6) expressing green fluorescent protein (GFP) from jellyfish (DNA plasmid of Av/GFP was kindly provided by Dr. Shailaga Shivprasad, University of Florida) revealed both AlMV CP and GFP by using antibodies specific for each protein in systemically infected leaves at 14 days post-inoculation (Fig. 4.7). However, during plant-to-plant passages, theamount of Av/GFP in the infected tissue gradually decreased, and after the third transfer, only Av/A4 was detectable (our unpublished results). This may be an advantage from an environmental safety point of view. With this approach, we could express an average of 50 ug/g fresh tissue of GFP. On the other hand, the use of wild-type TMV for the production of foreign proteins by co-inoculation has been unsuccessful, since the construct containing the foreign protein was outcompeted by wild-type TMV. An important component of this system, AlMV CP, is unique in its ability to encapsidate the genomic RNAs of unrelated viruses into infectious particles in the infected host. This unique ability of AlMV CP can be exploited to engineer hybrid vectors that specifically target selected crop species.

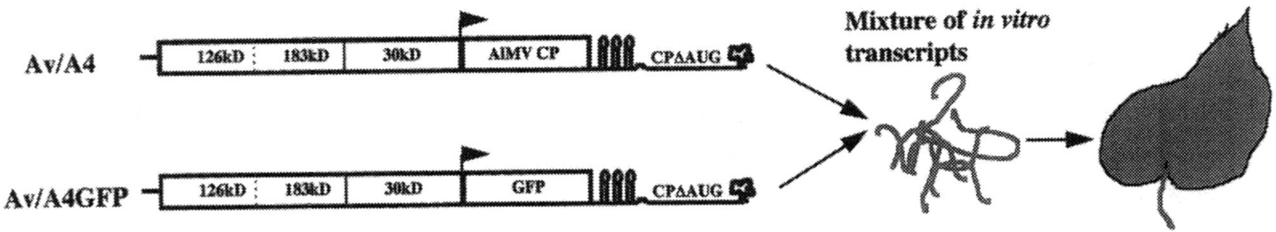

Figure 4.6: Schematic representation of virus-virus complementation for producing foreign proteins. The first component of this system is Av/A4, which expresses AlMV CP and supports systemic spread of virus. The second component is Av/A4GFP, which expresses foreign protein and depends on the Av/A4 component for systemic movement. To express the GFP in systemically infected tissue, *in vitro* synthesized transcripts of Av/A4 and Av/A4GFP were mixed and inoculated into *Nicotiana benthamiana* plants.

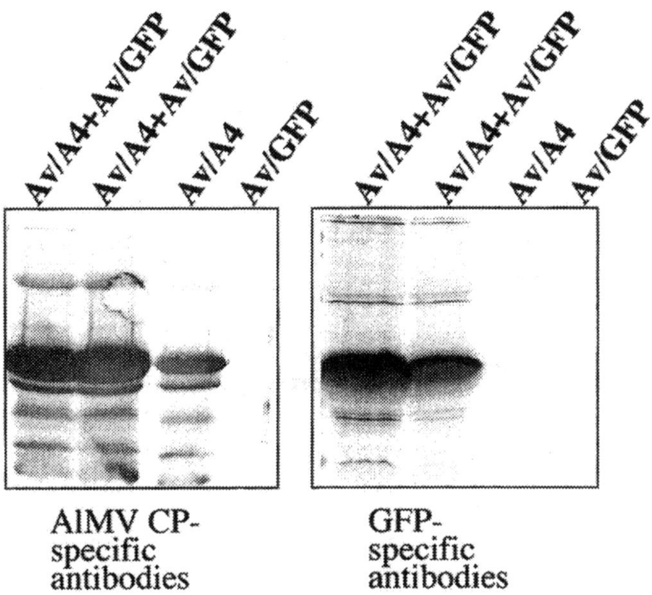

Figure 4.7: Western analysis of expression of recombinant GFP in *Nicotiana benthamiana* plants inoculated with Av/A4 and Av/A4GFP. Protein extracts from systemically infected leaves were separated by electrophoresis on a 12% SDS-polyacrylamide gel, transferred to a nitrocellulose membrane and reacted with protein-specific antibodies. AlMV CP-specific antibodies recognized the expected size protein (24.0 kD) in plants inoculated with Av/A4 or with the mixture of Av/A4 and Av/A4GFP. GFP-specific antibodies for recognized protein only in extracts from plants inoculated with the mixture of Av/A4 and Av/A4GFP. GFP-specific antibodies did not react with any proteins in plants inoculated with only Av/A4. Neither AlMV CP- nor GFP-specific antibodies reacted with any protein in extracts from plants inoculated with Av/GFP only, suggesting the lack of systemic movement.

SUMMARY

Since the first infectious cDNA clone of a plant RNA virus (brome mosaic virus) obtained in early 1980s, infectious cDNAs of numerous plant RNA viruses have been cloned. The availability of these clones has led to significant progress in the field of molecular plant virology, resulting in the development of virus-resistant transgenic plants and plant RNA viruses engineered to express heterologous sequences. The use of virus-based vectors has greatly contributed to studies of viral pathogenesis, host-pathogen interactions, virus replication, cell-to-cell and long distance movement of virus. Although plant virus-based vectors originally emerged as tools to study different aspects of virus pathogenesis, they have since found increasing use as vectors for expression of recombinant proteins of medical importance. In the last several years, a number of plant virus vectors have been engineered and used as a safe and inexpensive system to express foreign sequences, including vaccine antigens based on short peptides. Immunological testing of these plant virus-based antigens has demonstrated great promise for the development of plant virus-based subunit vaccines that may soon enter the clinic. Plant virus vectors also have been successfully used to express enzymes that have the appropriate biological activity. Most recently, these vectors have been used to study the functional activity of individual plant genes under *in vivo* conditions. Functional studies of individual plant genes may, in turn, lead to the discovery of loci that govern traits such as resistance to pathogens and productivity. The ease of genetic manipulation and the rapidity of expression time underlines the value of plant virus vectors as tools to provide high through-output screening of multiple genes.

REFERENCES

1. MASON, H.S., LAM, D.M.-K., ARNTZEN, C.J., Expression of hepatitis B surface antigen in transgenic plants, *Proc. Natl. Acad. Sci. USA*, 1992, **89**, 11745-11749.
2. KAPUSTA, J., MODELSKA. A., FIGLEROWICZ, M., PNIEWSKI, T., LETELLIER, M., LISOWA, O., YUSIBOV, V., KOPROWSKI, H., PLUCIENNICZAK, A., LEGOCKI, A., A plant derived edible vaccine against hepatitis B virus, *FASEB J.*, 1999, **13**, 1796-1799.
3. HAQ, T.A., MASON, H., CLEMENTS, J.D., ARNTZEN, C.J., Oral immunization with a recombinant bacterial antigen produced in transgenic plants, *Science*, 1995, **268**, 714-716.
4. MCGARVEY, P.B., HAMMOND, J., DIENELT, M.M., HOOPER, D.C., FU, Z.F., DIETZSCHOLD, B., KOPROWSKI, H., MICHAELS, F.H., Expression of the rabies virus glycoprotein in transgenic tomatoes, *Bio/Technology*, 1995, **13**, 1484-1487.
5. MASON, H.S., BALL, J.M., SHI, J.J., JIANG, X., ESTES, M.K., ARNTZEN, C.J., Expression of Norwalk virus capsid protein in transgenic tobacco and potato and its oral immunogenicity in mice, *Proc. Natl. Acad. Sci. USA*, 1996, **93**, 5335-5340.
6. WELTER, L.M., MASON, H.M., LU, W., LAM. D.M.-K., WELTER, M.W., Effective immunization of piglets with transgenic potato plants expressing a truncated TGEV S protein, *in* Vaccines: New Technologies & Applications, Cambridge Healthtech Institutes, 1996.

7. MA, S., ZHAO, D., YIN, A., MUKHERJEE, R., SINGH, B., QIN, H., STILLER, C.R., JEVNIKAR, A.M., Transgenic plants expressing autoantigens fed to mice induce oral immune tolerance, *Nature Med.*, 1997, **3**, 793-796.
8. PARMENTER, D.L., BOOTHE, J.G., VAN ROIJEN, G.J.H., YEUNG, E.S., MOLONEY, M.M., Production of biologically active hirudin in plant seeds using oleosin partitioning, *Plant Mol. Biol.*, 1995, **29**, 1167-1180.
9. CRAMER, C.L., WEISSENBORN, D.L., OISHI, K.K., RADIN, D.N., High-level production of enzymatically active human lysosomal proteins in transgenic plants, *in* Transgenic Plants: A Production System for Industrial and Pharmaceutical Proteins (M.R.L. Owen and J. Pen, eds.), Wiley, Chichester, 1996, pp. 299-310.
10. HIGO, K., SAITO, Y., HIGO, K., Expression of a chemically synthesized gene for human epidermal growth factor under the control of cauliflower mosaic virus 35S promoter in transgenic tobacco, *Biosci. Biotechnol. Biochem.*, 1993, **57**, 1477-1481.
11. TACKET, C.O., MASON, H.S., LOSONSKY, G., ESTES, M., LEVINE, M.M., ARNTZEN, C.J., Human immune responses to a novel Norwalk virus vaccine delivered in transgenic potatoes, *J. Infect. Dis.*, 2000, **182**, 302-305.
12. WEIMER, J.L., Alfalfa mosaic, *Phytopathology*, 1931, **21**, 122.
13. BOL, J.F., VLOTEN-DOTING, L., JASPARS, E.M., A functional equivalence of top component a RNA and coat protein in the initiation of infection by alfalfa mosaic virus, *Virology*, 1971, **46**, 73-85.
14. VAN DER VOSSEN, E.A., NEELEMAN, L., BOL, J.F., Early and late functions of alfalfa mosaic virus coat protein can be mutated separately, *Virology*, 1994, **202**, 891-903.
15. YUSIBOV, V., LOESCH-FRIES, L.S., N-terminal basic amino acids of alfalfa mosaic virus coat protein involved in the initiation of infection, *Virology*, 1995, **208**, 405-407.
16. YUSIBOV, V., LOESCH-FRIES, L.S., Functional significance of three basic N-terminal amino acids of alfalfa mosaic virus coat protein, *Virology*, 1998, **242**, 1-5.
17. BOL, J. F., Alfalfa mosaic virus and ilarviruses: Involvement of coat protein in multiple steps of the replication cycle. (Review, 100 refs.), *J. Gen. Virol.*, 1999, **80**, 1089-1102.
18. DE GRAAFF, M., MAN IN'T VELD, M.R., JASPARS, E.M., *In vitro* evidence that the coat protein of alfalfa mosaic virus plays a direct role in the regulation of plus and minus RNA synthesis: Implications for the life cycle of alfalfa mosaic virus, *Virology*, 1995, **208**, 583-589.
19. JASPARS, E.M., Plant viruses with a multipartite genome, *Adv. Virus Res.*, 1974, **19**, 37-149.
20. LOESCH-FRIES, L.S., JARVIS, N.P., KRAHN, K.J., NELSON, S.E., HALL, T.C., Expression of alfalfa mosaic virus RNA4 cDNA transcripts *in vitro* and *in vivo*, *Virology*, 1985, **146**, 177-187.
21. NEELEMAN, L., VAN DER KUYL, A.C., BOL, J.F., Role of alfalfa mosaic virus coat protein gene in symptom formation, *Virology*, 1991, **181**, 687-693.
22. NEELEMAN, L., VAN DER VOSSEN, E.A., BOL, J.F., Infection of tobacco with alfalfa mosaic virus cDNAs sheds light on the early function of the coat protein, *Virology*, 1993, **196**, 883-887.

23. VAN DER KUYL, A.C., NEELEMAN, L., BOL, J.F., Complementation and recombination between alfalfa mosaic virus RNA3 mutants in tobacco plants, *Virology*, 1991,**183**, 731-738.
24. VAN DER KUYL, A.C., NEELEMAN, L., BOL, J.F., Role of alfalfa mosaic virus coat protein in regulation of the balance between viral plus and minus strand RNA synthesis, *Virology*, 1991, **185**, 496-499.
25. BOL, J.F., VAN VLOTEN-DOTING, L., JASPARS, E.M.J., A functional equivalence of top component *a* RNA and coat protein in the initiation of infection by alfalfa mosaic virus, *Virology*, 1971, **46**, 73-85.
26. YUSIBOV, V.M., KUMAR, A., NORTH, A., JOHNSON, J.E., LOESCH-FRIES, L.S., Purification, characterization, crystallization and preliminary X-ray analysis of *E. coli* expressed alfalfa mosaic virus coat protein, *J. Gen. Virol.*, 1995, **77**, 567-573.
27. YUSIBOV, V., MODELSKA, A., STEPLEWSKI, K., AGADJANYAN, M., WEINER, D., HOOPER, C., KOPROWSKI, H., Antigens produced in plants by infection with chimeric plant viruses immunize against rabies virus and HIV-1, *Proc. Natl. Acad. Sci. USA,* 1997, **94**, 5784-5788.
28. SAITO, T., YAMANAKA, K., OKADA, Y., Long-distance movement and viral assembly of tobacco mosaic virus mutants, *Virology*, 1990, **176**, 329-336.
29. ISHIKAWA, M., MESHI, T., MOTOYOSHI, F., TAKAMATSU, N., OKADA, Y., *In vitro* mutagenesis of the putative replicase genes of tobacco mosaic virus, *Nucleic Acids Res.*, 1986, **14**, 8291-8308.
30. MESHI, T., WATANABE, Y., SAITO, T., SUGIMOTO, A., MAEDE, T., OKADA, Y., Function of 30 kD protein of tobacco mosaic virus: Involvment in cell-to-cell movement and dispensability for replication, *EMBO J.*, 1987, **6**, 2557-2563.
31. HUNTER, T.R., HUNT, T., KNOWLAND, J., ZIMMERN, D., Messenger RNA for the coat protein of tobacco mosaic virus, *Nature*, 1976, **260,** 759-760.
32. BRUENING, G., BEACHY, R.N., SCALLA, R., ZAITLIN, M., *In vitro* and *in vivo* translation of the ribonucleic acids of a cowpea strain of tobacco mosaic virus, *Virology*, 1976, **71**, 498-517.
33. BEACHY, R.N., ZAITLIN, M., BRUENING, G., ISRAEL, H.W., A genetic map for the cowpea strain of TMV, *Virology*, 1976, **73**, 498-507.
34. SMITH, G.P., Filamentous fusion phage: novel expression vectors that display cloned antigens on the virion surface, *Science*, 1985, **228**, 1315-1317.
35. GREENWOOD, J, WILLIS, A.E., PERHAM, R.N., Multiple display of foreign peptides on a filamentous bacteriophage. Peptides from Plasmodium falciparum circumsporozoite protein as antigens, *J. Mol. Biol.*, 1991, **220**, 821-827.
36. MASTICO, R.A., TALBOT, S.J., STOCKLEY, P.G., Multiple presentation of foreign peptides on the surface of an RNA-free spherical bacteriophage capsid, *J. Gen. Virol.*, 1993, **74**, 541-548.
37. BURKE, K.L., DUNN, G., FERGUSON, M., MINOR, P.D., ALMOND, J.W., Antigen chimeras of poliovirus as potential new vaccines, *Nature*, 1988, **332**, 81-82.
38. DEDIEU, J.F., RONCO, J., VAN DER WERF, S., HOGLE, J.M., HENIN, Y., GIRARD, M., Poliovirus chimeras expressing sequences from the principal neutralization domain of human immunodeficiency virus type 1, *J. Virol,* 1992, **66**, 3161-3167.

39. ARNOLD, G.F., RESNICK, D.A., LI, Y., ZHANG, A., SMITH, A.D., GEISLER, S.C., JACOBO-MOLINA, A., LEE, W., WEBSTER, R.G., ARNOLD, E., Design and construction of rhinovirus chimeras incorporating immunogens from polio, influenza, and human immunodeficiency viruses, *Virology*, 1994, **198**, 703-708.
40. HAMAMOTO, H., SUGIYAMA, Y., NAKAGAWA, N., HASHIDA, E., MATSUNAGA, Y., TAKEMOTO, S., WATANABE, Y., OKADA, Y., A new tobacco mosaic virus vector and its use for the systemic production of angiotensin-I-converting enzyme inhibitor in transgenic tobacco and tomato, *Bio/Technology*, 1993, **11**, 930-932.
41. FITCHEN, J., BEACHY, R.N., HEIN, M.B., Plant virus expressing hybrid coat protein with added murine epitope elicits autoantibody response, *Vaccine*, 1995, **13**, 1051-1057.
42. TURPIN, T.H., REINI, S.J., CHAROENVIT, Y., HOFFMAN, S.L., FALLARME, V., GRILL, L.K., Malarial epitopes expressed on the suface of recombinant tobacco mosaic virus. *Bio/Technology*, 1995, **13**, 53-57.
43. STACZEK, J., BENDAHMANE, M., GILLELAND, L.B., BEAHY, R.N., GILLELAND, H.E. Jr., Immunization with a chimeric tobacco mosaic virus containing an epitope of outer membrane protein F of *Pseudomonas aeruginosa* provides protection against challenge with *P. aeruginosa*, *Vaccine*, 2000, **18**, 2266-2274.
44. USHA, R., ROHLL, J.B., SPALL, V.E., SHANKS, M., MAULE, A.J., JOHNSON, J.E., LOMONOSOFF, G.P., Expression of an animal virus antigenic site on the surface of a plant virus particle, *Virology*, 1993, **197**, 366-374.
45. PORTA, C., SPALL, V.E., LOVELAND, J., JOHNSON, J.E., BARKER, P.J., LOMONOSSOFF, G., Development of cowpea mosaic virus as a high-yielding system for the presentation of foreign peptides, *Virology,* 1994, **202**, 949-955.
46. MCLAIN, L., PORTA, C., LOMONOSSOFF, G., DURRANI, Z., DIMMOCK, N.J., Human immunodeficiency virus type 1-neutralizing antibodies raised to a glycoprotein 41 peptide expressed on the surface of a plant virus, *AIDS Res. & Human Retroviruses*, 1995, **11**, 327-34.
47. DALSGAARD, K., UTTENTHAL, A., JONES, T.D., XU, F., MERRYWEATHER, A., HAMILTON, W.D.O., LANGEVELD, J.P.M., BOSHUIZEN, R.S., KAMSTRUP, S., LOMONOSSOFF, G.P., PORTA, C., VELA, C., CASAL, J.I., MELOEN, R.H., RODGERS, P.B., Plant-derived vaccine protects target animals against a viral disease, *Nature Biotechnol.*, 1997, **15**, 248-252.
48. MODELSKA, A., DIETZSCHOLD, B., FLEYSH, N., FU, Z.F., STEPLEWSKI, K., HOOPER, C., KOPROWSKI, H., YUSIBOV, V., Immunization against rabies with plant-derived antigen, *Proc. Natl. Acad. Sci. USA*, 1998, **95**, 2481-2485.
49. BELANGER, H., FLEYSH, N., COX, S., BARTMAN, B., DEEPALI, D., TRUDEL, M., KOPROWSKI, H., YUSIBOV, V., Human respiratory syncytial virus vaccine antigen produced in plants, *FASEB J.,* 2000, **14**, 2323-2328.
50. JOELSON, T., AKERBLOM, L., OXEFELT, L., STRANDBERG, B., TOMENIUS, K., MORRIS, T.J., Presentation of a foreign peptide on the surface of tomato bushy stunt virus, *J. Gen. Virol.*, 1997, **78**, 1213-1217.
51. DIETZSCHOLD, B., GORE, M., MARCHADIER, D., NIU, H.-S., BUNSCHOTEN, H.M., OTVOS, L., WUNNER, W.H., ERTL, H.C.J., OSTERHAUS, A.D.M.E., KOPROWSKI, H., Structural and immunological characterization of a linear virus-neutralizing epitope of the rabies virus glycoprotein and its possible use in a synthetic vaccine, *J. Virol.*, 1990, **64**, 3804-3809.

52. SHIVPRASAD, S., POGUE, G.P., LEWANDOWSKI, D.J., HIDALGO, J., DONSON, J., GRILL, L.K., DAWSON, W.O., Heterologous sequences greatly affect foreign gene expression in tobacco mosaic virus-based vectors, *Virology*, 1999, **255**, 312-23.
53. SPITSIN, S., STEPLEWSKI, K., FLYESH, N., BELANGER, H., MICHIEVA, T., SHIVPRASAD, S., DAWSON, W., KOPROWSKI, H., YUSIBOV, V., The coat protein alfalfa mosaic virus supports the long distance movement of TMV deficient in production of TMV coat protein, *Proc. Natl. Acad. Sci. USA*, 1999, **96**, 2549-2553.
54. TASCHNER, P.E.M., VAN DER KUYL, A.C., NEELEMAN, L., BOL, J.F., Replication of incomplete alfalfa mosaic virus genome in plants transformed with viral replicase genes, *Virology*, 1991, **181**, 445-450.
55. McCORMICK, A.A., KUMAGAI, M.H., HANLEY, K., TURPEN, T.H., HAKIM, I., GRILL, L.K., TUSE, D., LEVY, S., LEVY, R., Rapid production of specific vaccines for lymphoma by expression of the tumor-derived single-chain Fv epitopes in tobacco plants, *Proc. Natl. Acad. Sci. USA*, 1999, **96**, 703-708.
56. KUMAGAI, M.H., DONSON, J., DELLA-CIOPPA, G., GRILL, L.K., Rapid, high-level expression of glycosylated rice alpha-amylase in transfected plants by an RNA viral vector, *Gene,* 2000, **245**, 169-174.
57. VERCH, T., YUSIBOV, V., KOPROWSKI, H., Expression and assembly of a full length monoclonal antibody in plants using a plant virus vector, *J. Immunol. Methods,* 1998, **220**, 69-75.

Chapter Five

TRANSCRIPTION FACTORS AND METABOLIC ENGINEERING: NOVEL APPLICATIONS FOR ANCIENT TOOLS

Edward L. Braun, Anusha P. Dias, Todd J. Matulnik and Erich Grotewold

Department of Plant Biology & Plant Biotechnology Center
The Ohio State University
Columbus, OH 43210

INTRODUCTION

Metabolic engineering involves the deliberate modification of biochemical networks with the objective of altering the accumulation of specific metabolites. A general prerequisite to engineering a specific metabolic pathway is a good biochemical knowledge of the processes that govern the production and accumulation of the desired product(s). Over the past few decades a wealth of information has been obtained regarding the biosynthesis of a large number of metabolites in plants, animals, fungi, and bacteria. Many of these compounds have economic importance, and substantial effort has gone into increasing their accumulation by genetic selection and biotechnology.

A favorite strategy to increase or alter product formation has been to overexpress enzymes assumed to correspond to rate-limiting steps of particular pathways utilizing transgenic approaches. This strategy has been employed to alter lignin composition in *Arabidopsis thaliana* by overexpression of ferulate-5-hydroxylase (F5H)[1] and increase phenylpropanoid accumulation in tobacco by overexpression of phenylalanine ammonia lyase (PAL).[2] These approaches are limited by flux considerations,[3] the intrinsic plasticity of plant metabolism,[4] the possible participation of enzymes in complexes or metabolic channels,[5] and homeostatic control of metabolic pathways.[6] Moreover, pathways entirely absent from particular cell types can not be manipulated in this fashion because the expression of a single gene is not sufficient for product formation in the absence of the other enzymes of the pathway.

The simultaneous introduction of multiple genes from a specific pathway can overcome these limitations, at least in principle. Nevertheless, this strategy creates additional technical challenges and may result in unpredictable levels of product accumulation or the silencing of the introduced genes.[7] However, in some cases, this strategy has been successful in engineering new metabolic pathways in crop plants.[8] The expression of multiple genes from a single promoter to create polyproteins that are then cleaved by proteases provides an attractive alternative.[9]

TRANSCRIPTIONAL CONTROL OF PLANT METABOLIC PATHWAY GENES

The limitations of the approaches to engineering plant secondary metabolism described above suggest that new strategies are necessary. The simultaneous activation of all genes of a particular pathway by expression of one or a few transcriptional activators provides an emerging technology that may largely overcome the limitations of most currently available methods (also reviewed by C. Martin[10]). However, just as detailed knowledge of the relevant biochemical pathways is necessary for conventional approaches to metabolic engineering, precise knowledge of the biosynthetic genes subject to transcriptional regulation is necessary for metabolic engineering with transcription factors. Here, we review known roles for transcriptional regulators in controlling the accumulation of selected phytochemicals, focusing upon specific examples with excellent prospects for this novel approach to metabolic engineering.

Phenylpropanoid Accumulation

Plants are specialized to channel carbon on demand, or in response to specific stimuli, from primary metabolism through phenylalanine to the phenylpropanoid pathway, for the biosynthesis of a large number of compounds including lignin and flavonoids. The biosynthesis of phenylpropanoids requires the efficient flow of carbon into the shikimate pathway for the biosynthesis of phenylalanine. This pathway, responsible for the biosynthesis of aromatic amino acids,[11] plays a major role in the production of precursors for aromatic compounds in plants. Chorismate, the end product of the shikimate pathway, is converted to quinones, indoles, and aromatic amino acids, which are the precursors of many secondary metabolites (Fig. 5.1).

The first committed step in the phenylpropanoid pathway is catalyzed by PAL, which converts phenylalanine into cinnamic acid.[12,13] PAL is probably the best-studied biosynthetic enzyme in plant phenolic metabolism, and it is typically encoded by small multigene families in the higher plants (*e.g.*, the three characterized *Arabidopsis PAL* genes[13]). Accumulation of *PAL* mRNA increases in response to many stimuli, such as UV light and pathogens.[12] Genes encoding PAL were among the first isolated and were shown to respond to elicitors from plant pathogens by changes in mRNA accumulation.

Transcriptional activation of genes encoding enzymes involved in phenylpropanoid metabolism (Fig. 5.1), such as PAL, 4-coumarate CoA ligase (4CL), and cinnamyl alcohol dehydrogenase (CAD), probably represents a key step in the regulation of these pathways. However, the transcription factors that specify temporal and spatial expression of these genes remain poorly characterized. The snapdragon (*Antirrhinum majus*) *AmMYB308* and *AmMYB330* genes constitute excellent candidates for transcription factors involved in the regulation of genes specific to the phenylpropanoid pathway.[14,15] In addition, there are a number of other transcription factors that have been identified based upon their ability to bind elements in the promoters of phenylpropanoid biosynthetic enzymes (see below, this section), although their contribution to the regulation of this pathway remains unclear.

The coordinate regulation of the *PAL*, *4CL*, and *CAD* genes in many plant species suggests that specific transcription factors activate the expression of these genes. The promoters of several genes in the phenylpropanoid pathway have been

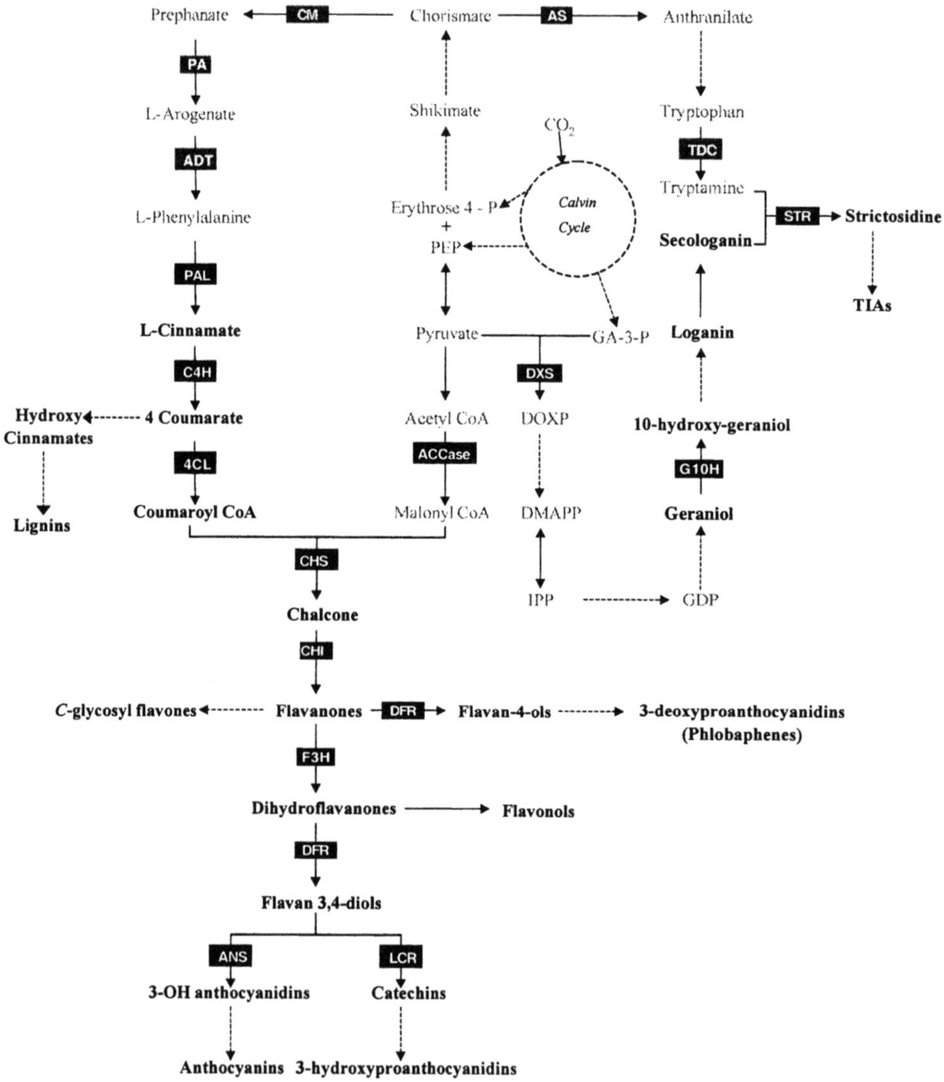

Figure 5.1: Schematic representation of the phenylpropanoid and terpenoid indole alkaloid biosynthetic pathways. Only core enzymes of each pathway are shown and their abbreviations are given in the text. The names of classes of intermediates or final compounds are indicated. Compounds and enzymes from primary metabolism are shown in gray.

dissected, and *cis*-acting regulatory elements important for expression have been identified. For example, members of the *PAL* gene family in parsley (*Petroselinum crispum*) contain the characteristic P-, A-, and L-boxes, also present in other phenylpropanoid biosynthetic genes.[16] Additional other *cis*-regulatory elements have been identified in genes of the pathway, including the G- and H-boxes. G-boxes (with the core consensus sequence CACGTG) mediate the response to various stimuli, including light and abscisic acid, and are recognized by bZIP proteins designated GBFs (G-box factors)[17] and potentially by other factors. [17] H-boxes, which are similar to the P-box identified by Sablowski and colleagues, [19] have the core consensus sequence CCTACC and are found in many phenylpropanoid biosynthetic genes. The G- and H-box regulatory elements in soybean (*Glycine max*) phenylpropanoid biosynthetic genes participate in the early response to pathogen attack. [20]

There are a number of candidate transcription factors that regulate the genes encoding enzymes involved in phenylpropanoid biosynthesis by recognizing H-boxes present in the promoters of these genes. The core consensus sequence of the H-box resembles the consensus binding site of plant R2R3 Myb transcription factors,[21-23] and the product of the snapdragon *AmMyb305* gene recognizes the H-box in the promoter of the tobacco (*Nicotiana tabacum*) *PAL2* gene. [19] This snapdragon Myb protein and a closely related paralog (AmMYB340) provide one of the first examples of transcriptional regulators that control structural genes from the phenylpropanoid pathway as well as flavonoid biosynthetic pathway, since they also activate genes encoding chalcone isomerase (CHI) and flavanone 3-hydroxylase (F3H) in tobacco.[24]

Despite the clear effects of AmMyb305 and AmMyb340 upon *PAL* mRNA accumulation,[24] the regulatory role of H-boxes is complex. Activation of transcription from an AmMYB305 binding site in tobacco protoplasts requires the presence of a G-box,[19] suggesting the need for additional factors *in vivo*. At least two different factors (KAP-1 and KAP-2) have been identified as H-box binding factors in the French bean (*Phaseolus vulgaris*),[25] suggesting that the R2R3 Myb proteins closely related to AmMYB305 and AmMYB340 are unlikely to be the only transcription factors that bind to H-boxes. An H-box binding factor designated BPF-1 (identified as a Box P-binding factor) has been purified from parsley, and transcription of this factor increases in response to elicitor treatment,[26] which also increases *PAL* mRNA accumulation. However, the correlation between *BPF*-1 and *PAL* mRNA accumulation is imperfect, and overexpression studies to establish that increased BPF-1 activity increases *PAL* transcription have not been conducted. Surprisingly, BPF-1 is closely related to the maize (*Zea mays*) IBP proteins[27] and the rice (*Oryza sativa*) RTBP1 protein,[28] which are single-repeat Myb homologs initially identified based upon their ability to interact with the initiator of the maize *Shrunken* gene and with plant telomeres, respectively. Thus, the biological role of these divergent Myb homologs remains unclear (also see below, section on *Indole Alkaloid Accumulation*).

In addition to Myb homologs, the bZIP protein G/HBF-1 has been identified in soybean,[29] based upon its ability to bind both G- and H-boxes from the *chs15* chalcone synthase promoter. G/HBF-1 is phosphorylated in response to elicitor treatment, and this enhances the binding to the soybean *chs15* promoter. It is unclear whether the apparent complexity of transcriptional regulation by H-boxes (or P-boxes) reflects the existence of multiple factors capable of binding H-boxes *in vivo*, heterogeneity among different plants studied, or differences among H-boxes examined in each study. However, the differences among studies make it difficult to use this information about transcriptional regulators to manipulate phenylpropanoid biosynthesis. From the standpoint of metabolic engineering, overexpression studies, such as those using the snapdragon R2R3 Myb proteins,[19,24] may provide the most useful information.

Flavonoid Accumulation

Flavonoids have multiple functions in plants, including roles as floral pigments for the attraction of pollinators, signalling molecules for beneficial microorganisms in the rhizosphere, and antimicrobial defense compounds.[30,31] In addition, flavonoids are emerging as important nutraceuticals because of their strong antioxidant properties, and several flavonoids have shown anti-tumor activities.[32] The regulation of flavonoid biosynthetic genes (which encode enzymes listed in Fig. 5.1) provides the best-described example of how plants control the expression of genes involved in secondary metabolite biosynthesis. [31] The conspicuous pigmentation provided by flavonoid-derived compounds has allowed the identification of mutants in a large number of structural and regulatory genes in this pathway in various model plants, such as maize (*Zea mays*), petunia (*Petunia hybrida*), and *Arabidopsis*.[33,34]

In maize, anthocyanin accumulation is controlled by two classes of regulatory proteins: a Myb-domain containing class (encoded by the *C1* and *Pl* genes[35,36]) and a basic helix-loop-helix (bHLH)-domain containing class (members of the *R/B* gene families[37]). Anthocyanin production requires interaction between a member of the Myb-domain C1/Pl family and a member of the bHLH-domain R/B family, [31] and the pattern of anthocyanin pigmentation in any particular plant part is controlled by the combinatorial, tissue-specific expression of these regulatory genes. Orthologs of the maize C1 and R regulators—homologs of C1 and R related to the maize proteins by the divergence of species rather than gene duplication[35]—have been identified in other plants, such as petunia and snapdragon.[31] These transcriptional regulators are exchangeable between monocots and higher dicots,[39] suggesting that it may be possible to use transcription factors from a single species for metabolic engineering in a variety of plants.

In addition to 3-hydroxy flavonoids and anthocyanins, maize and its close relatives like sorghum accumulate 3-deoxy flavonoids and derived pigments, which include the phlobaphenes. A single known genetic factor (*P*) controls 3-deoxy flavonoid and phlobaphene biosynthesis in maize.[40-42] *P* encodes a protein with a Myb domain more than 80% identical to the Myb domain of C1,[43] although evolutionary analyses suggest that the divergence between the genes encoding these transcription factors was fairly ancient.[44] In maize floral organs, *P* regulates the accumulation of a subset of mRNAs encoding flavonoid biosynthetic genes, namely *C2* (a chalcone synthase, CHS) and *A1* (dihydroflavonol 4-reductase, DFR). In sharp contrast, C1 is able to regulate the accumulation of mRNAs for *F3H*, *A2* (anthocyanidin synthase, ANS; an enzyme that has also been designated leucoanthocyanidin dioxygenase), *Bz1* (UDP-glucose flavonoid 3-O-glucosyltransferase), and *Bz2* (a glutathione *S*-transferase involved in the vacuolar localization of anthocyanins), as well as *C2* and *A1*,[23] when R or B are present. The regulatory activity of P is independent of the presence of the bHLH proteins R or B,[21,23,45] despite the fact that P and C1 recognize identical elements in the promoter of the *A1* gene.[21]

INDOLE ALKALOID ACCUMULATION

Alkaloids are classified in distinct groups based upon their core nitrogen-containing skeletons. This classification has a logical biochemical basis, since alkaloids with distinct skeletons are often derived from different primary metabolites, such as orinithine/arginine (nicotine and tropane alkaloids), tyrosine (isoquinoline alkaloids), or tryptophan (indole alkaloids).[46] We focus on the indole alkaloids in this review, because substantial progress has been made in elucidating the transcriptional regulation for genes encoding various enzymes involved in their biosynthesis.[47-49]

The regulation of tryptophan biosynthesis has been predicted to be more complex in plants than in other groups of organisms, such as the fungi and bacteria, since a large number of important compounds, including the plant hormone auxin, are derived from tryptophan or its precursors.[50] Further stressing the participation of *Myb* genes in plant metabolic control, an R2R3 Myb transcription factor designated ATR1[47] (altered tryptophan regulation) was identified that positively regulates accumulation of the *Arabidopsis ASA1* mRNA, which encodes the α subunit of anthranilate synthase (AS).[51] Since evolutionary analyses indicate that the *R2R3 Myb* gene family underwent a dramatic expansion early in the evolution of the land plants,[44] the existence of an R2R3 Myb protein that activates transcription of tryptophan biosynthetic enzymes is consistent with the prediction that plants regulate this pathway in unique ways (also see below, section on *Bioinformatics and Predicting Transcription Factor Function*).

Accumulation of the *ASA1* transcript increases substantially in response to wounding or infiltration with *Pseudomonas syringae*,[51] further stressing the unique patterns of regulation that genes encoding tryptophan biosynthetic enzymes exhibit in plants. However, any potential role of the ATR1 Myb protein in the induction of *ASA1* in response to specific environmental stimuli remains unclear. Indeed, any possible role of this Myb homolog or related paralogs in the regulation of other indole-derived secondary metabolites from *Arabidopsis*, such as the indolic phytoalexin camalexin,[52,53] has not been examined.

The greatest progress in elucidating the role of transcriptional regulation in the control of indole alkaloid accumulation has been made for compounds related to the *Vinca* alkaloids (vincristine and vinblastine). These are terpenoid indole alkaloids (TIAs) with important roles as chemotherapeutics in the treatment of cancer, which has generated substantial interest in their biosynthesis. TIAs have been identified from the Apocynaceae, Loganiaceae, and Rubiaceae, which are members of the Gentianales.[54] Extensive biosynthetic studies for TIAs have been conducted in the Madagascar periwinkle (*Catharanthus roseus*) to establish the identity of enzymes involved in their biosynthesis (Fig. 5.1).

The indole moiety of TIAs is derived from tryptophan after conversion to tryptamine by the enzyme tryptophan decarboxylase (TDC),[55] while the terpene moiety is derived from the iridoid glucoside secologanin.[55] Secologanin is derived from geraniol, synthesized by the chloroplast localized deoxyxylulose phosphate pathway.[56] Strictosidine synthase (STR) catalyzes the stereospecific condensation of tryptamine and secologanin, resulting in the formation of strictosidine, from which all other TIAs are derived.[55] Accumulation of these phytochemicals is induced by exposure to elicitors or jasmonic acid,[57] consistent with a role for TIAs in defense against pathogens and predators. However, the biosynthesis of TIAs appears to be complex, since the enzymes necessary for their biosynthesis are localized in distinct cell types, suggesting that intercellular translocation of intermediates is necessary.[58] Thus, the enzymes involved in TIA biosynthesis are subject to developmental control and respond to environmental stimuli, suggesting complex transcriptional control for the genes encoding them.

Members of the AP2/EREBP family of transcription factors, a diverse group of DNA-binding proteins thus far identified only in the plants[59] (and chlorophyte algae, based upon a *Chlamydomonas reinhardtii* EST encoding an AP2 homolog, accession number AW720749), have a role in the regulation of genes encoding TIA biosynthetic enzymes.[49,57] Two distinct periwinkle AP2-domain transcription factors, ORCA2 and ORCA3 (octadecanoid-derivative responsive *Catharanthus* AP2-domain proteins), have been identified as regulators of TIA biosynthetic genes. Both regulators show increased accumulation in response to jasmonic acid,[49,57] suggesting that they play a role in the induction of TIA accumulation by this plant hormone.

ORCA2 was identified in a yeast one-hybrid screen by using the jasmonic acid- and elicitor-response element from the periwinkle *STR* gene,[57] and ORCA2 is capable of trans-activating a fusion of the *STR* promoter to a GUS reporter gene. ORCA3 was identified by screening periwinkle cell lines containing T-DNA activation tags for lines with elevated TDC activity.[49] ORCA3 overexpression increased mRNA accumulation for TIA biosynthetic genes, including *TDC*, *STR*, and *D4H* (desacetoxyvindoline 4-hydroxylase), and genes necessary for biosynthesis of TIA precursors (the anthranilate synthase α subunit [AS] and the deoxyxylulose phosphate pathway enzyme D-1-deoxyxylulose 5-phosphate synthase [DXS]). ORCA3 binds to elements in the promoters of *TDC* and *SPR* and is capable of activating transcription from these promoters. However, ORCA3 overexpression does not result in TIA accumulation, because the gene encoding geraniol 10-hydroxylase (G10H), necessary for synthesis of loganin, was not induced in cells overexpressing ORCA3.[49] A third AP2 protein (ORCA1) was identified by the yeast one-hybrid screen by using the periwinkle *STR* promoter,[57] although this protein could not transactivate in plant cells from this promoter element.

The induction of genes encoding enzymes involved in both early (TDC and STR) and late (D4H) steps of TIA biosynthesis (see Fig. 5.1) by a single transcription factor is surprising, since these enzymes appear to be localized to different cell types in periwinkle plants.[58] This suggests that additional regulatory factors exist and that these factors show differential regulation *in planta* and in suspension cultures (or after exposure to elicitors). Indeed, a number of additional candidate transcription factors that may play a role in the regulation of TIA biosynthesis have been identified. For example, the promoter of the periwinkle *STR* gene contains several regions that are recognized by the tobacco GT-1 protein,[60] a member of a small family of plant transcription factors characterized by the trihelix DNA-binding domain.[61] One of the promoter regions recognized by GT-1 is capable of driving transcription of a reporter gene in response to elicitor treatment in transgenic tobacco,[60] suggesting that GT-factors may play a role in the regulation of TIA biosynthetic genes. A periwinkle BPF-1 homolog induced by elicitor in a jasmonic acid-independent manner has been shown to bind a novel element in the *STR* promoter, raising the possibility that this divergent class of Myb homologs play a role in the regulation of TIA biosynthetic genes.[62] Given the variety of biological processes that BPF-1 homologs have been implicated in (see above, section on *Phenylpropanoid Accumulation*), it seems critical to establish the roles that these divergent Myb proteins play in plants.

SUMMARY - PHYTOCHEMICAL ACCUMULATION

The regulation of phytochemical accumulation by transcription factors is clearly complex, with distinct types of transcription factors contributing to the regulation of various enzymes involved in both primary and secondary metabolism. This complexity may pose substantial problems for metabolic engineering. Indeed, differences between suspension cultures and regulation *in planta* (*e.g.*, for TIA biosynthesis) suggest the existence of differentially expressed and uncharacterized transcriptional regulators. Despite problems, there are good prospects for metabolic engineering of many pathways (see below, next section). Better understanding of the roles played by these regulators will allow the use of transcription factors to change the accumulation of secondary metabolites.

METABOLIC ENGINEERING BY USING PLANT TRANSCRIPTION FACTORS

Clearly, transcriptional regulators provide unique tools to modify plant traits.[10] Plant transcription factors often control several genes of a particular pathway in a coordinate fashion, providing an opportunity to activate entire biochemical pathways by expressing single transgenes. Despite the apparent power of this strategy, few examples of the successful use of transcription factors to manipulate plant metabolism have been described (Table 5.1). One reason for this is our limited knowledge of the transcriptional control of plant metabolic pathways. Nevertheless, the few reported examples suggest that regulatory factors will become a favorite strategy to manipulate plant metabolism.

Since the cloning of the first plant transcription factor more than a decade ago,[35] the regulators of anthocyanin biosynthesis have been extensively studied. One of the most intriguing findings is the exchangeability of these factors among plant species. Ectopic expression of the bHLH factors Lc (a member of the maize R gene family) or DELILA (the snapdragon ortholog of R) in transgenic tobacco or *Arabidopsis* plants resulted in a dramatic increase in anthocyanin production.[63,64] These and other studies[39] suggest that for pathways conserved in different plants, such as the anthocyanin pathway, orthologous regulators are likely to be exchangeable. Indeed, this may allow exceptional flexibility in the selection of transcription factors and plant species used for metabolic engineering.

Recently, the two main branches of maize flavonoid biosynthesis have been activated independently by ectopic expression of R2R3 Myb transcription factors in maize Black Mexican Sweet (BMS) cultured cells.[65] Expression of P resulted in the accumulation of 3-deoxy flavonoids and *C*-glycosyl flavones, similar to the compounds controlled by the *P* gene in maize floral organs.[42] In addition, the expression of P in BMS cells also resulted in an unexpected increase in the accumulation of the phenylpropanoids ferulic and chlorogenic acids,[65] suggesting an additional role of *P* in the control of phenylpropanoid biosynthesis. Consistent with this observation, a *PAL* transcript showed increased accumulation in response to the ectopic expression of P in a subsequent study.[66] Simultaneous expression of C1 and R in BMS cells resulted in a high level of anthocyanin accumulation,[65] and the induction of many mRNAs encoding flavonoid biosynthetic enzymes was readily

Table 5.1: Examples of successful metabolic engineering of phytochemical accumulation using transcription factors.

Regulator	Genes Controlled	Pathway Controlled[a]	Evidence Derived from:[b]
Zm-C1 (R2R3 Myb) & Zm-R (bHLH)	*C2* (CHS), *F3H*, *A1* (DFR), *A2* (ANS), *BZ1*, *BZ2*	Anthocyanins (+)	Loss-of-function mutants[126] and gain-of-function[77] analyses
Zm-P (R2R3 Myb)	*C2* (CHS), *A1* (DFR)	Phlobaphenes (+) *C*-Glycosyl flavones (+)	Loss-of-function mutants[53] and gain-of-function[77] analyses
Am-MYB308 (R2R3 Myb)	*C4H*, *4CL*, *CAD*	Phenylpropanoids (-) Lignin (-)	Gain-of-function analyses[c] [24]
Am-MYB330 (R2R3 Myb)	*4CL*	Phenylpropanoids (-) Lignin (-)	Gain-of-function analyses[c] [24]
Cr-ORCA3 (AP2)	*ASα*, *TDC*, *DXS*, *CPR*, *STR*, *D4H*, (not *G10H*)	Terpenoid indole alkaloids[d] (+)	Gain-of-function analyses[61]

[a] The specific phytochemicals that show altered accumulation and the nature of that change (activation [+] or repression [-]).
[b] Evidence that a particular transcriptional regulator controls a given pathway is provided by the mutant phenotypes (loss-of-function mutants) and increased accumulation of particular compounds when the regulator is overexpressed in plants or plant cells (gain-of-function evidence).
[c] Gain-of-function analyses were carried out in a different species of plant from the plant encoding the regulator.
[d] Terpenoid indole alkaloid biosynthesis was observed only when loganin was added to the cell cultures, probably reflecting the absence of increased *G10H* expression.

detectable.[66] Although P and C1+R expression in BMS cells resulted in the accumulation of 3-deoxy- and 3-hydroxy-flavonoids, respectively, neither *CHI*[67] (chalcone flavanone isomerase), mRNA accumulation, nor CHI enzyme activity was detectable in BMS cells expressing P or C1+R,[65] despite the participation of CHI in the biosynthesis of both types of flavonoids (Fig. 5.1). These findings indicate that it may not be necessary to have all the genes that play a role in specific pathways coordinately increased to dramatically increase flux by the corresponding regulators.

Since the induction of anthocyanin accumulation requires the simultaneous expression of both C1 and R, a chimeric factor (CRC) with the activities of both proteins was created.[66] Surprisingly, expression of P or CRC in BMS cells resulted in the induction of many more genes than those previously expected to be regulated by P or C1+R.[66] These genes encode glutathione *S*-transferases, MRP transporters, at least one cytochrome P450, and several other proteins predicted to participate in the modification and subcellular localization of the phytochemicals controlled by P and C1+R.[66] Thus, ectopic expression of transcription factors also provides a novel way to investigate currently unknown functions played by regulatory proteins (also see below, section on *Strategies For Using Plant Transcription Factors*).

Novel functions for transcription factors also can be identified by expression in a heterologous system, demonstrated by studies of the snapdragon AmMYB308 and AmMYB340 R2R3 Myb proteins in tobacco. These paralogs exhibit 94% identity within their DNA binding domains,[68] suggesting that they have similar DNA-binding activities. However, they are expressed in different regions of the snapdragon plant, with *AmMYB308* normally expressed throughout the plant and *AmMYB340* expressed only in flowers. When expressed from the *CaMV* 35S promoter in tobacco, the transgenic plants show reduced growth and patches of white, dead cells on mature leaves, in sharp contrast to the normal brown phenotype of senescent tobacco.[15] HPLC analysis of methanol-soluble extracts from these transgenic plants showed a reduced accumulation for a number of phenolics, with esters of caffeic and ferulic acid esters reduced 70-98%. Flavonoids were also affected, accumulating to levels about half of those in untransformed plants. Significant reduction in the accumulation of mRNAs encoding phenylpropanoid biosynthetic enzymes C4H, 4CL, and CAD was also observed. Lignin formation, which requires phenylpropanoid precursors, was also affected and showed an estimated total reduction of more than 50% in mature tissues and almost 20% in younger tissues. This finding is especially interesting for paper production, as lignin removal from wood pulp is currently achieved by environmentally-challenging chemical treatments.

A particularly elegant application of metabolic engineering to the identification of transcription factor function is provided by the identification of the TIA biosynthesis regulator encoded by *ORCA3* in periwinkle.[49] The strategy used corresponds essentially to metabolic engineering accomplished by overexpressing random genes, and can be applied to any metabolic pathway for which a convenient screen can be devised. To examine the regulation of TIA biosynthesis, the ability of TDC to detoxify tryptophan analogs such as 4-methyltryptophan was exploited, allowing the identification of TDC overproducing cell lines that were generated by the transformation of periwinkle cells with a T-DNA activation tag. This allowed the identification of a single transcription factor (ORCA3) that increased the expression of genes involved in both secondary metabolism and the biosynthetic pathways for precursors (see above, section on *Indole Alkaloid Accumulation*). However, ORCA3 overexpression is not sufficient for accumulation of TIAs in periwinkle cells, since at least one gene encoding an enzyme necessary for biosynthesis of the terpene moiety (G10H) is not induced by expression of this transcription factor (see below, section on *Limitations Of Using Transcription Factors*). A similar activation tag strategy has resulted in the identification of an *Arabidopsis* Myb homolog (*pap1-D*) capable of inducing anthocyanin accumulation throughout the plant when overexpressed (results of J. Borewitz, *et al.*, described by R. A. Dixon and C. L. Steel[69]).

The use of transcription factors to manipulate plant traits need not be limited to altering the accumulation of specific phytochemicals. For example, the expression of transcription factors could be used to alter stress resistance. One such experiment involved the *Arabidopsis* DREB1A AP2-domain transcription factor, which recognizes the DRE (dehydration response element) found in the promoters of several dehydration- and cold-stress inducible genes. When DREB1A was expressed in *Arabidopsis* plants from the promoter of the *rd29A* gene, which contains DRE elements itself and is induced by various stress conditions, plants highly tolerant to drought, salt, and freezing were obtained.[70] Expression of DREB1A from the *rd29A* promoter produced minimal negative effects on the plant in non-stress conditions, while expression of the same gene from the constitutive *CaMV* 35S promoter produced many undesirable effects. However, it is unclear how an activation loop in the *rd29*A::DREB1A plants was prevented, since *rd29A* is regulated by DREB1A.[70] Regardless of the specific features of the promoter used, this study indicates that other important plant traits can be manipulated by the ectopic expression of transcription factors.

IDENTIFICATION OF NOVEL TRANSCRIPTIONAL REGULATORS

Although the utilization of transcription factors as tools to engineer plant metabolism provides an attractive alternative to more classical approaches, there are formidable obstacles to the application of this approach outside of those examples that we list in Table 5.1, since there is almost no information about the role of transcriptional regulation in other plant metabolic pathways. Biochemical methods aimed at characterizing factors that interact with specific *cis*-acting DNA-regulatory elements in the promoters of genes encoding enzymes of particular pathways have been extensively used. However, these approaches are restricted by the low level of expression for many regulatory proteins and the limitations imposed by the specific cell types where particular phytochemicals accumulate.

The screening for mutants in regulatory genes has been helpful when the pathways controlled yielded visible intermediates or final products, as is the case for flavonoids[33] or the tryptophan pathway.[47] However, secondary metabolites are defined as compounds that are not essential for viability[71] so mutations in most secondary metabolic pathways will have relatively modest phenotypic effects. This, together with the functional redundancy characteristic of higher plants, makes the identification of mutants in regulators of secondary metabolism a low-payoff approach. This assertion is dramatically exemplified by the enormous effort invested in identifying DNA insertions (T-DNA or transposons) in 36 *Arabidopsis R2R3 Myb* genes that yielded few detectable phenotypes in the large variety of conditions tested.[72] However, the large number of distinct transcriptional regulators present in plants (see below, next section) suggests that the identification of specific candidate transcription factors for use in metabolic engineering represents a necessary initial step for this strategy. The growing interest in plant genomics is likely to provide substantial information that can be used to identify regulators that represent candidates for further study.

Bioinformatics and Predicting Transcription Factor Function

Transcriptional regulators may represent as much as 10% of the genes present in plant genomes (based upon analyses of the complete sequence of chromosome 4 from *Arabidopsis*[73]). Transcription factors have been classified into families based on the presence of conserved DNA-binding domains, with many of the families of transcriptional regulators that have been identified in the animals having plant homologs.[74] However, the cellular processes associated with particular classes of regulatory proteins in plants and animals often differ. For example, plant MADS-box proteins control a number of homeotic functions,[75] largely performed by homeodomain proteins in animals, while plant homeodomain proteins participate in functions associated with meristem identity and boundary establishment.[76] In addition to these broadly distributed groups of transcription factors, groups apparently specific to plants, such as the AP2 proteins[59] and trihelix factors,[61] also have been identified.

Plants also express members of transcription factor families with broad phylogenetic distributions that exhibit substantial qualitative differences from their homologs in animals and fungi. For example, animals express a handful of Myb-domain transcription factors, with the best characterized proteins—those related to the vertebrate c-Myb proto-oncoprotein—having three Myb-homologous direct repeats.[77] Although similar factors are present in plants,[78-80] the majority of the plant *Myb* gene family encodes proteins derived from the three repeat proteins by loss of the first repeat.[78] Thus, the majority of plant Myb genes encode proteins characterized by the presence of only two Myb repeats directly related to the second and third repeats in the vertebrate c-Myb protein (designated R2 and R3).[44,79,81] This gene family underwent a striking expansion 250-550 million years ago,[44] early in evolution of land plants, resulting in the presence of more than 150 paralogous *Arabidopsis* genes encoding R2R3 Myb proteins (Wenning, Braun and Grotewold, unpublished observation).

It seems likely that the expansion of gene families encoding transcription factors, such as the *R2R3 Myb* gene family, was directly associated with the evolution of novel functions that these factors regulate. Additional transcription factor families, such as the AP2 and the MADS-box families, underwent similar expansions. This suggests that one may be able to make functional inferences from their patterns of evolution. Analyses of genes encoding MADS-box proteins from the ferns *Ceratopteris pteroides* and *Ophioglossum pedunculosum* revealed that many of the MADS-box proteins involved in floral homeotic functions arose by gene duplication after the divergence of ferns and seed plants.[82] Likewise, the majority of characterized R2R3 Myb proteins control plant-specific functions thought to have arisen during the colonization of the land or the development of the vascular system.[44,79,81] One might infer that transcription factor genes that diverged prior to these periods might encode proteins with distinctive functions.

For example, phylogenetic analyses indicate the existence of deep-branching R2R3 Myb homologs related to the *Arabidopsis* AtMyb1 protein[83] and are characterized by the absence of a leucine insertion in the R2 Myb repeat[78] that is present in typical R2R3 Myb proteins from plants. This atypical class arose before the origin of land plants, since a member of this group is present in the chlorophyte alga *C. reinhardtii* (an EST with accession number AV396320). Unlike typical R2R3 Myb proteins, these atypical Myb proteins represent poor candidates for regulators of secondary metabolite accumulation. These differences emphasize the importance of carefully differentiating among distinct groups of homologous transcription factors, such as the typical R2R3 Myb proteins, the atypical (AtMyb1-like) R2R3 Myb proteins, the three repeat Myb homologs,[78-80] and the distantly related BPF-1/IBP-like Myb homologs.[26-28,62]

Evolutionary analyses can provide information suggesting that a regulator is unlikely to be involved in certain processes by providing evidence that a specific subfamily arose prior to the evolution of the process. Indeed, focusing upon gene duplications correlated with the evolution of specific phenotypes (such as the accumulation of a distinctive phytochemical) as candidates for regulators of that phenotype seems reasonable. Recent progress in angiosperm phylogeny[84] suggests that it may soon be possible to infer the divergence times for many different groups of plants fairly accurately, allowing this strategy to be applied to plants that accumulate interesting metabolites. However, the identification of recent duplications in genes encoding transcription factors must consider the patterns of duplication for other genes in the organism under consideration. Thus, the identification of more than two closely related genes would be necessary for plants that have undergone a genome duplication (such as maize[85]). Preliminary analyses of this type have been conducted,[86] revealing six groups of maize *R2R3 Myb* genes that underwent more than one gene duplication in the past 40 million years.

One of these groups of closely related maize *R2R3 Myb* genes contains at least nine closely-related paralogs. This group includes the maize *P* gene and has been designated the proline-to-alanine clade,[44] because of a specific amino acid substitution at a slowly evolving position that characterizes the group. Since the regulatory activity of *P* (activation of a subset of flavonoid biosynthetic genes resulting in phlobaphene pigment accumulation[23]) has been demonstrated only in maize and sorghum thus far,[87] the recent expansion of this group is consistent with an association between gene duplications and the evolution of novel regulatory functions. Some members of the proline-to-alanine clade exhibit a surprising degree of sequence divergence, suggesting a higher rate of evolution.[44] Establishing the functions of these divergent proteins will be informative, since many paralogs with different functions exhibit evolutionary rate differences.[88] Changes in the ratio of synonymous to nonsynonymous differences may prove useful for this, since functional changes should alter the accumulation of nonsynonymous rather than synonymous changes. However, correlations between the rate of evolution at synonymous and nonsynonymous sites[89] and the saturation of synonymous divergence[44] may complicate this type of analysis.

To determine whether a similar correlation was apparent for AP2-domain regulators of TIA biosynthesis, we used the ORCA proteins as queries for database searches. ORCA2 and ORCA3 are each other's top hit in BLAST searches, and both proteins recognize the same *Arabidopsis* protein (an uncharacterized ORF designated At2g44840) as their top hit in another organism (Braun and Grotewold, unpublished observations), suggesting a gene duplication after the divergence of the Gentianales and Brassicales. Searches indicated that ORCA1 is closely related to the *Arabidopsis* DREB2A protein,[57] consistent with the functional differences between ORCA1 and ORCA2/3. However, this relationship may be more complex, since ORCA2, ORCA3, and At2g44840 exhibit similar synonymous and nonsynonymous distances from each other (Braun and Grotewold, unpublished observations), consistent with a gene duplication prior to the divergence of the Gentianales and Brassicales. Resolving the timing of these gene duplications and establishing whether these genes exhibit differences in evolutionary rate will prove interesting.

Although the identification of genes encoding transcription factors with novel functions may be facilitated by identifying recently expanded gene clades or genes exhibiting evolutionary rate differences, functional inferences will require additional data. If transcription factors that exhibit a close phylogenetic relationship also exhibit similar regulatory functions, the use of evolutionary information will be powerful. Functional analyses of orthologous genes encoding MADS-box and R2R3 Myb proteins have indicated that functions are often conserved in distantly related plant taxa.[31, 90] Closely related paralogous transcription factors may also have similar regulatory functions. For example, the conserved domains of both MADS-box proteins[90] and R2R3 Myb proteins (Wolfe, Braun and Grotewold, unpublished observations) with related functions are often closely related phylogeneticaly.

As predicted by a correlation between phylogeny and function, the maize C1 and petunia AN2 regulators of anthocyanin biosynthesis cluster together in a larger group that also includes P, which may constitute a "phenylpropanoid group." Consistently, other studies positioned the snapdragon AmMYB330 and AmMYB308 in a clade sister to that containing C1.[91] Surprisingly, the snapdragon AmMYB305 and AmMYB340 proteins appear to be distantly related to this phenylpropanoid group,[91] suggesting some independent recruitment of genes to phenylpropanoid regulation. A major advantage of rigorous phylogenetic analyses relative to simple database searches (*e.g.*, BLAST) is the existence of well-accepted methods to evaluate support for specific clades, such as bootstrap or jackknife resampling[92,93] or the Bremer support index.[94] However, these rigorous analyses reveal that many phylogenetic associations among paralogous transcription factor groups (including the phenylpropanoid group) are poorly supported,[44] and researchers should be aware of artifacts driven by factors such as differences in evolutionary rates or amino acid composition.[95] The correlation between evolutionary relationships and regulatory functions of transcription factor genes requires further evaluation.

Genomic and cDNA sequence data may provide rich sources of information regarding the functions of specific transcriptional regulators in addition to the phylogenetic relationships that can be inferred. For example, differences in the composition of synonymous positions of plant genes (such as third codon position GC content) and intron size may have important functional correlates.[96] The identification of transcription factors with distinctive domain organization or evidence for recombination between relatively distantly related transcription factor genes might also represent excellent methods to identify candidate transcriptional regulators. The continued accumulation of sequence data and functional information from plant genomics should improve the selection of criteria for the identification of candidate transcription factors and suggest additional criteria that can be employed to this end.

Overall, we are optimistic regarding the use of evolutionary genomic analyses to identify candidates for transcription factors that regulate the accumulation of specific phytochemicals. For example, preliminary analyses suggest that another member of the proline-to-alanine clade has regulatory activities that overlap with those of P (Dias and Grotewold, unpublished observations), consistent with our assertion that the association between phylogeny and function is sufficiently strong to be useful for functional prediction. However, the use of sequence analyses should be viewed as a means to identify candidates for specific regulatory functions, to be followed with experiments (such as gain-of-function experiments) to evaluate the regulatory function of the candidate transcription factors in a direct fashion (see below, next section).

STRATEGIES FOR USING PLANT TRANSCRIPTION FACTORS

If a transcription factor controlling a particular metabolic pathway is known, the regulator can be expressed in plants or cultured plant cells from constitutive or inducible promoters, and the accumulation of the metabolite can be evaluated (Fig. 5.2). Some of the basic steps to follow after the identification of candidate transcription factors controlling particular metabolic pathways, to the characterization of gene products or phytochemicals that accumulate in cultured plant cells expressing the regulator are detailed below. This or similar strategies have been successfully applied in the examples we list in Table 5.1.

The Value of Gain-of-Function Approaches

The gain-of-function approach can be used to establish functions of particular regulators. The gene for the transcription factor is expressed in the plant or cultured plant cells under a constitutive or inducible promoter. After selection and propagation of the plant or plant cells expressing the regulator, several strategies are possible to determine which particular pathway the factor controls in the absence of an obvious phenotype. Differentially accumulated chemicals can be quantified with HPLC or GC-MS. This type of approach allowed the description of novel regulatory roles played by the maize *P* gene.[65] Alternatively, other methods to evaluate differential gene expression of possible target genes are available (Fig. 5.2), such as the Curagen GeneCalling technology. GeneCalling was efficiently applied to demonstrate that the ectopic expression of P or C1+R (the CRC chimera) induces or represses a large number of genes in maize BMS cells.[66] While this technology is expensive, the rapid development of microarrays for many model plants[97] provides a good alternative. Indeed, smaller arrays focusing on known metabolic pathways may suffice in many cases.

Comparison of mRNAs from cells expressing the transcriptional activator to control cells will allow the identification of differentially expressed genes, allowing the metabolic pathway(s) that are controlled to be established. This would permit subsequent biochemical confirmation by looking for the accumulation of the pathway products. These data, together with information on the temporal and tissue-specific expression of the regulator, can facilitate the prediction of loss-of-function phenotypes for the regulatory gene, closing the circle in this analysis (Fig. 5.2). Such data will provide substantial information regarding the criteria for the computational identification of candidate transcription factor genes.

This approach does not establish whether differentially expressed genes are directly or indirectly activated or repressed by the corresponding regulators. Fusing transcriptional regulators to the glucocorticoid receptor (GR) and comparing the activation of the putative regulated genes in the presence and absence of dexamethasone (DEX) and a protein synthesis inhibitor, such as cycloheximide, can identify genes that represent direct targets of specific transcriptional regulators. Similar studies have already been used to identify direct targets of the floral homeotic gene AP3.[98] Alternatively, one could examine the binding of the transcription factor to relevant promoters *in vitro* and examine the activation of reporter gene fusions *in planta*, similar to the studies performed with many of the proteins listed in Table 5.1. However, experiments of that type are relatively difficult to apply on a large scale, unlike the use of GR fusions where the mRNAs that accumulate in the presence of both DEX and cycloheximide can be assessed using microarrays or the Curagen GeneCalling technology.

While today is the time of plant genomics, tomorrow's goal will be relating the expression of particular transcription factors to the accumulation of specific proteins or enzymatic activities. Altering enzyme activities represents the ultimate goal of metabolic engineering with transcription factors. Detailed comparisons of mRNA accumulation and protein synthesis in the budding yeast *Saccharomyces cerevisiae* have shown profound differences in the translatability of specific mRNAs,[99] suggesting that differences exist in other groups of organisms as well. In fact, the point mutation that originally defined the *Arabidopsis ATR1* gene (allele *atr1D*) generated a stop codon in an upstream open reading frame, suggesting that ATR1 is subject to translational control.[47] Although few proteomic studies have been conducted in plants,[100] the comparison of isogenic cell lines distinguished only by ectopic expression of transcriptional regulators provided an opportunity to conduct a pilot proteomic study. The patterns of ^{35}S-methionine-labeled proteins obtained from BMS cells expressing P or C1+R were compared by two-dimensional polyacrylamide electrophoresis, and proteins induced by these regulators were revealed (Fig. 5.3). The number of proteins that P or C1+R induce in these analyses[101] is closer to the number of genes expected to be regulated by these factors[65] and significantly lower than the number estimated by using the Curagen GeneCalling technology.[66] These apparent differences may reflect translational regulation or differences in the sensitivity of the methods used.

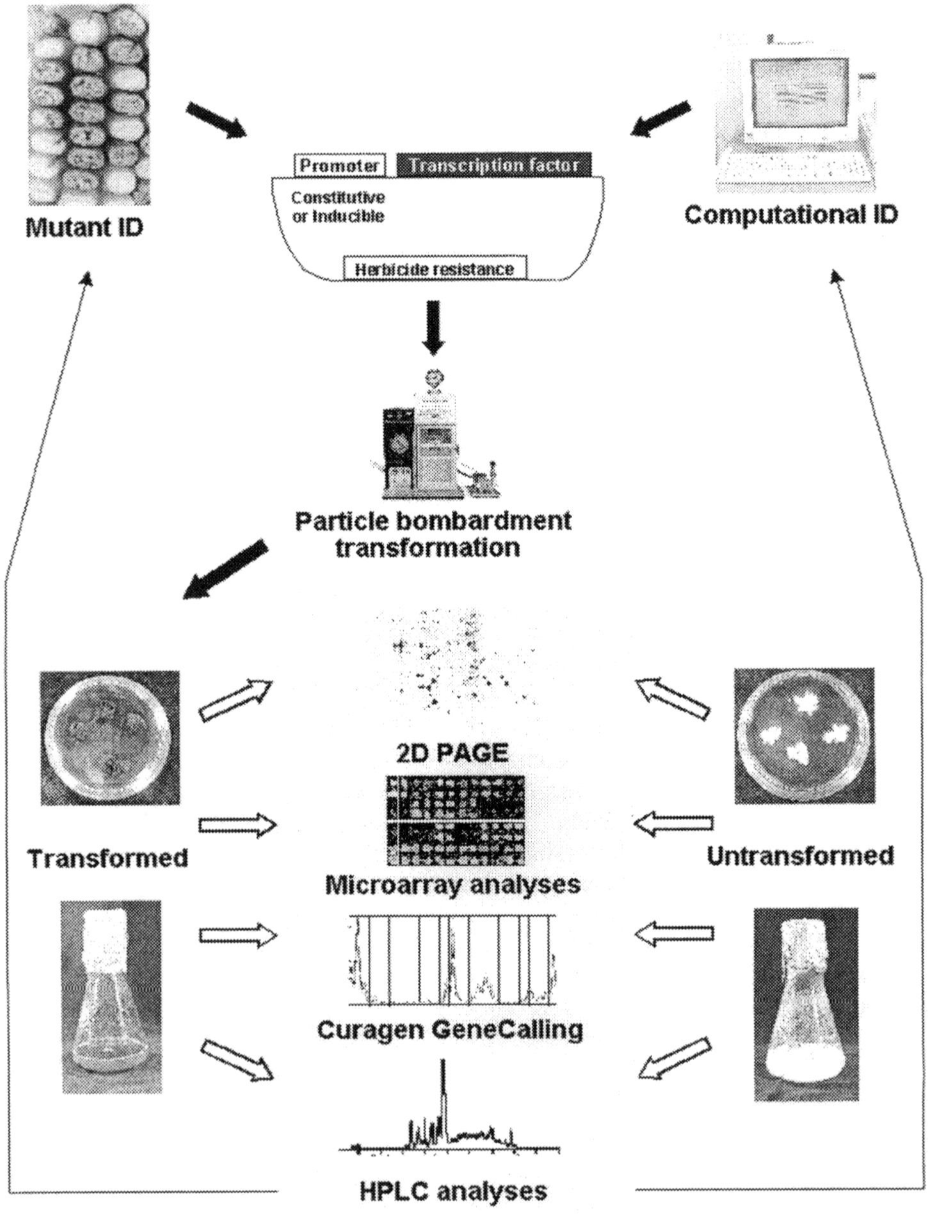
Mutant ID
Promoter
Transcription factor
Constitutive
or inducible
Herbicide resistance
Computational ID
Particle bombardment
transformation
2D PAGE
Transformed
Untransformed
Microarray analyses
Curagen GeneCalling
HPLC analyses

Figure 5.2: General scheme suggested for the engineering of plant metabolic pathways using transcription factors. Transcription factors either characterized by mutant analysis (mutant ID) or deduced from sequence information (computational ID) arc cloned under constitutive or inducible plant promoters. The maize ear shows the phenotype of the *Ac/Ds* transposable element insertions in the *R* (*r-m3* allele, spotted aleurone), or *P* (*P-vv* allele, variegated pericarp) genes. The transcription factors are introduced into plant cells (BMS cells are shown in this case) using the general method of particle bombardment. Transgenic plant cell lines (calli expressing C1+R from the *CaMV* 35S promoters are shown) are selected for herbicide resistance and HPLC or other methodologies are used to investigate the accumulation of different phytochemicals. Alternatively, or in addition, specific genes or proteins induced by the regulators can be identified using a number of available technologies. This information can be used for transcription factors of unknown function to predict function and help identify loss-of-function mutations or to improve the criteria used for computational identification (represented by the thin feedback arrows).

Limitations of Using Transcription Factors

The main limitation of the use of transcription factors for plant metabolic engineering is our knowledge of regulators that control plant metabolic pathways. We propose here several new approaches to determine the function of plant regulators that should complement other more classical methods. These studies will reveal the specificity of transcription factors under various conditions. Maintenance of this specificity will determine the future usefulness of transcription factors for metabolic engineering.

A second limitation is that a single transcription factor might not be sufficient to induce all genes of a given metabolic pathway. For example, the maize regulatory factor C1 is incapable of activating transcription in the absence of the co-activator R.[31] This is overcome either by expressing both regulators together[65] or by generating the CRC chimera.[66] The dogma has been that all genes in a particular pathway are coordinately activated. This, however, may turn out not to be the general case. The maize P or C1+R regulators activate all known flavonoid biosynthetic genes, with the exception of *CHI*.[65,66] Preliminary experiments indicate that maize *CHI1* is regulated by mechanisms different from the other genes in the pathway (Grotewold, unpublished observations). In higher dicots (often called eudicots[84]), C1 orthologs only activate "late genes" in the pathway, and not the early

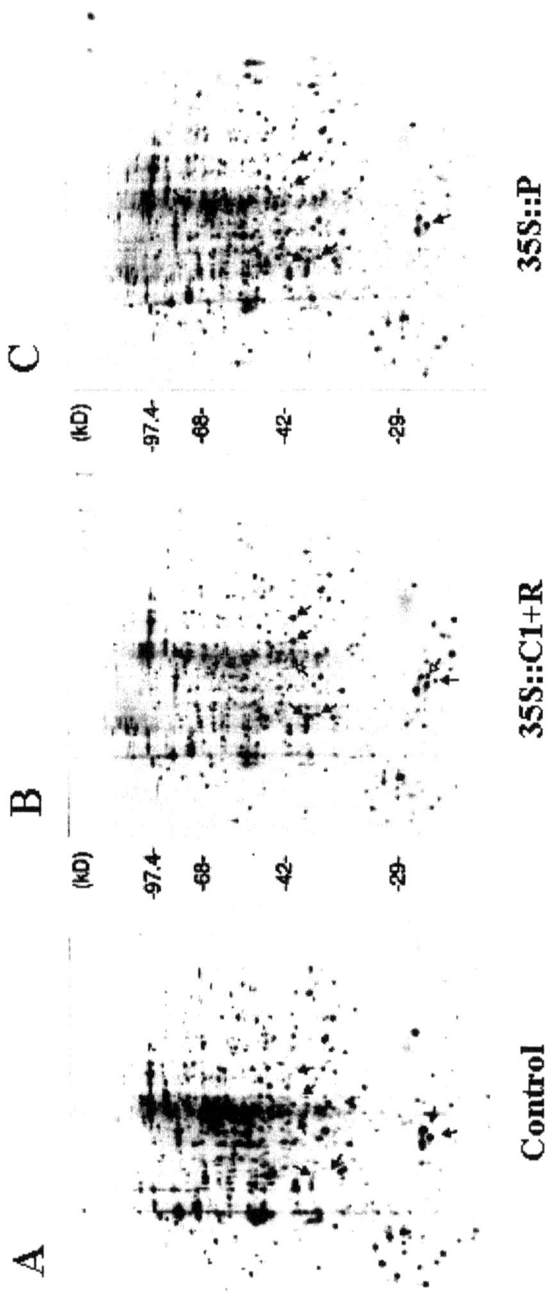
A
B
C
(kD)
97.4
68
42
29
Control
35S::C1+R
35S::P

Figure 5.3: Proteomic analysis of maize BMS callus cells expressing P and C1+R. The results of 2-D PAGE analyses for maize BMS callus cells (**A**); cells transformed with 35S::C1+R (**B**); or cells transformed with 35S::P (**C**). Arrows in (**B**) and (**C**) indicate protein spots absent in the control (**A**) line. Arrows indicate the positions of the relevant spots in the untransformed lines. Empty arrows in (**B**) correspond to protein spots seen only in the 35S::C1+R lines.

ones,[31] such as *chs* or *chi* (Fig. 5.1). Variation in the specific genes activated in different eudicots suggests that functionally-relevant changes in the promoters of flavonoid biosynthetic genes occur on short evolutionary time scales. The evolutionary flexibility of promoters for other genes involved in secondary metabolism is unclear, but we expect substantial variation to exist. Despite this variation, these regulators are capable of inducing anthocyanin accumulation in a number of eudicots examined (see above, section on *Flavonoid Accumulation*).

Just as the expression of C1+R resulted in the activation of a subset of flavonoid biosynthetic genes in maize BMS cells, overexpression of ORCA3 in periwinkle cells induced only a subset of TIA biosynthetic enzymes[49] (see above, section on *Indole Alkaloid Accumulation*). Cells overexpressing ORCA3 failed to accumulate TIAs unless externally fed with loganin,[49] demonstrating that failure to induce all of the genes participating in a process can be problematic. However, engineered cells expressing part of a biosynthetic pathway may allow the conversion of relatively inexpensive precursors to a more valuable end product.

The use of transcription factors to manipulate plant metabolism typically depends upon the use of constitutive or inducible promoters that drive high levels of expression of the regulator. It is conceivable that the over-expression of these transcription factors could result in non-specific effects associated with an increased binding of the regulator to DNA elements in the promoters of other genes, "squelching" other cellular factors or components of the transcriptional machinery.[102] However, studies that use maize cells stably or transiently expressing P or C1+R argue that these factors retain their biological specificity, even when over-expressed.[31,65,66] Thus, over-expression of transcription factors does not necessarily lead to non-specific effects. Nevertheless, ectopic expression of *P* in *Arabidopsis* plants resulted in phenotypes (greatly reduced plant size and pigment accumulation) that we have been unable to explain.[103] However, neither the function of the *Arabidopsis P* ortholog (encoded by *AtMyb12* and its close relatives, based upon our phylogenetic reconstructions[44,78]) nor the specific ways in which the regulatory function of the maize *P* gene differs from that of its orthologs in other species have been established. Thus, it is premature to conclude that the phenotype conferred by ectopic expression of P in *Arabidopsis* reflects non-specific transcriptional activation. Expression of C1 in *Arabidopsis* has no obvious phenotypic effect unless expressed together with the co-activator R, which results in anthocyanin accumulation,[63] suggesting that the biological specificity of C1 is maintained.

CONCLUSIONS

The plant kingdom provides a vast source of compounds with important biological activities, and ambitious projects to exploit this diversity have been undertaken recently. The time when plants and cultured plant cells will be routinely used as factories to produce compounds with importance to medicine and agriculture has yet to come, but important progress in the area of plant metabolic engineering has recently been made. Transcription factors are emerging as important tools for these processes, as they allow the activation of entire pathways with just one or a small number of transgenes. Clearly, one of the main challenges in the future will be establishing the regulation of important plant metabolic pathways.

ACKNOWLEDGEMENTS

We are grateful to Vince de Luca for critical reading of this manuscript and to Johan Memelink and his colleagues for sharing unpublished information. We thank our colleagues at the Ohio State Plant Biotechnology Center and Dan Crawford for many helpful interactions and we thank the editors for comments that improved this manuscript. We are extremely grateful to Ben Bowen for helpful interactions, suggestions, and inspiration while conducting much of the work described in this review. Our work is supported in part by grants from the United States Department of Agriculture and the National Science Foundation.

REFERENCES

1. MEIER, K., SHIRLEY, A.M., CUSUMANO, J.C., BELL-LELONG, D.A., CHAPPLE, C., Lignin monomer composition is determined by the expression of a cytochrome P450-dependent monooxygenase in *Arabidopsis*, *Proc. Natl. Acad. Sci. USA*, 1998, **95**, 6619-6623.
2. HOWLES, P.A., SEWALT, V.J.H., PAIVA, N.L., ELKIND, Y., BATE, N.J., LAMB, C., DIXON, R. A., Overexpression of L-phenylalanine ammonia-lyase in transgenic tobacco plants reveals control points for flux into phenylpropanoid biosynthesis, *Plant Physiol.*, 1996, **112**, 1617-1624.
3. NIEDERBERGER, P., PRASAD, R., MIOZZARI, G., KACSER, H., A strategy for increasing an *in vivo* flux by genetic manipulations, *Biochem. J.*, 1992, **287**, 473-479.

4. AP REES, T., Prospects of manipulating plant metabolism, *Trends Biotech.*, 1995, **13,** 375-378.
5. WINKEL-SHIRLEY, B., Evidence of enzyme complexes in the phenylpropanoid and flavoniod pathways, *Physiol. Plant.*, 1999, **107**, 142-149.
6. FRIDLYAND, L.E., BACKHAUSEN, J.E., SCHEIBE, R., Homeostatic regulation upon changes of enzyme activities in the Calvin cycle as an example for general mechanisms of flux control, *Photosyn. Res.*, 1999, **61,** 227-239.
7. LEECH, M.J., MAY, K., HALLARD, D., VERPOORTE, R., DE LUCA, V., CHRISTOU, P., Expression of two consecutive genes of a secondary metabolic pathway in transgenic tobacco: Molecular diversity influences levels of expression and product accumulation, *Plant Mol. Biol.*, 1998, **38**, 765-774.
8. YE, X., AL-BABILI, S., KLOTI, A., ZHANG, J., LUCCA, P., BEYER, P., POTRYKUS, I., Engineering the provitamin A (β-carotene) biosynthetic pathway into (carotenoid-free) rice endosperm, *Science*, 2000, **287**, 303-305.
9. BECK VON BODMAN, S., DOMIER, L.L., FARRAND, S.K., Expression of multiple eukaryotic genes from a single promoter in *Nicotiana*, *Biotechnology*, 1995, **13**, 587-591.
10. MARTIN, C., Transcription factors and the manipulation of plant traits, *Curr. Opin. Biotech.*, 1996, **7**, 130-138.
11. WEAVER, L.M., HERRMANN, K. M., Dynamics of the shikimate pathway in plants, *Trends Plant Sci.*, 1997, **2**, 346-351.
12. KUHN, D.N., CHAPPELL, J., BOUDET, A., HAHLBROCK, K., Induction of phenylalanine ammonia-lyase and 4-coumarate:CoA ligase mRNAs in cultured plant cells by UV or fungal elicitor, *Proc. Natl. Acad. Sci. USA*, 1984, **81,** 1102-1106.
13. WANNER, L.A., LI, G., WARE, D., SOMSSICH, I.E., DAVIS, K.R., The phenylalanine ammonia-lyase gene family in *Arabidopsis thaliana, Plant Mol. Biol.*, 1995, **27**, 327-338.
14. TAMAGNONE, L., MERIDA, A., PARR, A., MACKAY, S., CULIANEZ-MACIA, F.A., The AmMYB308 and AmMYB330 transcription factors from *Antirrhinum* regulate phenylpropanoid and lignin biosynthesis in transgenic tobacco, *Plant Cell*, 1998, **10**, 135-154.
15. TAMAGNONE, L., MERIDA, A., STACEY, N., PLASKITT, K., PARR, A., CHANG, C.-F., LYNN, D., MAXWELL DOW, J., ROBERTS, K., MARTIN, C., Inhibition of phenolic acid metabolism results in precocious cell death and altered cell morphology in leaves of transgenic tobacco plants, *Plant Cell*, 1998, **10**, 1801-1816.
16. LOGEMANN, E., PARNISKE, M., HAHLBROCK, K., Modes of expression and common structural features of the complete phenylalanine ammonia-lyase gene family in parsley, *Proc. Natl. Acad. Sci. USA*, 1995, **92**, 5905-5909.
17. MENKENS, A.E., SCHINDLER, U., CASHMORE, A.R., The G-box: A ubiquitous regulatory element in plants bound by the GBF family of bZIP proteins, *TIBS*, 1995, **20**, 506-516.

18. DE PATER, S., PHAM, K., MEMELINK, J., KIJNE, J., RAP-1 is an *Arabidopsis* MYC-like R protein homologue, that binds to G-box sequence motifs, *Plant Mol. Biol.*, 1997, **34**, 169-174.
19. SABLOWSKI, R.W.M., MOYANO, E., CULIANEZ-MACIA, F.A., SCHUCH, W., MARTIN, C., BEVAN, M., A flower-specific Myb protein activates transcription of phenylpropanoid biosynthetic genes, *EMBO J.*, 1994, **13**, 128-137.
20. WEISSHAAR, B., JENKINS, G. I., Phenylpropanoid biosynthesis and its regulation, *Curr. Opin. Plant Biol.*, 1998, **1**, 251-257.
21. SAINZ, M.B., GROTEWOLD, E., CHANDLER, V.L., Evidence for direct activation of an anthocyanin promoter by the maize C1 protein and comparison of DNA binding by related Myb domain proteins, *Plant Cell*, 1997, **9**, 611-625.
22. WILLIAMS, C.E., GROTEWOLD, E., Differences between plant and animal Myb domains are fundamental for DNA-binding and chimeric Myb domains have novel DNA-binding specificities, *J. Biol. Chem.*, 1997, **272**, 563-571.
23. GROTEWOLD, E., DRUMMOND, B., BOWEN, B., PETERSON, T., The Myb-homologous *P* gene controls phlobaphene pigmentation in maize floral organs by directly activating a flavonoid biosynthetic gene subset, *Cell*, 1994, **76**, 543-553.
24. MOYANO, E., MARTINEZ-GARCIA, F., MARTIN, C., Apparent redundancy in *myb* gene function provides gearing for the control of flavonoid biosynthesis in *Antirrhinum* flowers, *Plant Cell*, 1996, **8**, 1519-1532.
25. YU, L.M., LAMB, C.J., DIXON, R. A., Purification and biochemical characterization of proteins which bind to the H-box *cis*-element implicated in transcriptional activation of plant defense genes, *Plant J.*, 1993, **3**, 805-816.
26. DA COSTA E SILVA, O., KLEIN, L., SCHMELZER, E., TREZZINI, G.F., HAHLBROCK, K., BPF-1, a pathogen-induced DNA-binding protein involved in the plant defense response, *Plant J.*, 1993, **4**, 125-135.
27. LUGERT, T., WERR, W., A novel DNA-binding domain in the *Shrunken* initiator-binding protein (IBP1), *Plant Mol. Biol.*, 1994, **25**, 493-506.
28. YU, E.Y., KIM, S.E., KIM, J.H., KO, J.H., CHO, M.H., CHUNG, I.K., Sequence-specific DNA recognition by the Myb-like domain of plant telomeric protein RTBP1, *J. Biol. Chem.*, 2000, **275**, 24206-24214.
29. DRÖGE-LASER, W., KAISER, A., LINDSAY, W.P., HALKIER, B.A., LOAKE, G.J., DOERNER, P., DIXON, R.A., LAMB, C., Rapid stimulation of a soybean protein-serine kinase that phosphorylates a novel bZIP DNA-binding protein, G/HBF, during the induction of early transcription-dependent defenses, *EMBO J.*, 1997, **16**, 726-738.
30. KOES, R.E., QUATTROCCHIO, F., MOL, J.N.M., The flavonoid biosynthetic pathway in plants: Function and evolution, *BioEssays*, 1994, **16**, 123-132.
31. MOL, J., GROTEWOLD, E., KOES, R., How genes paint flowers and seeds, *Trends Plant Sci.*, 1998, **3**, 212-217.
32. DI CARLO, G., MASCOLO, N., IZZO, A.A., CAPASSO, F., Flavonoids: Old and new aspects of a class of natural therapeutic drugs, *Life Sci.*, 1999, **65**, 337-353.
33. DOONER, H.K., ROBBINS, T.P., JORGENSEN, R.A., Genetic and developmental control of anthocyanin biosynthesis, *Annu. Rev. Genet.*, 1991, **25**, 173-199.

34. SHIRLEY, B.W., Flavonoid biosynthesis: 'New' functions for an 'old' pathway, *Trends Plant Sci.*, 1996, **1**, 377-382.
35. PAZ-ARES, J., GHOSAL D., WEINLAND, U., PETERSON, P.A., SAEDLER, H., The regulatory *c1* locus of *Zea mays* encodes a protein with homology to *myb* proto-oncogene products and with structural similarities to transcriptional activators, *EMBO J.*, 1987, **6**, 3553-3558.
36. CONE, K.C., COCCIOLONE, S.M., BURR, F.A., BURR, B., Maize anthocyanin regulatory gene *pl1* is a duplicate of *c1* that functions in the plant, *Plant Cell*, 1993, **5**, 1795-1805.
37. LUDWIG, R., HABERA, L.F., DELLAPORTA, S.L., WESSLER, S.R., *Lc*, a member of the maize *R* gene family responsible for tissue-specific anthocyanin production, encodes a protein similar to transcriptional activators and contains the *myc*-homology region, *Proc. Natl. Acad. Sci. USA*, 1989, **86**, 7092-7096.
38. FITCH, W.M., Distinguishing homologous from analogous proteins, *Systematic Zool.*, 1970, **19**, 99-113.
39. QUATTROCCHIO, F., WING, J.F., LEPPEN, H.T.C., MOL, J.N.M., KOES, R.E., Regulatory genes controlling anthocyanin pigmentation are functionally conserved among plant species and have distinct sets of target genes, *Plant Cell*, 1993, **5**, 1497-1512.
40. STYLES, E.D., CESKA, O., Genetic control of 3-hydroxy- and 3-deoxy-flavonoids in *Zea mays*, *Phytochemistry*, 1975, **14**, 413-415.
41. STYLES, E.D., CESKA, O., The genetic control of flavonoid synthesis in maize, *Can. J. Genet. Cytol.*, 1977, **19**, 289-302.
42. STYLES, E.D., CESKA, O., Pericarp flavonoids in genetic strains of *Zea mays*, *Maydica*, 1989, **34**, 227-237.
43. GROTEWOLD, E., ATHMA, P., PETERSON, T., Alternatively spliced products of the maize *P* gene encode proteins with homology to the DNA-binding domain of Myb-like transcription factors, *Proc. Natl. Acad. Sci. USA*, 1991, **88**, 4587-4591.
44. RABINOWICZ, P.D., BRAUN, E.L., WOLFE, A.D., BOWEN, B., GROTEWOLD, E., Maize *R2R3 Myb* genes: Sequence analysis reveals amplification in higher plants, *Genetics*, 1999, **153**, 427-444.
45. GROTEWOLD, E., Does P protein require a partner, as C1 protein does?, *Maize Genet. Coop. News*, 1995, **69**, 32.
46. HASHIMOTO, T., YAMADA, Y., Alkaloid biogenesis: Molecular aspects, *Annu. Rev. Plant Physiol.*, 1994, **45**, 257-285.
47. BENDER, J., FINK, G.R., A Myb homologue, ATR1, activates tryptophan gene expression in *Arabidopsis*, *Proc. Natl. Acad. Sci. USA*, 1998, **95**, 5655-5660.
48. MENKE, F.L.H., PARCHMANN, S., MUELLER, M. J., KIJNE, J. W., MEMELINK, J., Involvement of the octadecanoid pathway and protein phosphorylation in fungal elicitor-induced expression of terpenoid indole alkaloid biosynthetic genes in *Catharanthus roseus*, *Plant Physiol.*, 1999, **199**, 1289-1296.
49. VAN DER FITS, L., MEMELINK, J., ORCA3, a jasmonate-responsive transcriptional regulator of plant primary and secondary metabolism, *Science*, 2000, **289**, 295-297.

50. RADWANSKI, E.R., LAST, R.L., Tryptophan biosynthesis and metabolism: Biochemical and molecular genetics, *Plant Cell*, 1995, **7**, 921-934.
51. NIYOGI, K.K., FINK, G., Two anthranilate synthase genes in *Arabidopsis*: Defense-related regulation of the tryptophan pathway, *Plant Cell*, 1992, **4**, 721-733.
52. TSUJI, J., JACKSON, E.P., GAGE, D.A., HAMMERSCHMIDT, R., SOMERVILLE, S., Phytoalexin accumulation in *Arabidopsis thaliana* during the hypersensitive reaction to *Psuedomonas syringae* pv *syringae, Plant Physiol.*, 1992, **98**, 1304-1309.
53. ZHAO, J., LAST, R.L., Coordinate regulation of the tryptophan biosynthetic pathway and indolic phytoalexin accumulation in *Arabidopsis*, *Plant Cell*, 1996, **12**, 2235-2244.
54. ROBERTS, M.F., STRACK, D., Biochemistry and physiology of alkaloids and betalains, *in* Biochemistry of Plant Secondary Metabolism (M. Wink, ed.) CRC Press LLC, Baton Rouge, 1999, pp. 358.
55. MEIJER, A.H., VERPOORTE, R., HOGE, J.H.C., Regulation of enzymes and genes involved in terpenoid indole alkaloid biosynthesis in *Catharanthus roseus*, *J. Plant Res.*, 1993, **3**, 145-164.
56. CONTIN, A., VAN DER HEIJDEN, R., LEFEBER, A.W.M., VERPOORTE, R., The iridoid glucoside secologanin is derived from the novel triose phosphate/pyruvate pathway in a *Catharanthus roseus* cell culture, *FEBS Lett.*, 1998, **434**, 413-416.
57. MENKE, F.L.H., CHAMPION, A., KIJNE, J.W., MEMELINK, J., A novel jasmonate- and elicitor-responsive element in the periwinkle secondary metabolite biosynthetic gene *Str* interacts with a jasmonate- and elicitor-inducible AP2-domain transcription factor, ORCA2, *EMBO J.*, 1999, **18**, 4455-4463.
58. ST-PIERRE, B., VAZQUEZ-FLOTA, F., DE LUCA, V., Multicellular compartmentation of *Catharanthus roseus* alkaloid biosynthesis predicts intercellular translocation of a pathway intermediate, *Plant Cell*, 1999, **11**, 887-900.
59. REICHMANN, J.L., MEYEROWITZ, E.M., The AP2/EREBP family of plant transcription factors, *Biol. Chem.*, 1998, **379**, 633-646.
60. PASQUALI, G., ERVEN, A.S.W., OUWERKERK, P.B.F., MENKE, F.L.H., MEMELINK, J., The promoter of the strictosidine synthase gene from periwinkle confers elicitor-inducible expression in transgenic tobacco and binds nuclear factors GT-1 and GBF, *Plant Mol. Biol.*, 1999, **39**, 1299-1310.
61. ZHOU, D.-X, Regulatory mechanism of plant gene transcription by GT-elements and GT-factors, *Trends Plant Sci.*, 1999, **4**, 210-214.
62. VAN DER FITS, L., ZHANG, H., MENKE, F.L.H., DENEKA, M., MEMELINK, J., A *Catharanthus roseus* BPF-1 homologue interacts with an elicitor-responsive region of the secondary metabolic biosynthetic gene *Str* and is induced by elicitor via a JA-independent signal transduction pathway, *Plant Mol. Biol.*, 2000, **44**, 675-685.
63. LLOYD, A.M., WALBOT, V., DAVIS, R.W., *Arabidopsis* and *Nicotiana* anthocyanin production activated by maize regulators *R* and *C1*, *Science*, 1992, **258**, 1773-1775.
64. GOODRICH, J., CARPENTER, R., COEN, E.S., A common gene regulates pigmentation pattern in diverse plant species, *Cell*, 1992, **68**, 955-964.

65. GROTEWOLD, E., CHAMBERLAIN, M. , ST. CLAIRE, G., SWENSON, J., SIAME, B.A., BUTLER, L.G., SNOOK, M., BOWEN, B., Engineering secondary metabolism in maize cells by ectopic expression of transcription factors, *Plant Cell*, 1998, **10**, 721-740.
66. BRUCE, W., FOLKERTS, O., GARNAAT, C., CRASTA, O., ROTH, B., BOWEN, B., Expression profiling of the maize flavonoid pathway genes controlled by estradiol-inducible transcription factors CRC and P, *Plant Cell*, 2000, **12**, 65-79.
67. GROTEWOLD, E., PETERSON T., Isolation and characterization of a maize gene encoding chalcone flavanone isomerase, *Mol. Gen. Genet.*, 1994, **242**, 1-8.
68. JACKSON, D., CULIANEZ-MACIA, F., PRESCOTT, A.G., ROBERTS, K., MARTIN C., Expression patterns of *myb* genes from *Antirrhinum* flowers, *Plant Cell*, 1991, **3**, 115-125.
69. DIXON, R.A., STEELE, C.L., Flavonoids and isoflavonoids - a gold mine for metabolic engineering, *Trends Plant Sci.*, 1999, **4**, 394-400.
70. KASUGA, M., LIU, Q., MIURA, S., YAMAGUCHI-SHINOZAKI, K., SHINOZAKI, K., Improving plant drought, salt and freezing tolerance by gene transfer of a single stress-inducible transcription factor, *Nature Biotechnol.*, 1999, **17**, 287-291.
71. KOSSEL., Über die chemische zusammensetzung der zelle. Verhandlungen der Berliner , 1891, 181-186.
72. MEISSNER, R.C., HAILING, J., COMINELLI, E., DENEKAMP, M., FUERTES, A., GRECO, R., KRANZ, H., PENFIELD, S., PETRONI, K., URZAINQUI, A., MARTIN, C., PAZ-ARES, J., SMEEKENS, S., TONELLI, C., WEISSHAAR, B., BAUMANN, E., KLIMYUK, V., JONES, J.J.D., PEREIRA, A., WISMAN, E., BEVAN, M., Function search in a large transcription factor gene family in *Arabidopsis*: Assessing the potential of reverse genetics to identify insertional mutations in R2R3 *MYB* genes, *Plant Cell*, 1999, **11**, 1827-1840.
73. THE EUROPEAN UNION ARABIDOPSIS GENOME SEQUENCING CONSORTIUM & THE COLD SPRING HARBOR, WASHINGTON UNIVERSITY IN ST. LOUIS AND PE BIOSYSTEMS ARABIDOPSIS SEQUENCING CONSORTIUM., Sequence and analysis of chromosome 4 of the plant *Arabidopsis thaliana*, *Nature*, 1999, **402**, 769-777.
74. LIU, L., WHITE, M.J., MACRAE, T.H., Transcription factors and their genes in higher plants, *Eur. J. Biochem.*, 1999, **262**, 247-257.
75. THEISSEN, G., BECKER, A., DI ROSA, A., KANNO, A., KIM, J.T., MÜNSTER, T., WINTER, K.-U., SAEDLER, H., A short history of MADS-box genes in plants, *Plant Mol. Biol.*, 2000, **42**, 115-149.
76. REISER, L., SANCHEZ-BARACALDO, P., HAKE, S., Knots in the family tree: Evolutionary relationships and functions of knox homebox genes, *Plant Mol. Biol.*, 2000, **42**, 151-166.
77. LIPSICK, J. S., One billion years of Myb, *Oncogene*, 1996, **13**, 223-235.
78. BRAUN, E. L., GROTEWOLD, E., Newly discovered plant *c-myb*-like genes rewrite the evolution of the plant *myb* gene family, *Plant Physiol.*, 1999, **121**, 21-24.

79. JIN, H., MARTIN, C., Multifunctionality and diversity within the plant *MYB*-gene family, *Plant Mol. Biol.*, 1999, **41**, 577-585.
80. KRANZ, H., SCHOLTZ, K., WEISSHAAR, B., c-MYB oncogene-like genes encoding three MYB repeats occur in all major plant lineages, *Plant J.*, 2000, **21**, 231-235.
81. MARTIN, C., PAZ-ARES, J., MYB transcription factors in plants, *Trends Genet.*, 1997, **13**, 67-73.
82. MÜNSTER, T., PAHNKE, J., DI ROSA, A., KIM, J.T., MARTIN, W., SAEDLER, H., THEISSEN, G., Floral homeotic genes were recruited from homologous MADS-box genes preexisting in the common ancestor of ferns and seed plants, *Proc. Natl. Acad. Sci. USA*, 1997, **94**, 2415-2420.
83. SHINOZAKI, K., YAMAGUCHI-SHINOZAKI, K., URAO, T., KOIZUMI, M., Nucleotide sequence of a gene from *Arabidopsis thaliana* encoding a *myb* homologue, *Plant Mol. Biol.*, 1992, **19**, 493-499.
84. SOLTIS, P.S., SOLTIS, D.E., CHASE, M.W., Angiosperm phylogeny inferred from multiple genes as a tool for comparative biology, *Nature*, 1999, **402**, 402-404.
85. GAUT, B.S., DOEBLEY, J.F., DNA sequence evidence for the segmental allotetraploid origin of maize, *Proc. Natl. Acad. Sci. USA*, 1997, **94**, 6809-6814.
86. BRAUN, E.L., GROTEWOLD, E., Diversification of the *R2R3 Myb* gene family and the segmental allotetraploid origin of the maize genome, *Maize Genet. Coop. Newsl.*, 1999, **73**, 26-27.
87. CHOPRA, S., BRENDEL, V., ZHANG, J., AXTELL, J., PETERSON T., Molecular characterization of a mutable pigmentation phenotype and isolation of the first active transposable element from *Sorghum bicolor, Proc. Natl. Acad. Sci. USA*, 1999, **96**, 15330-15335.
88. GU, X., Statistical methods for testing functional divergence after gene duplication, *Mol. Biol. Evol.*, 1999, **16**, 1664-1674.
89. MOUCHIROUD, D., GAUTIER, C., BERNARDI, G., Frequencies of synonymous substitutions in mammals are gene-specific and correlated with frequencies of non-synonymous substitutions, *J. Mol. Evol.*, 1995, **40**, 107-113.
90. SMYTH, D., A reverse trend - MADS functions revealed, *Trends Plant Sci.*, 2000, **5**, 315-317.
91. ROMERO, I., FUERTES, A., BENITO, M.J., MALPICA, J.M., LEYVA, A., PAZ-ARES, J., More than 80 *R2R3-MYB* regulatory genes in the genome of *Arabidopsis thaliana*, *Plant J.*, 1998, **14**, 273-284.
92. FELSENSTEIN, J., Confidence limits on phylogenies - an approach using the bootstrap, *Evolution*, 1985, **39**, 783-791.
93. FARRIS, J.S., ALBERT, V.A., KALLERSJO, M., LIPSCOMB, D., KLUGE, A.G., Parsimony jackknifing outperforms neighbor joining, *Cladistics*, 1996, **12**, 99-124.
94. BREMER, K., The limits of amino acid sequence data in angiosperm phylogenetic reconstruction, *Evolution*, 1988, **42**, 795-803.
95. PHILIPPE, H., LAURENT, J., How good are deep phylogenetic trees?, *Curr. Opin. Genet. Dev.*, 1998, **8**, 616-623.

96. CARELS, N., BERNARDI, G., Two classes of genes in plants, *Genetics*, 2000, **154**, 1819-1825.
97. SOMERVILLE, C., SOMERVILLE, S., Plant functional genomics, *Science*, 1999, **285**, 380-383.
98. SABLOWSKI, R.W.M., MEYEROWITZ, E.M.,A Homolog of *NO APICAL MERISTEM* is an immediate target of the floral homeotic genes *APETALA3/PISTILLATA*, *Cell*, 1998, **92**, 93-103.
99. FUGE, E.K., BRAUN, E.L., WERNER-WASHBURNE, M., Protein synthesis in long-term stationary-phase cultures of *Saccharomyces cerevisiae*, *J. Bacteriol.*, 1994, **176**, 5802-5813.
100. THIELLEMENT, H., BAHRMAN, N., DAMERVAL, C., PLOMION, C., ROSSIGNOL, M., SANTONI, V., DE VIENNE, D., ZIVY, M., Proteomics for genetic and physiological studies in plants, *Electrophoresis*, 1999, **20**, 2013-2026.
101. GROTEWOLD, E., Ectopic expression of P and R+C1 induce few new proteins, *Maize Genet. Coop. Newsl.*, 1999, **73**, 23-24.
102. PTASHNE, M., How eukaryotic transcriptional activators work, *Nature*, 1988, **335**, 683-689.
103. RABINOWICZ, P. D., MA, H., GROTEWOLD, E., Consequences of the ectopic expression of the Myb-domain protein P, *Maize Genet. Coop. Newsl.*, 1997, **71**, 21-22.

Chapter Six

DESIGNER BASIC/LEUCINE-ZIPPER PROTEINS AS REGULATORS OF PLANT DETOXIFICATION GENES

Amir Rahbar[1,2], Christopher Johnson[1,3], and Jonathan Arias[1,2,3]

[1]*Center for Agricultural Biotechnology*
University of Maryland Biotechnology Institute

[2]*Dept. of Chemistry and Biochemistry*
University of Maryland, College Park

[3]*Program in Molecular and Cell Biology*
University of Maryland, College Park

INTRODUCTION

Biotic and industrial activities result in environmental accumulation of xenobiotic organic toxins and mutagens that can subsequently be taken up and transformed by plants. Yet, xenobiotic residues often retain significant latent mutagenic or toxic activities after being transformed and stored in the plant vacuole or apoplastic space.[1] Furthering our understanding of plant detoxification processes may, thus, positively impact bioremediation and agriculture. In terms of bioremediation, it is anticipated that this information could be useful in creating plants with improved uptake and storage of toxic environmental pollutants and endogenous phytochemicals. In terms of agriculture, crops with enhanced xenobiotic metabolism could be made safer for animal and human consumption given that consumption of the plant can result in later release of latent xenobiotic residues.

Sequence-specific DNA-binding proteins of the conserved basic/leucine-zipper (bZIP) family of transcription factors can mediate transcription of nuclear genes, a number of which are implicated in protective metabolic responses to environmental stress (*e.g.*, UV light and xenobiotic chemical stress) and infection in plants. Current efforts to characterize plant detoxification processes include understanding how these bZIP regulatory proteins modulate the rate of transcription of specific protective genes. Ultimately, the goal is to improve the metabolism of toxic phytochemicals by potentiating the activity of individual rate-limiting transcription factors and, thus, the expression of their target genes. In this review, we discuss the contribution to these transcriptional regulatory processes of plant bZIP factors, with emphasis on recent investigations of TGA1a, a tobacco bZIP factor that governs xenobiotic stress-activated transcription of detoxification genes.

XENOBIOTIC DETOXIFICATION IN PLANTS

Plant protection against xenobiotic organic toxins and mutagens is largely established by differential expression of a range of rate-limiting detoxification genes. Enzymes encoded by these nuclear genes are responsible for different phases of drug metabolism (Table 6.1).[2-4] During initial reactions of phase I metabolism, xenobiotics are transformed through oxidation, reduction, or hydrolysis to more reactive intermediates, largely through the action of cytochrome p450 mono-oxygenases.[2] In most cases, these intermediates are then covalently conjugated to peptides, sugars, or sulfur groups through the action of phase II enzymes. The protective role that these phase II enzymes confer upon the cell largely derives from the fact that the resultant conjugates exhibit higher aqueous solubility than the parental xenobiotic. Consequently, the modified xenobiotic becomes a substrate for phase III enzymes, including ABC- or GS-X-type transporters and specific ATPases, which serve to transport these compounds for excretion or storage.[3, 4]

Table 6.1: Overview of Xenobiotic Metabolism.

XENOBIOTIC METABOLISM				
Transformation	→	**Conjugation**	→	**Elimination**
Phase I		**Phase II**		**Phase III**
cytochrome P-450		**glutathione *S*-transferase**		**GS-X transporters**
NAD(P)H:oxido-reductase		**glutathione reductase**		**ABC-transporters**
		glutathione peroxidase		
quinone reductase		**glycosyltransferase**		
lipoxygenase		**sulfotransferase**		

Plant glutathione-S-transferase (GST) isoenzymes have been extensively studied regarding their roles in herbicide detoxification and defense against microbial pathogens. Cellular exposure to xenobiotic stress stimulates the rate of transcription of GST and other phase II enzyme-encoding genes. Changes in the activity of GST and other phase II enzymes are strongly correlated, in a number of cases, with cellular resistance to xenobiotic stress and infection.[5-7] All known GSTs have two regions consisting of a thioredoxin domain and a more divergent four-helix bundle domain.[2, 8] The amino-terminal thioredoxin domain contains the principal determinants for reduced glutathione (GSH) binding, whereas the carboxyl-terminal domain provides the specificity determinants for the second (*i.e.*, xenobiotic) substrate. GSTs lower the pKa of the sulfhydryl group of GSH bound at the thioredoxin domain from 9.0 in aqueous solution to 6.5. Consequently, the fundamental chemistry is thought to occur spontaneously through a nucleophilic attack by GS^- in the active site on a nearby electrophilic reaction center, which is itself bound on the other domain of the GST enzyme.[2]

In animals, yeast, and plants, transcriptional regulation of GST genes has been shown to involve the activities of different bZIP transcription factors. The hallmark of this class of factors is the basic/leucine zipper domain, a structural motif that consists of a region of about 25 basic amino acids that directly contacts DNA.[9] Adjacent to this region lies an α-helical sequence of leucine heptad repeats, termed the leucine zipper, which confers dimerization between monomeric bZIP polypeptides. The number of repeats in the zipper varies between 3 to 6, with the seventh residue of the heptad being leucine. Molecular models propose that the leucine zipper positions a diverging pair of α-helical basic regions to make sequence specific contacts with a dyad symmetrical DNA target site in the major groove.[9-11] Although molecular models describing the structure of the bZIP protein-DNA complex have been proposed, the molecular basis for bZIP binding specificity remains unsolved.

Functional studies with bZIP factors have led to the elucidation of distinct rate-limiting steps in their transcription activity (Fig. 6.1). As noted above, the basic residue-rich domain of bZIP transcription factors is essential for *cis*-element binding through DNA major groove contacts. Since all known bZIP factors bind DNA as dimers or oligomers, combinatorial dimerization among these factors can contribute to their DNA-binding specificity (Fig. 6.1-A).[12-16] Some bZIP factors form homodimers with themselves or heterodimers with other bZIP factors, or both. One example is the parsley bZIP factor CPRF4a that forms a homodimer with itself and a heterodimer with CPRF1.[17] Another example is the wheat bZIP protein EMBP-1 that has been shown to heterodimerize with osZIP-1, osZIP-2a, and osZIP-2b factors.[18] Dimerization specifically contributes to the DNA-binding activities of plant bZIP factors, as evidenced by the heterodimer of EMBP-1 and osZIP-1 that binds to the *Em-1a* element, whereas the heterodimer with osZIP-2a or osZIP-2b lacks binding to this element.

Another rate-limiting step is the nuclear import of bZIP transcription factors once they are assembled as dimers in order to regulate the expression of target genes (Fig. 6.1-B). Post-translational modification or binding to cytoplasmic proteins can modulate the nuclear concentration of specific bZIP factors, in some cases by altering the activity of nuclear localization signals (NLS) in these factors. Thus, phosphorylation by casein kinase II (CKII) increases parsley bZIP factor CPRF-2's rate of nuclear import.[19] Indeed, phosphorylation of CPRF-2, stimulated by light, is thought to promote exposure of its NLS or to liberate this bZIP factor from a cytosolic anchoring protein. Light also induces nuclear import of members of a class of bZIP proteins termed G-box binding factors (GBFs) that activate plant gene expression through the G-box motif.[20, 21]

Additional biochemical mechanisms can affect the activity of a bZIP factor upon nuclear import. These include stimulus-responsive interactions with nuclear mediator proteins (Fig. 6.1-C). These mediator proteins can function as either co-activators or co-repressors, respectively, to activate or attenuate *trans*-activation, or affect DNA binding (Fig. 6.1-D) by a bZIP factor. One example of a putative co-repressor is the *Arabidopsis thaliana* COP1 protein. COP1 does not appear to bind directly DNA but instead binds the bZIP factor HY5, thus inhibiting the latter's DNA-binding activity.[22] Specifically, whereas COP1 is localized in the nucleus in

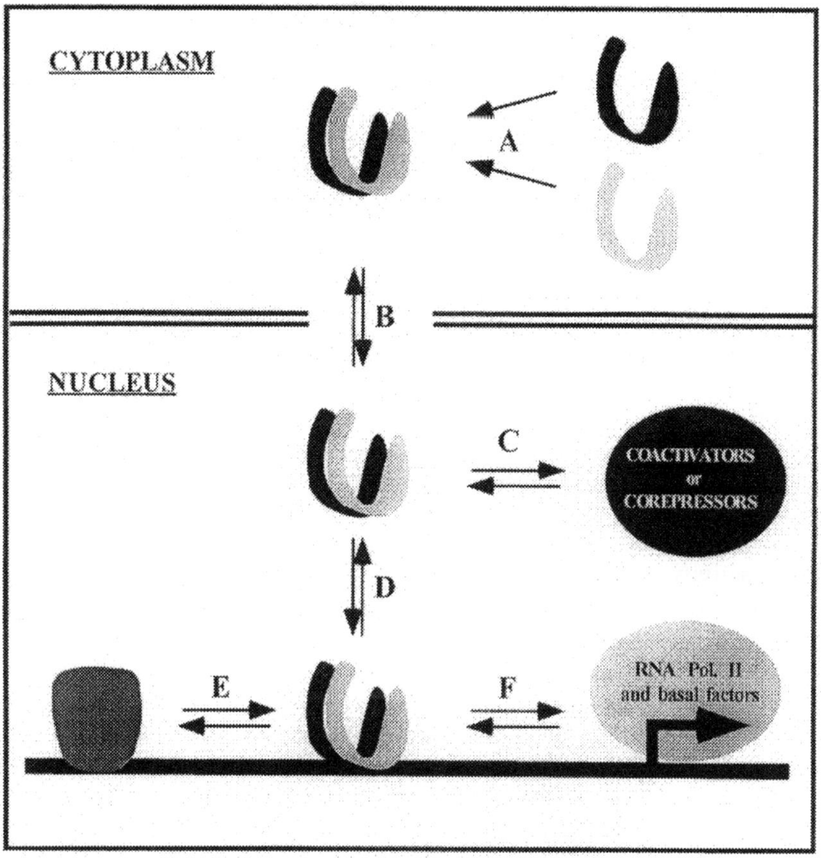

Figure 6.1: Rate-limiting steps in transcription involving bZIP factors: (A) dimerization between the same or different bZIP factor; (B) nuclear import; (C) interactions with mediator proteins, such as co-activators or co-repressors; (D) DNA-binding; (E) interaction with other DNA-binding proteins, including chromatin, and, (F) through interactions with basal transcription factors or RNA polymerase II.

the dark, COP1 releases HY5 and leaves the nucleus in light-irradiated cells, allowing HY5 to bind to its specific DNA-binding site. Also, from recent studies of pathogenesis-related (PR) gene expression in plants, the NPR1/NIM1 protein may function as a co-activator, as it associates with the *Arabidopsis* bZIP factor TGA2, thereby increasing the latter's binding to the *as-1 cis*-element.[23] A plant protein that may have characteristics of both a co-activator and co-repressor is VP1 of maize. VP1 is a key regulator of seed maturation and germination, and mutations in this protein cause premature germination (vivipary). VP1 increases the DNA-binding affinity of the bZIP protein EMBP-1 and potentiates transcription through its promoter binding site(s).[24] In contrast to its role as an activator, VP1 also represses transcription of the α-amylase gene during seed maturation, presumably through its interaction with an as yet unidentified transcription factor.[25]

Cross-talk within and among different bZIP transcription factors in animal systems is also important in modulating gene expression. Although plant bZIP factors have not been as well studied in this regard, it is thought that they function in a similar manner (Fig. 6.1-E). Experiments with *Arabidopsis* EREBP protein suggest that cross-coupling between it and other bZIP factors may be important in gene regulation during plant stress responses.[26] This cross-coupling is not limited only to interactions between bZIP factors, given that transcription from the wheat prolamin LMWG-1D1 gene is synergistically potentiated by the bZIP factor SPA and ESBF1, a member of the Dof family of transcriptional regulators.[27] Also, VP1 is thought to associate loosely with DNA in a non-specific manner in order to modulate bZIP factor activity. Another example is Histone H1, which has been shown to increase promoter binding of EMBP-1 *in vitro*.[28] The *Arabidopsis* OBP1 protein, a Dof factor, binds to promoter AAGG motifs and to adjacent bound members of the TGA family of bZIP proteins. This protein-protein interaction enhances the binding activity of bZIP and Dof transcription factors with their cognate promoter elements.[29]

Stimulus-induction of cellular protein kinase or phosphatase activities can affect the phosphorylation state and, consequently, DNA-binding activity of certain plant bZIP factors. One example is G/HBF-1, a novel soybean bZIP protein, which binds both G-box and adjacent H-box elements in the soybean chalcone synthase promoter. G/HBF1 is rapidly phosphorylated at key serine residues *in vivo* in response to fungal cell wall elicitor, resulting in enhanced DNA-binding activity.[30] The *Arabidopsis* bZIP factor GBF1 also has enhanced DNA-binding activity upon phosphorylation by CKII.[21] In contrast, HBP-1a, a member of the HBP-1a/GBF family of bZIP factors, is phosphorylated *in vitro* by a nuclear kinase at serine residues, resulting in a loss of DNA-binding activity.[31] In contrast, in maize endosperm, only the non-phosphorylated form of Opaque-2 has been shown to bind DNA with high affinity. Though the specific kinases and phosphatases that act upon Opaque-2 are unknown, their activities are thought to be regulated by light and dark.[32]

Once bound to DNA, bZIP factors may interact directly with RNA polymerase II or other basal transcription factors (Fig. 6.1-F). Although this type of regulatory interaction has not yet been directly established with bZIP factors of plants, the strong overall functional conservation of basal transcription factors of plants, animals, and yeast suggests that this regulation is likely.

TRANSCRIPTIONAL CONTROL OF XENOBIOTIC DETOXIFICATION BY bZIP FACTORS

Evidence from a number of eukaryotic model systems increasingly indicates that xenobiotic activation of GST and other phase II protective genes occurs largely through stimulus-induced changes in the activities of bZIP-type transcription factors. In yeast and animals, up-regulation of transcription by xenobiotic stress occurs through electrophile response elements (EREs) and antioxidant response elements (AREs),[33, 34] respectively, which are found in the promoters of these genes. For example, in yeast, the bZIP factor yAP-1 acts through the ERE present in detoxification genes, such as TRX2 and GSH1, and ATP-binding cassette transporter genes, such as YCF1, PDR5 and SNQ2.[35] Under oxidative stress conditions during which reactive oxygen species (ROS) are made, yAP-1 activity is increased. Specifically, the carboxy-terminal cysteine-rich domain (CRD), containing three conserved cysteines, gets cleaved, resulting in constitutive nuclear localization of yAP-1, increased DNA binding, and subsequent high levels of activity. It has been shown recently that the three cysteine residues through which regulation occurs are important for sensing the redox state of the cell.[35] Preliminary evidence seems to indicate that the regulation of yAP-1 occurs through modifying its interaction with another cellular protein in the cytoplasm, as opposed to direct post-translational modification of yAP-1.

Xenobiotic regulation of transcription in animals differs somewhat from that of yeast. For example, the xenobiotic-activated bZIP factor NRF-2 binds AP-1-like elements, or AREs, found in animal phase II genes such as NAD(P)H:quinone oxidoreductase (NOQ_1), which encodes an enzyme involved in detoxification of the cell and protection against oxidative stress.[36] However, the mechanism of action is different from that of yAP-1. NRF-2 is located in the cytoplasm where it is co-associated with the cytoskeletal protein Keap1.[37] This NRF-2-Keap1 complex could be a type of sensor for electrophiles or ROS antioxidants, given that under oxidative stress conditions the NRF-2-Keap1 association is disrupted, resulting in NRF-2 localization to the nucleus where it induces expression of NOQ_1 and other protective enzymes. Because NRF-2 homodimers do not appear to bind AREs,[37] it is more likely that this factor heterodimerizes with some other protein, and that this protein complex then activates transcription.[38]

The NRF-2 mechanism of action is also made more complicated by the fact that activation of NRF-2 is phosphorylation dependent. Protein kinase C (PKC) has been shown to phosphorylate NRF-2, resulting in nuclear import of the phosphorylated form of NRF-2.[39] Although both oxidative stress and phosphorylation have been shown to localize NRF-2 to the nucleus, the actual mechanism could be a combination of both, given that oxidative stress has been shown to increase cellular protein kinase signaling pathways. Specifically, under oxidative stress conditions, the PKC pathway may be activated, resulting in phosphorylation of NRF-2, which then disassociates from the cytoplasmic Keap1 protein and enters the nucleus where it binds the ARE and *trans*-activates its target genes.[39]

TGA1a: A XENOBIOTIC STRESS-ACTIVATED bZIP FACTOR OF TOBACCO

Although key structural and regulatory features of xenobiotic activated bZIP factors of animals and yeast have been elucidated, comparatively less is known about their functional counterparts in plants. Xenobiotic stress, induced by auxin herbicides, such as 2,4-dichlorophenoxyacetic acid (2,4-D), or their biologically "inactive" analogs, has been shown to stimulate transcription in plants through *as-1*-type *cis*-elements.[40, 41] Recent work has shown that these elements make important contributions to regulating the expression of plant genes involved in detoxification and host defense. Thus, elucidating the molecular basis of the *as-1*-dependent transcriptional response pathway in plants will be important for advancing basic knowledge of protective processes to develop effective strategies to manipulate endogenous protective genes.

Homologous *cis*-acting elements termed *as-1*, *ocs*, or *nos* are comprised of a pair of tandem head-to-tail or inversely oriented TGAC motifs that are separated by 8-9 bases.[41] These promoter regulatory elements are generally located just upstream of the TATA box, where they confer transcription in response to plant defense hormones, wounding, and xenobiotic stress. Intriguingly, the presence of these elements in promoter DNA of plant nuclear and viral genes suggests opposing roles in disease and host-defense.[41-43] Structurally and functionally, *as-1* elements resemble AREs and EREs of animals and yeast, described above.[44, 45] Both *as-1* and these related *cis*-elements contribute to the transcriptional regulation of the GST supergene family, members of which are implicated in cellular detoxification.[43, 46, 47] Other plant genes that contain functional *as-1*-type promoter elements encode for pathogenesis-related (PR) proteins that are thought to confer protection against infection and disease.[7, 48, 49]

As-1 is regulated by the activity of one or more bZIP proteins, termed "TGA factors", that selectively bind this element. Outside of their conserved DNA-binding domains, these plant TGA factors show little overall homology to animal and yeast bZIP proteins, such as NRF-2 or yAP-1, mentioned above. Multiple and distinct genes for TGA factors have been identified in both the same and different plant species. The best studied of these TGA factors to date is TGA1a of tobacco (Fig. 6.2). TGA1a mediates *as-1*-dependent transcription and is activated by xenobiotic stress.[50-52] The biological significance of this response is suggested by observations that *TGA1a* transcripts and protein are preferentially co-expressed in root-tip meristem cells with *GNT35*, an *as-1*-regulated GST gene whose expression is enhanced by xenobiotic stress.[53]

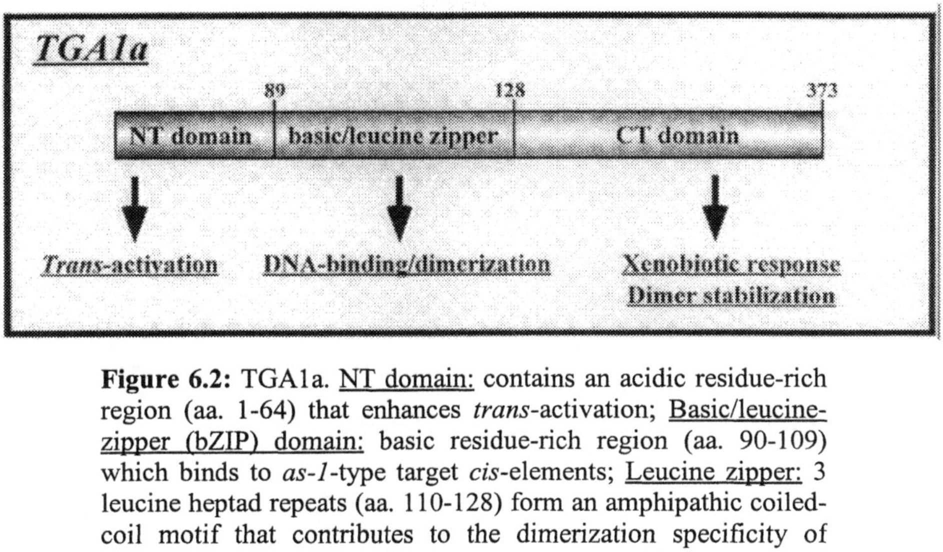

Figure 6.2: TGA1a. NT domain: contains an acidic residue-rich region (aa. 1-64) that enhances *trans*-activation; Basic/leucine-zipper (bZIP) domain: basic residue-rich region (aa. 90-109) which binds to *as-1*-type target *cis*-elements; Leucine zipper: 3 leucine heptad repeats (aa. 110-128) form an amphipathic coiled-coil motif that contributes to the dimerization specificity of TGA1a; CT domain: mediates response to xenobiotic stress and stabilizes TGA1a homodimer.

TGA1a was cloned in 1989 by Katagiri *et al.* by using concatamers of the *as-1*-like wheat *hex-1* element to probe a random primed cDNA library made from tobacco leaf mRNA.[54] The full-length protein of 373 amino acids with a molecular mass of ~41 kDa has three functional domains. The first domain is the amino-terminus (NT), which consists of both an acidic region (residues 5-52) and a basic region (residues 70-82). The second domain is a leucine zipper region (residues 86-142) consisting of three heptad repeats. The third domain is the carboxyl-terminus (CT, residues 143-373) that has a glutamine-rich region (residues 244-332) (Fig. 6.3). Dimerization, necessary for *as-1* binding by TGA1a, is thought to occur through coiled-coil interactions between adjacent bZIP domains.[9] The CT domain of TGA1a also promotes dimerization by this factor and mediates its response to

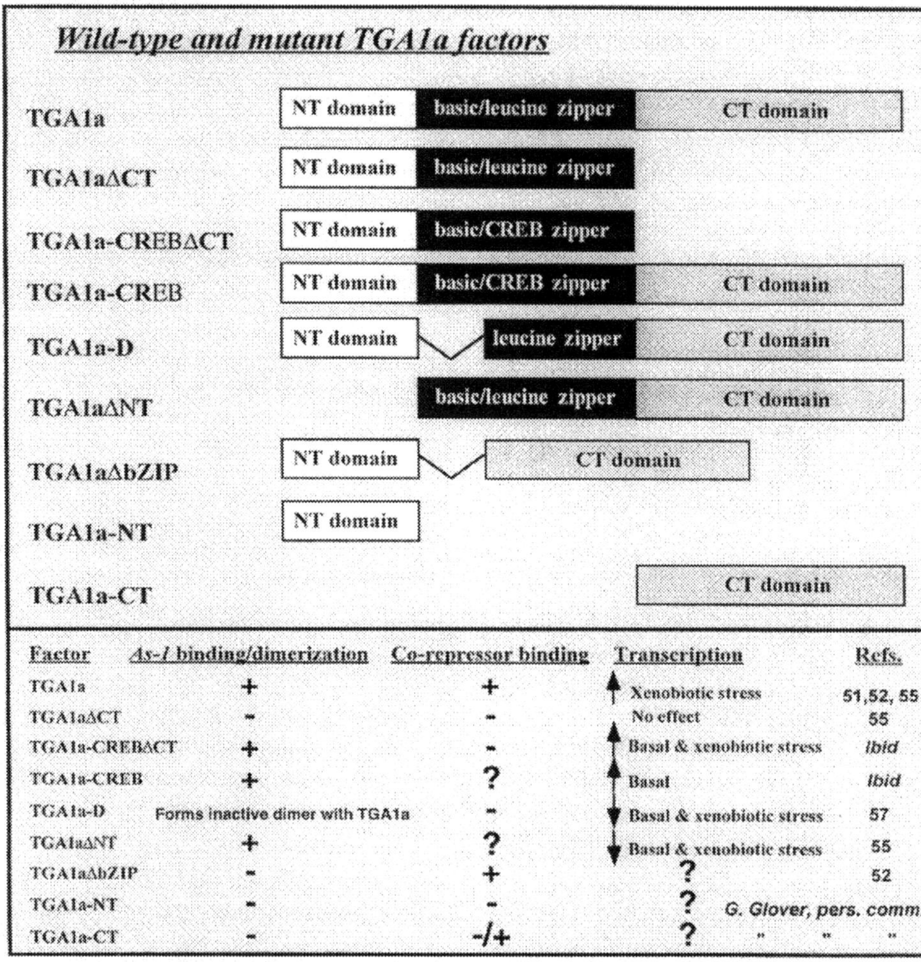

Factor	*As-1* binding/dimerization	Co-repressor binding	Transcription	Refs.
TGA1a	+	+	↑ Xenobiotic stress	51,52, 55
TGA1aΔCT	-	-	No effect	55
TGA1a-CREBΔCT	+	-	↑ Basal & xenobiotic stress	*Ibid*
TGA1a-CREB	+	?	↑ Basal	*Ibid*
TGA1a-D	Forms inactive dimer with TGA1a	-	↓ Basal & xenobiotic stress	57
TGA1aΔNT	+	?	↓ Basal & xenobiotic stress	55
TGA1aΔbZIP	-	+	?	52
TGA1a-NT	-	-	?	*G. Glover, pers. comm.*
TGA1a-CT	-	-/+	?	" " "

Figure 6.3: Organization and activities of wild-type and mutant TGA1a factors. TGA1a: wild-type factor; TGA1aΔCT: deletion of CT domain; TGA1a-CREBΔCT: same as TGA1aΔCT, but has CREB leucine zipper; TGA1a-CREB: same as TGA1a, but has CREB leucine zipper; TGA1a-D: internal deletion of basic domain residues; TGA1aΔNT: deletion of NT domain; TGA1aΔbZIP: deletion of bZIP domain; TGA1a-NT: NT domain alone; TGA1a-CT: CT domain alone. (+): active, (-) no activity detected, (?) not determined; Arrows: ↑ enhanced activity; ↓ decreased activity.

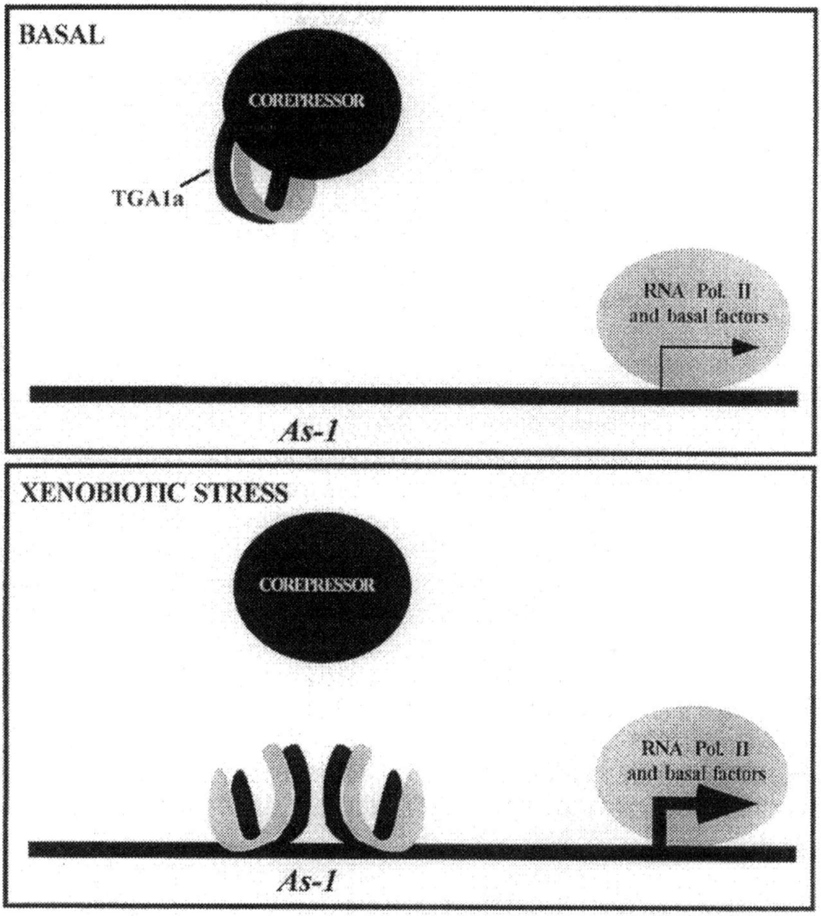

Figure 6.4: Model of basal and xenobiotic stress-activated transcription through TGA1a and its cognate *as-1* element. Basal transcription: TGA1a is sequestered in the nucleus by the 120 kDa co-repressor protein. Under these conditions, the rate of transcription through *as-1* is minimal. Xenobiotic stress: Exposure to xenobiotic stress induces dissociation of TGA1a from the co-repressor protein. Subsequent occupancy of *as-1* by TGA1a results in an enhanced rate of gene expression, as indicated by the broad arrow at the start site of transcription.

xenobiotic stress.[52] Gain-of-function studies of TGA1a have shown that the CT domain mediates this response through an intramolecular regulatory mechanism involving stress-reversible, CT domain-mediated repression of the NT activation domain.[52] Recent evidence indicates further that this regulatory mechanism involves a nuclear protein of 120 kDa, which binds to TGA1a through synergistic contributions with its NT activation and CT regulatory domains.[55] Xenobiotic stress induces a rapid and transient change in this nuclear protein's binding to TGA1a, with kinetics that are inversely correlated with changes in the rate of *as-1*-dependent transcription. Moreover, xenobiotic stress was also found to markedly stimulate the *as-1* binding activity of TGA1a.

These findings, and those of others, have been placed in the context of a working model of how *as-1*-dependent transcription is regulated by TGA1a (Fig. 6.4). Under basal conditions (*i.e.*, no stimulus), nuclear TGA1a is stably associated with a 120 kDa co-repressor protein that blocks the ability of TGA1a to bind *as-1*. Concomitantly, a TGA factor with little or no *trans*-activation potential, such as TGA2.2,[56] may occupy *as-1* and inhibit basal transcription while preserving future access by TGA1a to this site. In response to xenobiotic stress, the TGA1a/co-repressor complex dissociates, thus allowing TGA1a to bind *as-1* and displace the less active TGA factor. The cumulative effect of these changes is to enhance transcription. Prolonged exposure to xenobiotic stress results in the re-association of TGA1a with the co-repressor protein and a concomitant decline in *as-1*- dependent transcription. These changes suggest the existence of biochemical pathways for reversing initial stimulatory events through TGA1a.

STRATEGIES FOR ALTERING THE EXPRESSION OF DETOXIFICATION GENES WITH MODIFIED PLANT bZIP FACTORS

Biochemical steps that contribute to the rate-limiting activities of specific plant bZIP factors are potentially attractive targets for manipulation to confer gain or loss of transcription functions. As in the case of animal bZIP factors, "dominant inhibitor" versions of plant bZIP factors have been shown to attenuate selectively target gene activity. Specifically, research on the structure and function of TGA1a and its sub-domains has led to the design of modified forms of TGA1a with distinct dimerization, co-repressor binding, and transcription activities (Fig. 6.3). For example, TGA1a-D, a bZIP factor that lacks the basic domain (Fig. 6.3), heterodimerizes with wild-type TGA1a and blocks its ability to bind *as-1*.[57] Transgenic expression of TGA1a-D in tobacco inhibits ASF-1, a nuclear protein activity that also binds *as-1*. Similar mutations in PG13, a homolog of TGA1a, inhibit nuclear ASF-1 activity and transcription *in vivo* through *as-1*.[58] Although mutations that affect dimerization and DNA-binding are by far the most commonly employed, an alternative strategy for making dominant inhibitors is exemplified by the use of a mutant version of the bZIP factor EMBP-1, which lacks the proline-rich activation domain. Transgenic expression of this mutant bZIP factor (ΔEMBP) results in abnormal vegetative development of tobacco plants, particularly of stems and petioles.[59]

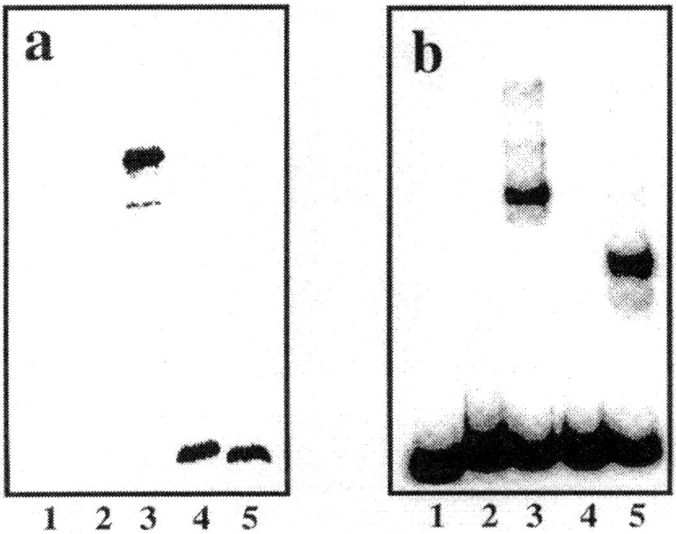

Figure 6.5: Differential binding of wild-type and mutant TGA1a factors to *as-1*. (a) Coupled transcription-translation reactions programmed with the following templates were fractionated by SDS-PAGE. [^{35}S]-labeled proteins from these reactions were subsequently revealed by fluorography: lane 1: no template; lane 2: vector DNA; lane 3: TGA1a; lane 4: TGA1aΔCT, and lane 5: TGA1a-CREBΔCT. (b) Gel-shift assays with an *as-1* oligonucleotide probe. Samples in (a) were incubated with labeled and fractionated by PAGE. The positions of free and protein-bound probe were revealed by autoradiography.

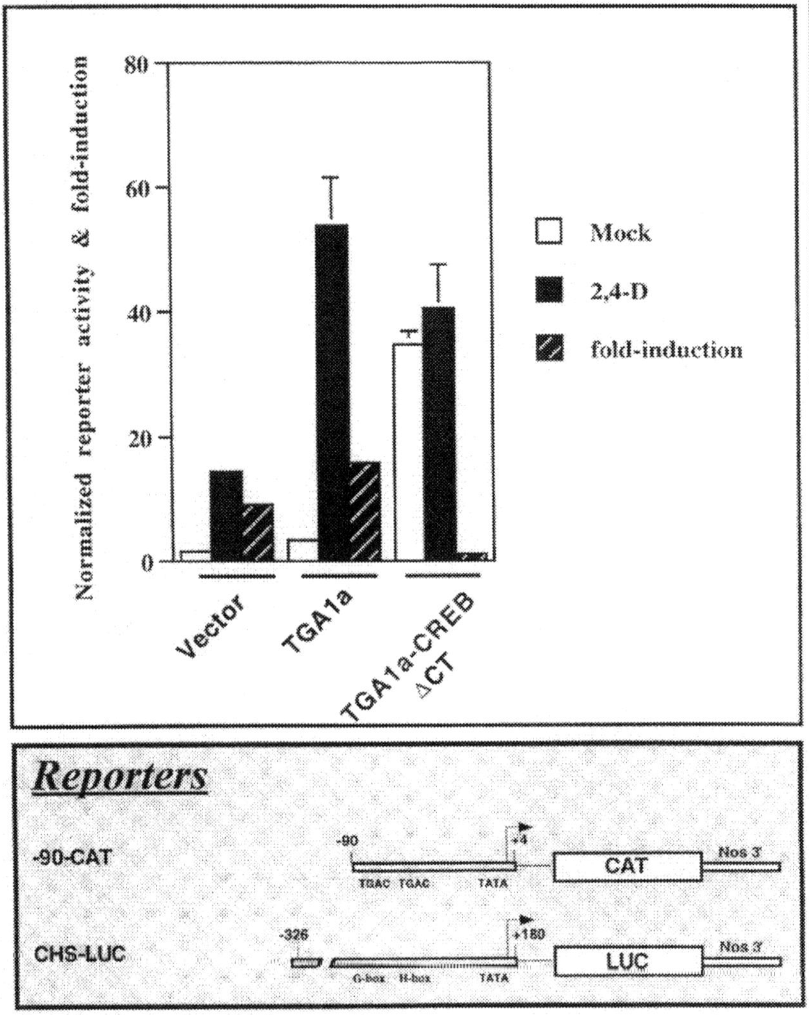
Normalized reporter activity & fold-induction
80
60
40
20
0
Mock
2,4-D
fold-induction
Vector
TGA1a
TGA1a-CREB
ΔCT
Reporters
-90-CAT
-90
+4
TGAC TGAC
TATA
CAT
Nos 3'
CHS-LUC
-326
+180
G-box H-box
TATA
LUC
Nos 3'

Figure 6.6: Transcription activities of wild-type and modified TGA1a. BY-2 tobacco protoplasts were transfected with the following effector plasmids: pMON 999 (vector), pMON-TGA1a (TGA1a) or pMON-TGA1a-CREBΔCT (TGA1a-CREBΔCT). Reporters: In these experiments, the following reporter genes were co-transfected into protoplasts to monitor: -90-CAT (chloramphenicol acetyltransferase) regulated by an *as-1* element and minimal promoter, and CHS-LUC (luciferase), regulated by a bean chalcone synthase promoter that is unresponsive to xenobiotic stress. Protoplasts were then incubated with 0.1% ethanol carrier (mock) or 100 μM 2,4-D (2,4-D) for 20h and assayed for reporter enzyme activities. To correct for differences in transfection efficiency, chloramphenicol acetyltransferase activity was normalized to that of luciferase. Results are the mean and standard error of 3 independent experiments. Fold-induction is the ratio of 2,4-D and mock values.

An alternative strategy to dominant negative forms of TGA factors is to inhibit or attenuate functions other than simple dimerization or DNA binding. Our prior work indicates that TGA1a preferentially enhances transcription through its cognate *as-1* element in response to xenobiotic chemical stress induced by 2,4-D.[52] The CT domain of TGA1a (residues 143-373) mediates this stimulus response through a mechanism involving stimulus-reversible repression of this factor's NT activation domain and through interactions with a 120 kDa nuclear co-repressor protein.[55] From these findings, we predicted that a modified TGA1a factor, termed TGA1aΔCT, which lacks the inhibitory CT domain, might show strong constitutive activity in governing *as-1*-dependent transcription. One complication in testing this modified factor, however, is that CT domain residues are also involved in dimer stabilization of TGA1a.[60] Thus, TGA1aΔCT was unable to dimerize efficiently through the leucine zipper, and, thus, bind DNA (Fig. 6.5). To address this limitation in assessing other CT domain functions, we substituted 3 heptad repeats from the leucine zipper of CREB, a mammalian bZIP factor, for the leucine zipper in TGA1aΔCT to create TGA1a-CREBΔCT (Fig. 6.3). UnlikeTGA1aΔCT, this modified factor was able to form a stable homodimer *in vitro* and bind *as-1* to a degree that is comparable to that of TGA1a (Fig. 6.5). As predicted, TGA1a-CREBΔCT did not bind the co-repressor protein and showed constitutive and enhanced basal and xenobiotic stress-induced transcriptional activity through *as-1* (Fig. 6.6). These results demonstrate that site-directed changes in regulatory sites of TGA1a can over-ride normal regulatory controls that govern basal repression of this bZIP factor and its response to xenobiotic stress.

CONCLUSIONS AND FUTURE DIRECTIONS

Animal and plant bZIP factors play key regulatory roles in stimulus-response pathways leading to changes in the rate of transcription of protective genes. In this review, we have illustrated how modifications of a plant bZIP factor can alter its response to stimulus. TGA1a is a xenobiotic stress-activated bZIP factor of tobacco that activates transcription through *as-1*-type *cis*-elements. TGA1a domains that govern basal and xenobiotic stress-induced transcription were identified and subsequently led to the development of a novel modified TGA1a factor that lacks normal regulatory control. Target genes that are regulated by TGA1a include those that encode for GST isoenzymes, which are implicated in xenobiotic detoxification. Investigations are underway to test whether modified forms of TGA1a and other TGA factors can stably affect the expression of these and other protective genes in plants.

ACKNOWLEDGEMENTS

This article was supported by the University of Maryland Biotechnology Institute and a grant to J. Arias from the National Science Foundation (MCB 9527364). The authors thank Marisela Morales and Elaine Rahbar for critical comments on the manuscript.

REFERENCES

1. SANDERMANN, H., Higher plant metabolism of xenobiotics: the 'green liver' concept, *Pharmacogenetics*, 1994, **4**, 225-241.
2. HAYES, J., PULFORD, D., The glutathione S-transferase supergene family: Regulation of GST and the contribution of the isoenzymes to cancer chemoprotection and drug resistance, *Crit. Rev. Biochem. Mol. Biol.*, 1995, **301**, 445-600.
3. GAILLARD, C., DUFAUD, A., TOMMASSINI, R., KREUZ, L., AMRHEIN, N., MARTINOIA, E., A herbicide antidote (safener) induces the activity of both the herbicide detoxifying enzyme and of a vacuolar transporter for the detoxified herbicide, *FEBS Lett.*, 1994, **352**, 219-221.
4. TOMMASINI, R., VOGT, E., SCHMID, J., FROMENTAU, M., AMRHEIN, N., MARTINOIA, E., Differential expression of genes coding for ABC transporters after treatment of *Arabidopsis thaliana* with xenobiotics, *FEBS Lett.*, 1997, **411**, 206-210.
5. DIXON, D., CUMMINS, I., COLE, D., EDWARDS, R., Glutathione-mediated detoxification systems in plants, *Curr. Opin. Plant Biol.*, 1998, **1**, 258-266.
6. MAUCH, F., DUDLER, R., Differential induction of distinct glutathione S-transferases of wheat by xenobiotics and by pathogen attack, *Plant Physiol.*, 1993, **102**, 1193-1201.

7. HAHN, K., STRITTMATTER, G., Pathogen-defense gene *prp-1* from potato encodes an auxin-responsive glutathione *S*-transferase, *Eur. J. Biochem.*, 1994, **22**, 619-626.
8. BABBITT, P., Reengineering the glutathione S-transferase scaffold: A rational design strategy pays off, *Proc. Natl. Acad. Sci. USA*, 2000, **97**, 10289-10300.
9. VINSON, C., SIGLER, P., MCKNIGHT, S., Scissors-grip model for DNA recognition by a family of leucine zipper proteins, *Science*, 1989, **246**, 911-916.
10. IZAWA, T., FOSTER, R., CHUA, N-H., Plant bZIP protein-DNA binding specificity, *J. Mol. Biol.*, 1993, **230**, 1131-1144.
11. PAOLELLA, D., YICHIN, L., FABIAN, M., SCHEPARTZ, A., Electrostatic mechanism for DNA bending by bZIP proteins, *Biochemistry*, 1997, **36**, 10033-10038.
12. SCHWECHHEIMER, C., BEVAN, M., The regulation of transcription factor activity in plants, *Trends Plant Sci.*, 1998, **3**, 378-382.
13. WEISS, M.A., ELLENBERGER, T., WOBBE, C.R., LEE, J.P., HARRISON, S.C., STRUHL, K., Folding transition in the DNA-binding domain of GCN4 on specific binding to DNA, *Nature*, 1990, **347**, 575-578.
14. MESHI, T., IWABUCHI, M., Plant transcription factors, *Plant Cell Physiol.*, 1995, **36**, 1405-1420.
15. LEE, K., Dimeric transcription factor families: It takes two to tango, but who decides on partners and venue?, *J. Cell Sci.*, 1992, **103**, 9-14.
16. BAXEVANIS. A., VINSON, C., Interactions of coiled coils in transcription factors: Where is the specificity?, *Curr. Opin. Genet. Dev.*, 1993, **3**, 278-285.
17. KIRCHER, S., LEDGER, S., HAYASHI, H., WEISSHAAR, B., SCHAFER, E., FROHNMEYER, H., CPRF4a, a novel plant bZIP protein of the CPRF family: Comparative analyses of light-dependant expression, post-translational regulation, nuclear import and heterodimerization, *Mol. Gen. Genet.*, 1998, **257**, 595-605.
18. NANTEL, A., QUATRANO, R., Characterization of three rice basic/leucine zipper factors, including two inhibitors of EmBP-1 DNA-binding activity, *J. Biol. Chem.*, 1996, **271**, 31296-31305.
19. KIRCHER, S., WELLMER, F., NICK, P., RUGNER, A., SCHAFER, E., HARTER, K., Nuclear import of the parsley bZIP transcription factor CPRF2 is regulated by phytochrome photoreceptors, *J. Cell Biol.*, 1999, **144**, 201-211.
20. TERZAGHI, W.B., BERTEKAP, Jr, R.L., CASHMORE, A.R., Intracellular localization of GBF proteins and blue light-induced import of GBF2 fusion proteins into the nucleus of cultured *Arabidopsis* and soybean cells, *Plant J.*, 1997, **11**, 967-982.
21. KLIMCZAK, L.J., SCHINDLER, U., CASHMORE, A.R., DNA binding activity of the *Arabidopsis* G-box binding factor GBF1 is stimulated by phosphorylation by casein kinase II from broccoli, *Plant Cell*, 1992, **4**, 87-98.
22. ANG, L.H., CHATTOPADHYAY, S., WEI, N., OYAMA, T., OKADA, K., BATSCHAUER, A., DENG, X.W., Molecular interaction between COP1 and HY5 defines a regulatory switch for light control of *Arabidopsis* development, *Mol. Cell*, 1998, **1**, 213-222.

23. DESPRES, C., DELONG, C., GLAZE, S., LIU, E., FOBERT, P.R., The *Arabidopsis* NPR1/NIM1 protein enhances the DNA binding activity of a subgroup of the TGA family of bZIP transcription factors, *Plant Cell*, 2000, **12**, 279-290.
24. HILL, A., NANTEL, A., ROCK, C.D., QUATRANO, R.S., A conserved domain of the viviparous-1 gene product enhances the DNA binding activity of the bZIP protein EmBP-1 and other transcription factors, *J. Biol. Chem.*, 1996, **271**, 3366-3374.
25. HOECKER, U., VASIL, I., MCCARTY, D., Integrated control of seed maturation and germination programs by activator and repressor functions of Viviparous-1 of maize, *Genes Dev.*, 1995, **9**, 2459-2469.
26. BUTTNER, M., SINGH, K.B., *Arabidopsis thaliana* ethylene-responsive element binding protein (AtEBP), an ethylene-inducible, GCC box DNA-binding protein interacts with an ocs element binding protein, *Proc. Natl. Acad. Sci. USA*, 1997, **94**, 5961-5966.
27. CONLAN, R., HAMMOND-KOSACK, M., BEVAN, M., Transcription activation mediated by the bZIP factor SPA on the endosperm box is modulated by ESBF-1 *in vitro*, *Plant J.*, 1999, **19**, 173-181.
28. SCHULTZ, T.F., SPIKER, S., QUATRANO, R.S., Histone H1 enhances the DNA binding activity of the transcription factor EmBP-1, *J. Mol. Biol.*, 1996, **271**, 25742-25745.
29. ZHANG, B., CHEN, W., FOLEY, R., BUTTNER, M., SINGH, K., Interactions between distinct types of DNA binding proteins enhance binding to *ocs* element promoter sequences, *Plant Cell*, 1995, **7**, 2241-2252.
30. DROGE-LASER, W., KAISER, A., LINDSAY, W.P., HALKIER, B.A., LOAKE, G.J., DOERNER, P., DIXON, R.A., LAMN, C., Rapid stimulation of a soybean protein-serine kinase that phosphorylates a novel bZIP DNA-binding protein, G/HBF-1, during the induction of early transcription-dependent defenses, *EMBO J.*, 1997, **16**, 726-738.
31. MESHI, T., MODA, I., MINAMI, M., OKANAMI, M., IWABUCHI, M., Conserved Ser residues in the basic region of the bZIP-type transcription factor HBP-1a(117): Importance in DNA binding and possible targets for phosphorylation, *Plant Mol. Biol.*, 1998, **36**, 125-136.
32. CICERI, P., GIANAZZA, E., LAZZARI, B., LIPPOLI, G., GENGA, A., HOSCHECK, G., SCHMIDT, R.J., VIOTTI, A., Phosphorylation of Opaque2 changes diurnally and impacts its DNA binding activity, *Plant Cell*, 1997, **9**, 97-108.
33. ITOH, K., ISHII, T., WAKABAYASHI, N., YAMAMOTO, M., Regulatory mechanisms of cellular responses to oxidative stress, *Free Rad. Res.*, 1999, **31**, 319-324.
34. COLEMAN, S., EPPING, E., STEGGERDA, S., MOYE-ROWLEY, W., Yap1p activates gene transcription in an oxidant-specific fashion, *Mol. Cell. Biol.*, 1999, **19**, 8302-8313.
35. KUGE, S., JONES, N., NOMOTO, A., Regulation of yAP-1 nuclear localization in response to oxidative stress, *Eur. Mol. Biol. J.*, 1997, **16**, 1710-1720.
36. VENUGOPAL, R., JAISWAL, A., Nrf1 and Nrf2 positively and c-Fos and Fra1 negatively regulate the human antioxidant response element-mediated expression of NAD(P)H:quinone oxidoreductase$_1$ gene, *Proc. Natl. Acad. Sci. USA*, 1996, **93**, 14960-14965.

37. ITOH, K., WAKABAYASHI, N., KATOH, Y., ISHII, T., IGARISHI, K., ENGEL, J., YAMAMOTO, M., Keap1 represses nuclear activation of antioxidant responsive elements by Nrf2 through binding to the amino-terminal Neh2 domain, *Gen. Dev.*, 1999, **13**, 76-86.
38. NGUYEN, T., HUANG, H.C., PICKETT, C., Transcriptional regulation of the antioxidant response element, *J. Biol. Chem.*, 2000, **275**, 15466-15473.
39. HUANG, H.C., NGUYEN, T., PICKETT, C., Regulation of the antioxidant response element by protein kinase C-mediated phosphorylation of NF-E2-related factor 2, *Proc. Natl. Acad. Sci. USA*, 2000, **97**, 12475-12480.
40. ULMASOV, T., OHMMIYA, A., HAGEN, G., GUILFOYLE, T. 1995. The soybean *GH2/4* gene that encodes a glutathione *S*-transferase has a promoter that is activated by a wide range of chemical agents, *Plant Physiol.*, 1995, **108,** 919-927.
41. BOUCHEZ, D., TOKUHISA, J., LLEWELLYN, D., DENNIS, E., ELLIS, J., The *ocs*-element is a component of the promoters of several T-DNA and plant viral genes, *EMBO J.*, 1989, **8**, 4197-4204.
42. LAM, E., BENFEY, P., GILMARTIN, P., FANG, R-X., CHUA, N-H., Site-specific mutations alter in vitro factor binding and change promoter expression pattern in transgenic plants., *Proc. Natl. Acad. Sci. USA*, 1989, **86**, 7890-7894.
43. DROOG, F. SPEK, A., VAN DER KOOY, A., DE RUYTER, A., HOGE, H., LIBBENGA, K., HOOYKAAS, P., VAN DER ZAAL, B., Promoter analysis of the auxin-regulated tobacco glutathione S-transferase genes *Nt103-1* and *Nt103-35*, *Plant Mol. Biol.*, 1995, **29**, 413-429.
44. FRILING, R., BERGELSON, S., DANIEL, V., Two adjacent AP-1-like binding sites form the electrophile element of the murine glutathione *S*-transferase Ya subunit gene, *Proc. Natl. Acad. Sci. USA*, 1992, **89**, 668-672.
45. WASSERMAN, W., FAHL, W., Functional antioxidant responsive elements, *Proc. Natl. Acad. Sci. USA*, 1997, **94**, 5361-5366.
46. CHEN, W., CHAO, G., SINGH, K., The promoter of a H_2O_2-inducible, *Arabidopsis* glutathione *S*-transferase gene contains closely linked OBF- and OBP1-binding sites, *Plant J.*, 1996, **10**, 955-966.
47. CHEN, W., SINGH, K., The auxin, hydrogen peroxide and salicylic acid induced expression of the *Arabidopsis GST6* promoter is mediated in part by an *ocs* element, *Plant J.*, 1999, **19,** 667-677.
48. LEBEL, E., HEIFETZ, P., THORNE, L., UKNES, S., RYALS, J., WARD, E., Functional analysis of regulatory sequences controlling *PR-1* gene expression in *Arabidopsis*, *Plant J.*, 1998, **16**, 223-233.
49. ZHOU, J., TRIFA, Y., SILVA, H., PONTIER, D., LAM, E., SHAH, J., KLESSIG, D.F., NPR1 differentially interacts with members of the TGA/OBF family of transcription factors that bind an element of the *PR-1* gene required for induction by salicylic acid, *Mol. Plant Microbe Interact.*, 2000, **13**, 191-202.
50. YAMAZAKI, K-I, KATAGIRI, F., IMASEKI, H., CHUA, N-H., TGA1a, a tobacco DNA-binding protein, increases the rate of initiation in a plant *in vitro* transcription system, *Proc. Natl. Acad. Sci. USA*, 1990, **87**, 7035-7039.

51. NEUHAUS, G., NEUHAUS-URL, G., KATAGIRI, F., SEIPEL, K., CHUA, N-H., Tissue-specific expression of *as-1* in transgenic tobacco, *Plant Cell*, 1994, **6**, 827-834.
52. PASCUZZI, P., HAMILTON, D., BODILY, K., ARIAS, J., Auxin-induced stress potentiates *trans*-activation by a conserved plant basic/leucine-zipper factor, *J. Biol. Chem.*, 1998, **273**, 26631-26637.
53. KLINEDINST, S., PASCUZZI, P., REDMAN, J., DESAI, M., ARIAS, J., A xenobiotic-stress-activated transcription factor and its cognate target genes are preferentially expressed in root tip meristems, *Plant Mol. Biol.*, 2000, **42**, 679-688.
54. KATAGIRI, F., LAM, E., CHUA, N-H., Two tobacco DNA-binding proteins with homology to the nuclear factor CREB, *Nature*, 1989, **240**, 727-730.
55. JOHNSON, C., GLOVER, G., ARIAS, J., Regulation of DNA binding and *trans*-activation by a xenobiotic stress-activated plant transcription factor, 2000, in press.
56. NIGGEWEG, R., THUROW, C., WEIGEL, R., PFITZNER, U., GATZ, C., Tobacco TGA factors differ with respect to interaction with NPR1, activation potential and DNA-binding properties., *Plant Mol. Biol.*, 2000, **42**, 775-788.
57. MIAO, Z., LAM, E., Construction of a trans-dominant inhibitor for members of the TGA family of transcription factors conserved in higher plants, *Plant J.*, 1995, **7**, 887-896.
58. RIEPING, M.,FRITZ, M., PRAT, S., GATZ, C., A dominant negative mutant of PG13 suppresses transcription from a Cauliflower mosaic Virus 35S truncated promoter in transgenic Tobacco plants, *Plant Cell*, 1994, **6,** 1087-1098.
59. ECKARDT, N., MCHENRY, L., GUILTINAN, M., Overexpression of ΔEMBP, a truncated dominant negative version of the wheat G-box binding protein EMBP-1, alters vegetative development in transgenic tobacco, *Plant Mol. Biol.*, 1998, **38**, 539-549.
60. KATAGIRI, F., SEIPEL, K., CHUA, N., Identification of a novel dimer stabilization region in a plant bZIP transcription activator, *Mol. Cell Biol.*, 1992, **12**, 4809-4816.

Chapter Seven

PROPERTIES AND METABOLIC ENGINEERING OF ALFALFA PHENYLPROPANOID PATHWAY *O*-METHYLTRANSFERASES

Richard A. Dixon[1], Fang Chen[1], Xian-Zhi He[1], Joseph P. Noel[2], and Chloe Zubieta[2]

[1]*Plant Biology Division*
Samuel Roberts Noble Foundation
2510 Sam Noble Parkway
Ardmore, OK 73401

[2]*Structural Biology Laboratory*
Salk Institute for Biological Studies
10010 North Torrey Pines Road
La Jolla, CA 92138

INTRODUCTION

Plants elaborate an enormous array of natural products, derived mainly from the phenylpropanoid, isoprenoid, and (various) alkaloid biosynthetic pathways. Many of these compounds function in interactions of plants with physical and biotic factors in their environments. In addition, plant natural products may have positive or negative impacts on human or animal health and nutrition. Such functions suggest approaches to crop improvement based upon manipulation of natural product profiles. Limitations to natural product pathway engineering include lack of knowledge of specific enzymatic steps, flux control points, or regulatory architecture of the biosynthetic pathways.

We have taken a multidisciplinary approach to understand and begin to manipulate several classes of biologically active natural products of the phenylpropanoid pathway, namely flavonoids, isoflavonoids, and lignin. Flavonoids are ubiquitous in higher plants, where they function as UV protectants,[1] defense compounds,[2] and stimulators of beneficial mutualistic interactions with microorganisms, insects, and other organisms.[3] Isoflavonoids are restricted primarily to the Leguminosae, where they function as antimicrobial phytoalexins.[4] Additionally, both flavonoids and isoflavonoids have received much recent attention in view of reports implicating them as dietary components that may protect against cardiovascular disease, cancer, and osteoporosis.[5-9] Lignin, a major structural polymer of secondarily thickened plant cell walls, inhibits digestibility of forage grasses and legumes, and thereby impacts performance of grazing ruminant animals.[10-12]

Alfalfa (*Medicago sativa* L), the world's major forage legume, has been chosen as the target for our studies on natural product pathways. Lignin, isoflavonoids, and condensed tannins (polymerized flavan 3,4-diols) all impact forage quality and utilization in forage legumes. Furthermore, the development of the closely related species *Medicago truncatula* as a model legume in the Noble Foundation's Plant Biology Division and through major US National Science Foundation and European Community funded programs, provides powerful genomics tools, such as large expressed sequence tag (EST) databases (http://www.ncgr.org/research/mgi/; http://www.noble.org/medicago/; http://www.tigr.org/tdb/mtgi/; http://www.sequence.toulouse.inra.fr/Mtruncatula.html) that are directly applicable to alfalfa because of the high degree of sequence conservation of genes between the two *Medicago* species. Thus, it is now possible to use both biochemical and molecular genetic approaches for understanding phenylpropanoid biosynthesis and other natural product pathways in *Medicago* species.

Phenylpropanoid natural products usually contain one or more hydroxyl groups, originating from the activity of microsomal cytochrome P450 monooxygenases[13] or soluble 2-oxoglutarate dependent dioxygenases.[14] Altering the hydrophobicity of phenylpropanoid compounds often causes major changes in their biological activity, or is a necessary step for subsequent structural modifications. A common modification to phenolic hydroxyl groups is *O*-methylation, catalyzed by members of a superfamily of small molecule *O*-methyltransferase enzymes (OMTs) that utilize S-adenosyl L-methionine (SAM) as methyl group donor.[15,16]

The first plant phenolic *O*-methyltransferase to be cloned was the so-called caffeic acid 3-*O*-methyltransferase (COMT) of lignin biosynthesis,[17] and this enzyme typifies a major class of related plant OMTs. Perhaps ironically, the appellation caffeic acid *O*-methyltransferase recently has been shown to be a misnomer, as the enzyme appears to function *in vivo* as a 5-hydroxyconiferaldehyde *O*-methyltransferase[18] and should, therefore, be renamed to reflect this activity. This is perhaps fortuitous, because the acronym COMT often caused confusion with catechol *O*-methyltransferase, known as COMT in the mammalian literature. Catechol OMT is more closely related structurally to caffeoyl CoA *O*-methyltransferase, another OMT involved in lignin biosynthesis.[19,20]

The properties, primary sequences, and phylogenetic relationships of members of the plant OMT superfamily have been reviewed.[15,16] The number of putative plant OMT sequences in the databases is rapidly increasing due to sequence depositions from a number of major EST sequencing projects, for example from Arabidopsis, rice and *Medicago truncatula*.

In this chapter, we describe the molecular characteristics of four alfalfa *O*-methyltransferases involved in the biosynthesis of flavonoids, isoflavonoids, and lignin. Genetic manipulation of the activities of three of these enzymes in transgenic alfalfa has been shown to have profound effects on pathway flux. These effects indicate useful strategies for crop improvement based on modification of OMT expression.

ISOFLAVONE SYNTHESIS AND *O*-METHYLATION IN ALFALFA

The 2-Hydroxyisoflavanone Synthase Cytochrome P450

Figure 7.1 provides a simplified scheme for isoflavonoid biosynthesis in alfalfa. The major isoflavonoids of alfalfa are the constitutively produced glucosyl malonate of the methylated isoflavone formononetin (found in roots and cell suspension cultures) and the inducible pterocarpan phytoalexin medicarpin. These isoflavonoids are derived from the ubiquitous flavonoid precursor flavanone, itself derived from isomerization of the chalcone produced from 4-coumaroyl CoA

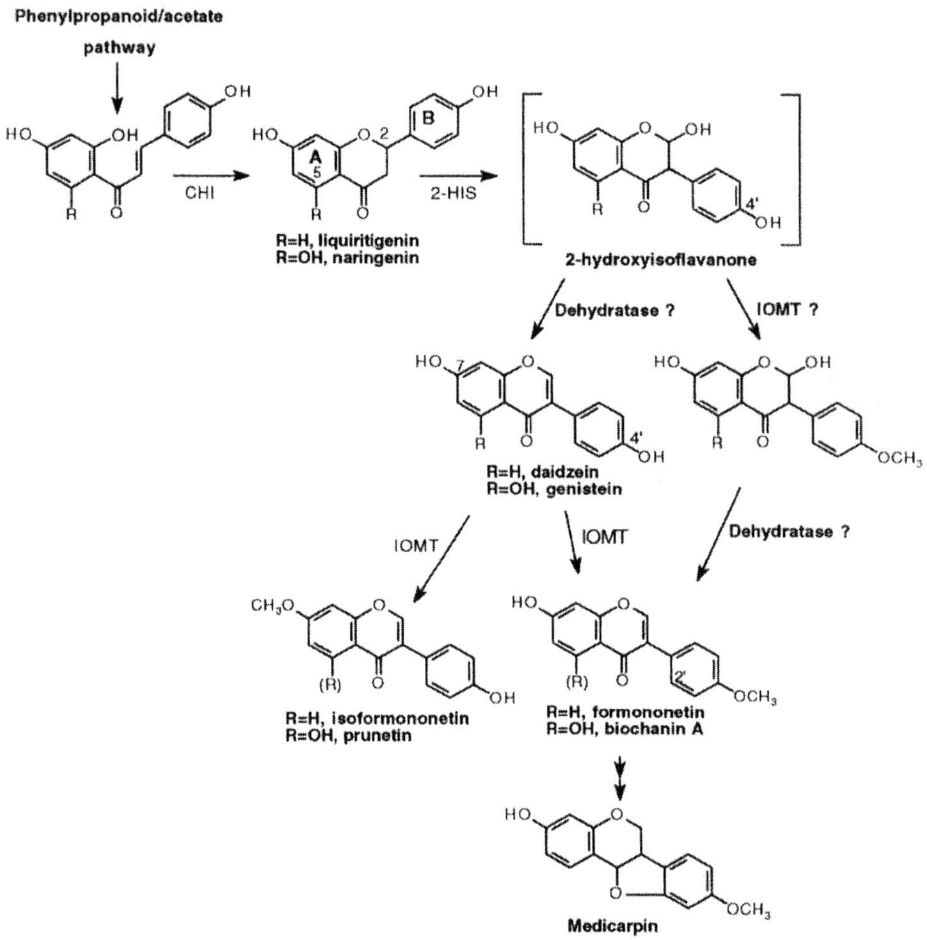
Phenylpropanoid/acetate pathway
CHI
R=H, liquiritigenin
R=OH, naringenin
2-HIS
2-hydroxyisoflavanone
Dehydratase ?
IOMT ?
R=H, daidzein
R=OH, genistein
IOMT
IOMT
Dehydratase ?
R=H, isoformononetin
R=OH, prunetin
R=H, formononetin
R=OH, biochanin A
Medicarpin

Figure 7.1. Formation and *O*-methylation of isoflavones. The flavanones naringenin and liquiritigenin are formed by the action of CHS and CHI, with loss of the 5-hydroxyl in liquiritigenin occurring by co-action of CHR with CHS. Aryl migration of the B-ring is catalyzed by 2-HIS leading, after dehydration, to daidzein or genistein. 4'-*O*-Methylation could hypothetically occur at two stages: after dehydration to daidzein, or before dehydration, at the level of the 2-hydroxyisoflavanone. Conversion of formononetin to medicarpin proceeds via 2'-hydroxylation, two successive reductions (catalyzed by IFR and vestitone reductase) and a ring closure step.[24]

and three molecules of malonyl CoA by chalcone synthase (CHS), a member of a superfamily of plant low molecular weight polyketide synthases.[21] In many legumes, the isoflavonoids predominantly lack the 5-hydroxyl (R = H in Fig. 7.1). Loss of this group is brought about by co-action of a polyketide reductase, trivially termed chalcone reductase (CHR), with CHS to produce 2',4,4'-trihydroxychalcone rather than naringenin chalcone, which is the normal product of the CHS polyketide synthase reaction.[22]

The genes encoding the first two enzymes specific for isoflavonoid biosynthesis in alfalfa (Fig. 7.1) recently have been isolated and characterized. The entry point enzyme into the isoflavonoid branch pathway, 2-hydroxyisoflavanone synthase (2-HIS, more loosely known as "isoflavone synthase"), is an unstable, membrane-bound cytochrome P450.[23,24] The enzyme 2-HIS catalyzes the 2-hydroxylation of flavanone, accompanied by migration of the B-ring from the 2- to the 3-position, to yield a 2-hydroxyisoflavanone. This intermediate is unstable, and will dehydrate, either spontaneously or by the action of a specific dehydratase enzyme,[25] to yield isoflavone. It is the possession of 2-HIS that distinguishes legumes and other isoflavonoid-producing species from the majority of plants that do not produce this class of compound. Thus, 2-HIS is the key for genetically introducing the isoflavonoid pathway into non-legume crop species, and it is briefly described here because of its intimate association with the subsequent B-ring methylation reaction of isoflavonoid biosynthesis.

We first cloned the 2-HIS gene by a functional genomics approach, based on its predicted expression pattern in soybean seeds (for production of the constitutive isoflavone phytoestrogens daidzein and genistein) and fungally infected hypocotyls (for production of the isoflavonoid phytoalexin glyceollin). Potential candidate ESTs from seed and infected hypocotyl libraries were tested for the correct expression pattern by RNA gel blot analysis, and then expressed in insect cells for functional identification of the 2-HIS activity producing the isoflavones daidzein or genistein from their corresponding flavanones.[26] Recently, 2-HIS has also been cloned from several other plants, including the legumes licorice, lupin, mung bean and *Medicago truncatula*, as well as from the non-legume sugar beet[27,28] (C.L. Steele and R.A. Dixon, unpublished results).

Soybean 2-HIS has been expressed in transgenic Arabidopsis (a crucifer) [28] (C.L Steele, C.-J. Liu and R.A. Dixon, unpublished results), which then accumulates genistein as a series of rhamnose and glucose-linked conjugates. This demonstrates that dehydration of the 2-hydroxyisoflavanone product of 2-HIS to the corresponding isoflavone can occur without the need for a separate dehydratase enzyme, and provides the first evidence that it will be possible to introduce health-promoting isoflavones into dietary crops.

Isoflavone 7-O-Methyltransferase Functions as a 2-Hydroxyisoflavanone 4'-O-Methyltransferase in vivo

The second enzymatic step in isoflavonoid biosynthesis in alfalfa is *O*-methylation of the B-ring 4'-hydroxyl group to yield formononetin (Fig. 7.1). However, all previous studies have identified only an enzyme capable of methylating the A-ring 7-hydroxyl of daidzein to yield isoformononetin,[24,29] a compound that is undetectable in alfalfa. We purified the isoflavone 7-OMT (IOMT) to homogeneity, obtained protein sequence information,[30] and subsequently cloned the corresponding gene.[31] Expression of the enzyme in *E. coli* confirmed the 7-position specificity of the enzyme *in vitro*.[31]

Transgenic alfalfa plants with strong constitutive expression of IOMT were produced by transformation with the alfalfa IOMT cDNA sequence under control of the cauliflower mosaic virus 35S promoter. IOMT expression in leaves of these plants is ectopic, in that the enzyme is not normally expressed in the aerial parts of the plant unless responding to infection or other stresses. Leaves of IOMT over-expressing plants have identical metabolite profiles to wild-type plants, and can not make isoflavonoids in the absence of induction of 2-HIS. If leaves of IOMT over-expressing plants are fed daidzein, it is converted to isoformononetin as catalyzed by IOMT *in vitro*. However, IOMT over-expressing plants exhibit a greater and more rapid accumulation of 4'-*O*-methylated isoflavonoids (including the phytoalexin medicarpin) than control lines following fungal infection or exposure to the abiotic elicitor copper chloride.[32] Thus, IOMT appears to be involved in the methylation of the 4'-position of isoflavonoids *in vivo*, and acts operationally as a rate-determining enzyme for isoflavonoid phytoalexin biosynthesis in alfalfa. The more rapid accumulation of medicarpin in IOMT over-expressing plants results in resistance to the foliar fungal pathogen *Phoma medicaginis*[32] (Fig. 7.2), providing direct evidence of a role for medicarpin in disease resistance.

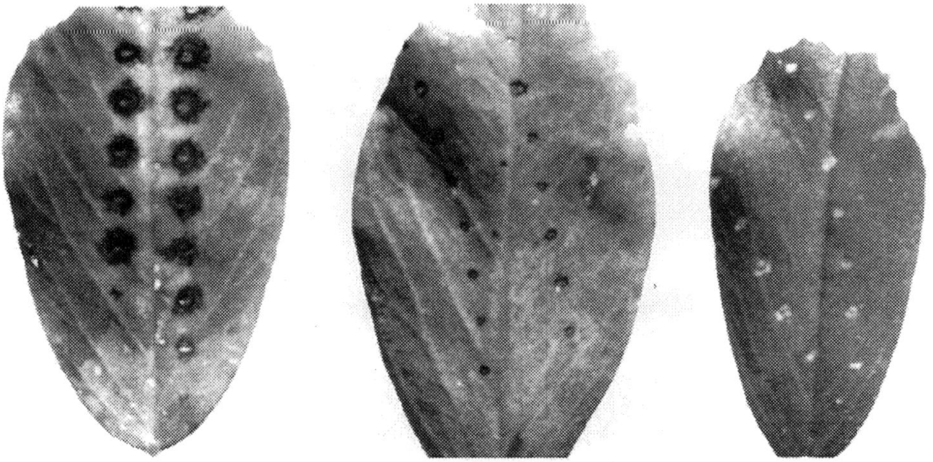

1 2 3

Figure 7.2. Disease resistance of transgenic alfalfa over-expressing isoflavone *O*-methyltransferase. Leaves of a control empty vector transformed line (1) and IOMT over-expressing line (2) were wounded on the leaves with a tracing wheel and then sprayed with a spore suspension of *Phoma medicaginis* (500,000 spore/mL). Leaves of the control line were also sprayed with water alone (3). The inoculated plants were then incubated in a growth chamber for 5 days.

Intuition would place 2-HIS, the entry point enzyme of the isoflavonoid pathway, as a major rate-determining step, rather than IOMT. The increased production of isoflavonoids resulting from IOMT over-expression can be explained by the fact that constitutive over-expression of IOMT leads to increased transcript levels of at least 7 other genes of medicarpin biosynthesis, including 2-HIS, following infection[32] (Fig. 7.3). However, these genes are not strongly expressed in uninfected IOMT over-expressing plants. Thus, genetic manipulation of IOMT provides a means to enhance an infection-inducible natural phytoalexin response without the need to genetically manipulate other enzymes of the pathway and without production of potentially phytotoxic phytoalexins in the absence of infection. The mechanism whereby IOMT over-expression co-induces other enzymes of the isoflavonoid pathway is currently unknown.

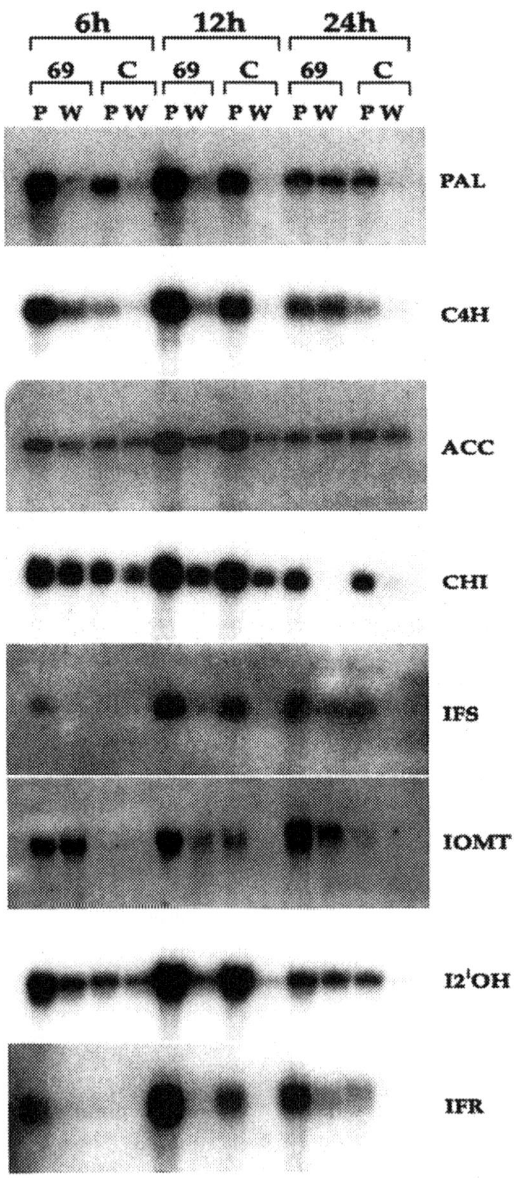

Figure 7.3. RNA gel blot analysis of isoflavonoid pathway transcripts in *Phoma* infected leaves of empty vector control (C) and IOMT over-expressing (linc 69) alfalfa. Total RNA was isolated from leaves at 6, 12 and 24 hr after infection with *P. medicaginis* (P) or treatment with water (W), and probed with labeled cDNA probes for alfalfa PAL, alfalfa C4H, alfalfa acetyl CoA carboxylase (ACC), alfalfa CHI, *Medicago truncatula* 2-HIS (IFS), alfalfa IOMT, *Medicago truncatula* isoflavone 2'-hydroxylase (I2'OH) and alfalfa isoflavone reductase (IFR). Blots were stripped and re-probed with a ribosomal RNA probe to confirm equal loading and transfer efficiency (data not shown).

The question arises as to why IOMT exhibits 7-position specificity *in vitro*, but appears to be involved in 4'-*O*-methylation of isoflavonoids *in vivo*. Our original hypothesis was that some kind of metabolic channeling, whereby IOMT associated with 2-HIS, resulted in altered regiospecificity, perhaps by allowing different spatial presentation of daidzein to the active site of the enzyme compared to that arising from free diffusion of the substrate from the cytoplasmic milieu. Alfalfa IOMT is an operationally soluble enzyme. However, alfalfa microsomes do contain immunodetectable IOMT, and can catalyze the formation of formononetin rather than isoformononetin from radiolabeled liquiritigenin *in vitro* (Fig. 7.1). Sub-cellular localization studies that use particle bombardment of IOMT-green fluorescent protein (GFP) fusion constructs in alfalfa have shown that IOMT can localize to the endoplasmic reticulum (ER) following elicitation of isoflavonoid biosynthesis (C.-J. Liu and R.A. Dixon, unpublished results). This is consistent with the proposed metabolic channel occurring on the surface of the ER.

The three dimensional crystal structure of IOMT recently has been solved at the 1.4 A level.[33] IOMT was co-crystallized with SAM and daidzein, but the solved structure revealed isoformononetin and S-adenosyl homocysteine in the active site. This suggests that the enzyme binds daidzein such that its 7-hydroxyl group is in close proximity to the methyl group of the methyl donor SAM. However, modeling studies indicate that 2,7,4'-trihydroxyisoflavanone, the unstable product of 2-HIS (Fig. 7.1), can fit into the active site of IOMT in such a way that the 4'-hydroxyl group is now facing the methyl group of SAM. Binding of 2,4',7-trihydroxyisoflavanone is thermodynamically more stable than that of daidzein.[33] Catalysis in IOMT proceeds through deprotonation of the 7-hydroxyl group on the A-ring of daidzein, or the 4'-hydroxyl group on the B-ring of 2,7,4'-trihydroxyisoflavanone, by histidine 257, which sets up subsequent attack by the hydroxyl anion on the methyl group of SAM. Because of the positive charge on the sulfur of SAM, the deprotonation step readily facilitates the transmethylation process.[33]

Based on the above modeling studies, 2,4',7-trihydroxyisoflavanone is likely to be the true substrate of IOMT *in vivo*. The formation of isoformononetin *in vitro* may, therefore, be of no physiological significance *in planta*. However, the association of IOMT with microsomes may be essential for 4'-*O*-methylation in isoflavonoid biosynthesis because close interaction of IOMT with 2-HIS would ensure that 4'-*O*-methylation take place prior to dehydration of the unstable trihydroxyisoflavanone. If dehydration took place first, formation and subsequent methylation of daidzein would lead to the metabolic "dead-end" product isoformononetin due to the 7-position specificity of IOMT with daidzein.

Because of this metabolic channeling, daidzein is not an intermediate in isoflavonoid biosynthesis in alfalfa. Daidzein does, however, accumulate to quite high levels in seeds of soybean, along with its 5-hydroxy derivative genistein.[34] The isoflavonoid phytoalexins of soybean are not methylated, and dehydration of 2,4',7-trihydroxyisoflavanone leads to daidzein, which, following infection, is converted to 2'-hydroxydaidzein by an inducible isoflavone 2'-hydroxylase (I2'H) cytochrome P450 during the biosynthesis of the prenylated pterocarpan phytoalexin glyceollin.[24] 2'-Hydroxydaidzein is the preferred substrate for the isoflavone reductase (IFR) of soybean.[35] Interestingly, the I2'H of alfalfa can not hydroxylate daidzein, requiring the presence of the 4'-*O*-methyl group for its activity. Thus, a combination of strict substrate specificity for the isoflavone hydroxylase, coupled to close physical association of the 2-hydroxyisoflavanone *O*-methyltransferase with 2-HIS, together allow for conversion of isoliquiritigenin to 2'-hydroxyformononetin, the immediate precursor for isoflavanone formation in the biosynthesis of the alfalfa phytoalexin medicarpin.

The protein most related to alfalfa isoflavone (2-hydroxyisoflavanone) *O*-methyltransferase at the amino acid sequence level (50.5 % identical) is 6a-hydroxymaackiain 3-*O*-methyltransferase from pea,[36] the 3-position of pterocarpans being equivalent to the 7-position of flavonoids/isoflavonoids. Other enzymes have been characterized that preferentially methylate flavonoids on the 7-position of the A-ring, but OMTs also exist with strict specificities for other positions on the flavonoid nucleus, for example the enzymes associated with the biosynthesis of polymethylated flavonols in *Chrysosplenium americanum*.[37-39]

The knowledge of the structure and detailed mechanism of isoflavone OMT, which includes information on the amino acid groups that constrain the substrate in the correct orientation within the active site[33] now makes it possible to modify the regiospecificity of flavonoid methylation by targeted site-directed mutagenesis.

LIGNIN *O*-METHYLTRANSFERASES AND THE GENETIC MANIPULATION OF LIGNIN QUANTITY AND COMPOSITION FOR FORAGE IMPROVEMENT

Lignin and Forage Digestibility

Lignin is a complex polymer of hydroxylated and methoxylated phenylpropane units, linked via oxidative coupling.[40] Dicotyledonous angiosperm lignins contain two major monomer species, termed guaiacyl (G) and syringyl (S) units (Fig. 7.4). G units are characterized by a single methoxyl substitution on the aromatic ring, whereas S units are di-methoxylated. The S and G units in lignin are joined through at least five different types of linkages. It has become increasingly apparent that lignin is not simply a random polymer of monolignol units, and lignin composition is developmentally programmed, both temporally and spatially.[41,42]

Lignin levels increase with progressive maturity in alfalfa stems. In addition, the lignin composition changes with advanced maturity towards a progressively higher S/G ratio.[43] Both lignin concentration and lignin methoxyl content (S/G ratio) have been negatively correlated with forage digestibility for ruminant animals. Small reductions in lignin content may have correspondingly larger impacts on digestibility.[44] Although a number of studies have linked decreased forage digestibility to increased S/G ratio as a function of increased plant maturity, other studies have questioned the effect of lignin composition on digestibility.[45] Our approach to lignin manipulation for improvement of forage quality has been to gain an understanding of the pathways leading to the synthesis of the monolignol building blocks in alfalfa, and to target key genes for transgenic modification.

O-Methylation Reactions as Part of a "Metabolic Grid" for Lignin Biosynthesis

The “text book” view of lignin biosynthesis is a metabolic grid, through which side chain reduction and ring hydroxylation/methylation can, in theory, occur at several different levels (Fig. 7.4). We have critically evaluated the evidence for the metabolic grid model, and recently proposed an alternative model for alfalfa, based on metabolic channeling, that is consistent with both biochemical and molecular genetic evidence.[46]

Our studies have centered on the *O*-methylation reactions of G and S lignin biosynthesis. We cloned the alfalfa caffeic acid *O*-methyltransferase (COMT) and caffeoyl CoA *O*-methyltransferase (CCOMT) genes believed to be involved in parallel pathways for monolignol methylation at the levels of free hydroxycinnamic acids and their corresponding CoA esters, respectively[17,43] (Fig. 7.4). For COMT

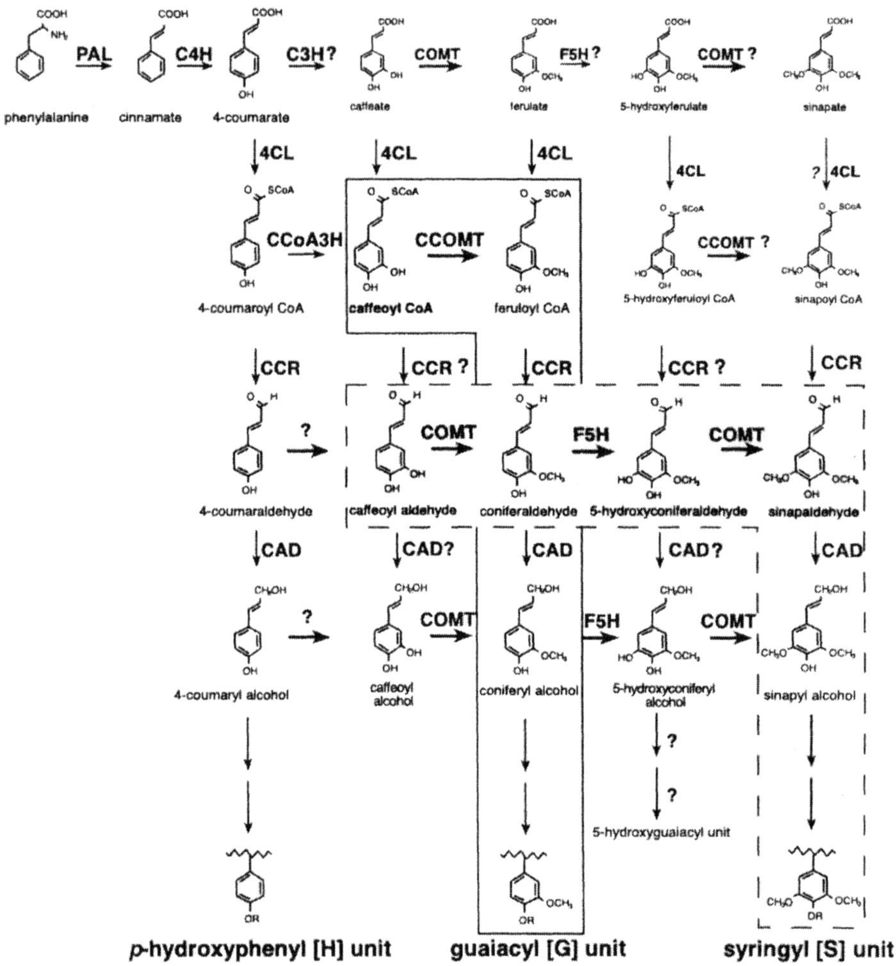

Figure 7.4. Biochemical pathways to monolignols. The complete metabolic grid of reactions is shown. Reactions shown in smaller type may not occur *in vivo*. The reactions shown in the solid boxed area seem the most likely route to G lignin *in vivo*. The reactions in the dashed boxed area represent those reactions consistent with both *in vivo* and *in vitro* evidence for being involved specifically in S lignin biosynthesis.

CCOMT to catalyze parallel pathways within a metabolic grid, it is necessary for both enzymes to be expressed in the same cells. Immunolocalization studies at the light and electron microscope levels demonstrated expression of both COMT and CCOMT in the cytoplasm of primary and secondary xylem parenchyma cells in alfalfa stems.[47] Both enzymes were localized also to phloem in mature internodes, where COMT was found in cytoplasm and vacuoles, whereas CCOMT was localized only to cytoplasm.

A study of developmental changes in OMT activities in successive stem internodes by using caffeic and 5-hydroxyferulic acids, their coenzyme A esters, and caffeoyl aldehyde, caffeoyl alcohol, 5-hydroxyconiferaldehyde, and 5-hydroxconiferyl alcohol revealed the unexpected finding that activities were highest against caffeoyl aldehyde and alcohol, and that these reactions were catalyzed by COMT.[48] Analysis of recombinant alfalfa COMT purified from *E. coli* confirmed that caffeoyl and 5-hydroxyferuolyl aldehydes and alcohols were kinetically favored substrates, and that caffeic acid is a poor substrate.[48] The properties of the enzymes are summarized in Table 7.1. If the metabolic grid functions at all *in vivo*, the reactions would most likely occur at the level of the aldehydes and alcohols, consistent with previous suggestions based on the substrate preferences of COMT from aspen[18] and ferulate 5-hydroxylase from Arabidopsis.[49]

Alfalfa microsomes, which contain immunodetectable COMT protein, can convert caffeoyl aldehyde to sinapaldehyde in the presence of SAM and NADPH (D. Guo and R.A. Dixon, unpublished results). This is consistent with the operation of a metabolic channel, involving an association of COMT with the membrane bound ferulate 5-hydroxylase (5FH, now shown to be a coniferaldehyde 5-hydroxylase), that carries out both methylation reactions of S lignin biosynthesis. Together with the transgenic modification data, these results suggest that the 3-methoxyl group of S lignin originates from COMT-catalyzed methylation of caffeoyl aldehyde, whereas the same position in G lignin is methylated by CCOMT, COMT, and possibly other enzymes.

Young alfalfa internodes express a novel form of COMT (COMT II), having preference for caffeic acid, but also active with hydroxycinnamoyl CoA esters and some flavonoids.[50] This apparently new *O*-methyltransferase was separated from the previously characterized COMT described above (COMT I) by anion exchange and hydrophobic interaction chromatography. COMT I, but not COMT II, is found in mature internodes. COMT II is not recognized by anti-(COMT I) serum. In addition to substrate preference, COMT II differs from COMT I in its native molecular weight, pH optimum, and its low Km for caffeic acid[50] (7. 1). COMT II may be, therefore, the true caffeic acid OMT *in vivo*.

Further OMTs are expressed at high level in stems of *Medicago truncatula* based on frequency of their transcripts in a stem library from which over 10,000 ESTs have been sequenced at the Noble Foundation (www.noble.org/medicago/).

These are candidates for providing the residual OMT activity we have observed both *in vitro* and *in vivo* following down-regulation of COMT and CCOMT.

Table 7.1: Properties of alfalfa O-methyltransferases[a] active against hydroxycinnamic acids and their coenzyme A esters.

Substrate	Vmax (nkat mg^{-1} protein)	Km (μM)	Vmax/Km
COMT I:			
Caffeic acid	5.9	59.5	0.1
5-Hydroxyferulic acid	4.4	6.8	0.6
Caffeoyl aldehyde	19.2	6.9	2.8
Caffeoyl alcohol	20.8	11.5	1.8
5-Hydroxyconiferaldehyde	5.6	1.8	3.1
5-Hydroxyconiferyl alcohol	5.3	3.8	1.4
COMT II:			
Caffeic acid	-	0.6	-
5-Hydroxyferulic acid	-	5.4	-
CCOMT:			
Caffeoyl CoA	0.71	4.3	0.17
5-Hydroxyferuolyl CoA	0.77	6.2	0.12

[a]The properties of COMT I and CCOMT were determined with recombinant protein expressed in *E. coli*, whereas kinetics of COMT II were determined with a partially purified preparation from alfalfa stem.

Genetic Modification of Lignin O-Methyltransferases in Alfalfa

We were able to down-regulate strongly COMT and CCOMT, either singly or in combination, in transgenic alfalfa.[51] The results provided striking evidence for involvement of COMT in the biosynthesis of S lignin, but questioned whether CCOMT was involved in S lignin biosynthesis. Thus, down-regulation of COMT completely eliminated S units, even in the condensed portion of the lignin polymer revealed as dimeric units following analysis by thioacidolysis, whereas down-regulation of CCOMT had no effect on S lignin[51] (Table 7.2). Reduction in either enzyme resulted in a fewer G units, but these were never eliminated, and the effect of simultaneous down-regulation of both enzymes was not additive, suggesting the possible involvement of other OMTs in lignin biosynthesis.

Table 7.2: Effects of down-regulation of COMT and CCOMT in transgenic alfalfa.

COMT to 4.5% of wild-type activity	**CCOMT to 3.5% of wild-type activity**
Lignin level reduced by 25%	Lignin level reduced by 15%
G units reduced by 25%	G units reduced by 25%
Total loss of S units in β-O-4-linked and dimeric fractions	No loss of S units
5-hydroxyguaiacyl units accumulate	No 5-hydroxyguaiacyl units
No increase in soluble precursors	Accumulation of caffeoyl-β-D-glucose
2 % increase in forage digestibility	6% increase in forage digestibility

CCOMT down-regulated alfalfa plants accumulated high levels of soluble caffeoyl β-D-glucose, which is not a substrate for CCOMT[51] (Table 7.2). It presumably arises from conjugation of free caffeic acid resulting from thioesterase catalyzed cleavage of caffeoyl CoA. Surprisingly, neither this compound nor any other monolignol precursor/conjugate accumulates in COMT down-regulate plants that are totally blocked in S lignin biosynthesis. This provides further indirect evidence against caffeic acid being a substrate for monolignol biosynthesis.

Field trials and digestibility studies have been conducted with transgenic alfalfa lines modified in expression of COMT and CCOMT. *In vivo* digestibility studies, in which samples were introduced into the rumens of fistulated steers, indicated significant increases in the rate and end-point of digestibility of forage from CCOMT down-regulated lines, with smaller improvements in digestibility in COMT down-regulated lines (D. Guo and R.A. Dixon, unpublished results). This result has been reproduced by using *in vitro* digestibility assays, and suggests that both lignin content and composition are important factors determining forage digestibility. The more condensed lignin resulting from removal of S units in COMT down-regulated lines would, on the basis of studies on paper pulping, be predicted to result in reduced digestibility, and the improvement in digestibility in COMT down-regulated lines suggests that a reduction in lignin levels can overcome this effect. The *in vitro* digestibility of tobacco plants modified in expression of COMT, phenylalanine ammonia-lyase (PAL), and cinnamate 4-hydroxylase (C4H) has also been studied.[12,52] Interestingly, down regulation of C4H completely eliminated S lignin, whereas PAL down-regulation led to similar reduction in lignin levels but to an increased S/G ratio.[52] Both situations resulted in a similar improvement in digestibility. Overall, these studies point to lignin content having a greater impact on digestibility than lignin composition.

ALFALFA CHALCONE 2'-*O*-METHYLTRANSFERASE AND ITS METABOLIC RELATION TO CHALCONE ISOMERASE

Methoxychalcone as a Nodulation Gene Inducer

A characteristic feature of leguminous plants is their ability to enter into a symbiotic relationship with nitrogen fixing bacteria of the genus *Rhizobium*. This process is initiated by bacterial recognition of flavonoid compounds released from legume roots.[3] This recognition leads to activation of Rhizobial nodulation (*nod*) genes that results in synthesis of lipochitooligosaccharide signal molecules that activate root hair curling and cortical cell divisions in the host plant and thereby initiate bacterial entry and root nodule formation.[53]

The most potent of the *Rhizobium meliloti nod* gene inducers released from alfalfa roots is 2'-methoxy, 4, 4'-dihydroxychalcone.[54] This compound is formed by direct methylation of 2', 4, 4'-trihydroxychalcone, the product of the co-action of CHS and CHR as described above, by a specific chalcone *O*-methyltransferase (Fig. 7.5). This enzyme has been purified previously from alfalfa cell cultures,[55] and its gene cloned.[56] Because the chalcone 2'-hydroxyl group has to be free for isomerization to the flavanone, the action of chalcone OMT commits chalcone to the biosynthesis of *nod* gene inducers and away from formation of flavonoids and protective isoflavonoids (Fig. 7.5). Chalcone OMT and chalcone isomerase are present in the same cell types in alfalfa roots,[56] and have similar Michaelis constants.[55] CHI is a catalytically near perfect enzyme, with the second order rate constant for flavanone synthesis approaching the diffusion controlled limit for entry of substrate to the active site (kcat/Km = $1.71 \times 10^8\ M^{-1}\ min^{-1}$). It is, therefore, important to understand how flux is controlled into the two mutually exclusive metabolic pathways originating from trihydroxychalcone.

Structural Biology of Alfalfa Chalcone Isomerase and Chalcone O-Methyltransferase

We have suggested previously that the potentially competing pathways might be able to "co-exist" if the enzymes were involved in discrete complexes, perhaps with the products of different members of the alfalfa *CHS* gene family.[57] We have not, however, been able to demonstrate direct physical interactions between alfalfa CHS and CHI, or CHS and chalcone OMT, by using the yeast two-hybrid system. (J.T. Reddy and R.A Dixon, unpublished results).

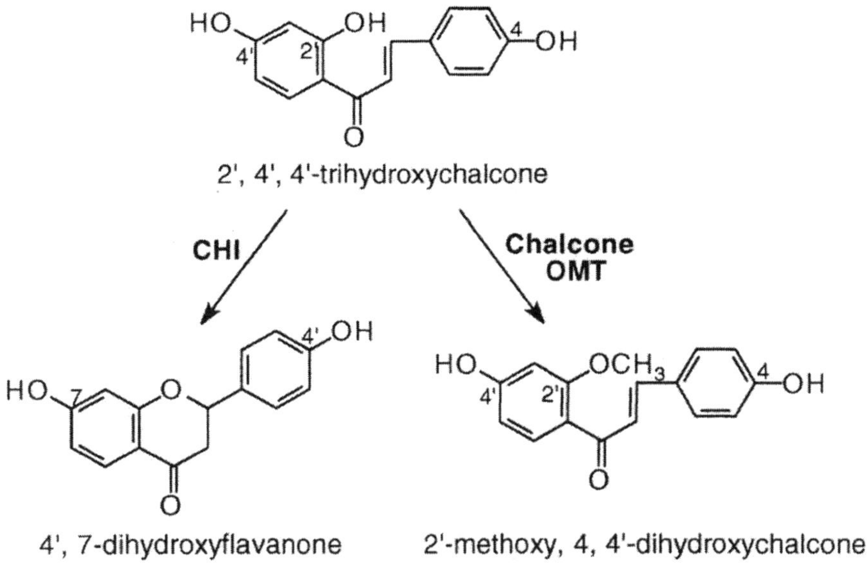

Figure 7.5. The reactions catalyzed by chalcone isomerase and chalcone 2'-*O*-methyltransferase in alfalfa.

In contrast, physical interactions do appear to occur between CHI and CHS in Arabidopsis,[58] a species that does not possess chalcone OMT. Arabidopsis CHI contains a C-terminal extension to its amino acid sequence that is not found in alfalfa CHI. This extension might be involved in targeting Arabidopsis CHI to a complex with CHS following some type of lipid modification to CHI.[58]

The three-dimensional crystal structures of alfalfa chalcone OMT and CHI recently have been described.[33,59] Alfalfa CHI contains an evolutionary novel protein fold that, to date, has not been observed in proteins from organisms other than plants.[59] Chalcone OMT has a catalytic site that bears close similarity to that of IOMT.[33] The fold of the SAM-binding domain is conserved throughout all classes of SAM-dependent methyltransferases.[16,60] Unique features of the plant OMTs include a second domain that is involved in dimerization of the 40 kDa subunits (Fig. 7.6) and that contributes to the substrate binding site. In chalcone OMT, the A-ring of chalcone is bound by the thioester moieties of methionines 189 and 329, with threonine 332 being within hydrogen bonding distance to the chalcone 4'-hydroxyl (Fig. 7.6). These interactions contribute towards ensuring that the 2'-hydroxyl group is correctly positioned for deprotonation by the putative catalytic base, histidine 278, followed by methylation by SAM in a reaction analogous to that catalyzed by IOMT.

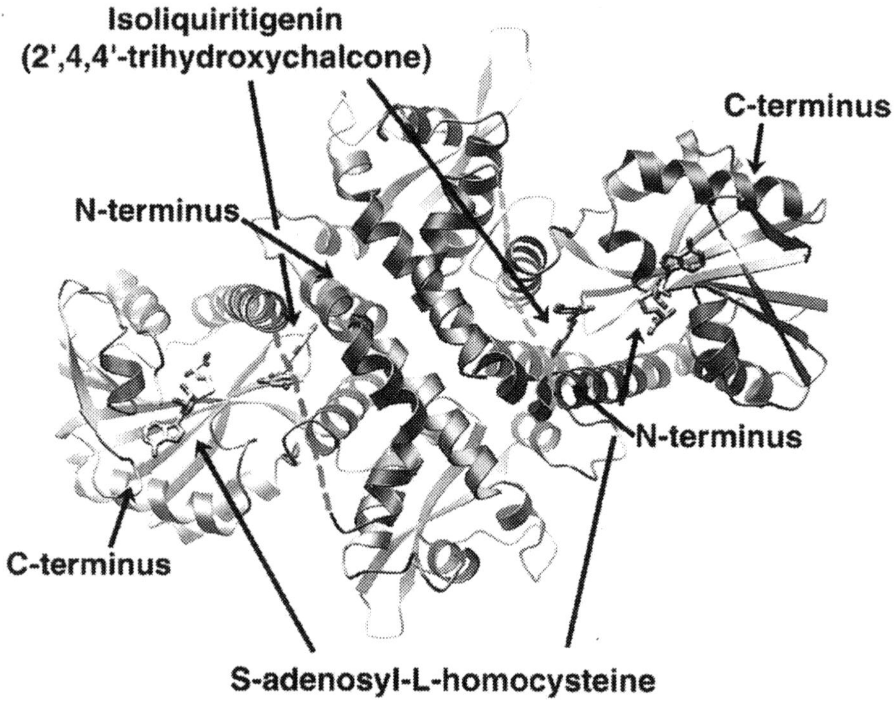

Figure 7.6. Ribbon representation of the alfalfa chalcone OMT homodimer with bound S-adenosyl L-homocysteine (SAH) and 2',4,4'-trihydroxychalcone molecules. From ref 33.

An understanding of the catalytic mechanisms of different, position-specific plant small molecule *O*-methyltransferases will serve as a foundation for understanding the molecular basis for regiospecificity in plant methylation reactions, and will facilitate the design of novel methyltransferases with altered kinetics and specificity by targeted site-directed mutagenesis. At the same time, the availability of the three-dimensional structure of alfalfa CHS,[61] in addition to the CHI and chalcone OMT structures referred to above, should lead to an understanding of how these enzymes might interact within the cell to facilitate channeling of metabolites into the biochemically and functionally competing *nod* gene inducer and phytoalexin pathways involved, respectively, in the attraction or repulsion of microorganisms. Because *nod* gene inducer levels can, under some conditions, present limitations to nodulation efficiency,[62] it will be interesting to evaluate the effects of over- or under-expression of chalcone OMT in transgenic plants.

CONCLUSIONS

Knowledge of the biochemistry and structural biology of *O*-methyltransferase enzymes involved in the biosynthesis of isoflavonoids, chalcones, and monolignols has improved our understanding of how the various branch pathways of phenylpropanoid biosynthesis are organized at the subcellular level. Genetic modification of the expression of these enzymes has already suggested methods to improve the disease resistance and forage quality of alfalfa. In addition, such genetic modification experiments have revealed that the reactions ascribed to specific OMTs, based on catalytic activity with supposed substrates *in vitro,* are not necessarily the preferred reactions for these enzymes *in vivo*. Thus, caffeic acid OMT should now be re-named hydroxycinnamaldehyde OMT, and isoflavone OMT should be re-named 2-hydroxyisoflavanone OMT. These types of observations, which may be more wide-spread within plant secondary metabolism, highlight the importance of combining phytochemistry, biochemistry, structural biology, and functional genomics in order to obtain a complete picture of how biosynthetic pathways function *in vivo.*

ACKNOWLEDGMENTS

Drs. Dianjing Guo, Kota Parvathi, and Chang-Jun Liu (Noble Foundation) were responsible for a significant portion of the work on the substrate specificity and genetic manipulation of alfalfa COMT and CCOMT, and the subcellular localization of IOMT. The work in Richard Dixon's laboratory was supported by The Samuel Roberts Noble Foundation and AgriBioTech, Inc. The work in Joseph Noel's laboratory was supported by the Salk Institute, the Samuel Roberts Noble Foundation, and NIH. We thank Cuc Ly, Darla Boydston and Paul Horton for assistance with graphics.

REFERENCES

1. LI, J., OU-LEE, T.M., RABA, R., AMUNDSON, R.G., LAST, R.L., Arabidopsis flavonoid mutants are hypersensitive to UV-B irradiation, *Plant Cell*, 1993, **5**, 171-179.
2. SNYDER, B.A., NICHOLSON, R.L., Synthesis of phytoalexins in sorghum as a site-specific response to fungal ingress, *Science*, 1990, **248**, 1637-1639.
3. PETERS, N.K., FROST, J.W., LONG, S.R., A plant flavone, luteolin, induces expression of *Rhizobium meliloti* nodulation genes, *Science*, 1986, **233**, 977-980.

4. DIXON, R.A., PAIVA, N.L., Stress-induced phenylpropanoid metabolism, *Plant Cell*, 1995, **7**, 1085-1097.
5. RICE-EVANS, C.A., MILLER, N.J., Antioxidant activities of flavonoids as bioactive components of food, *Biochem. Soc. Trans.*, 1996, **24**, 790-794.
6. HOLLMAN, P.C.H., HERTOG, M.G.L., KATAN, M.B., Role of dietary flavonoids in protection against cancer and coronary heart disease, *Biochem. Soc. Trans.*, 1996, **24**, 785-789.
7. DRAPER, C.R., EDEL, M.J., DICK, I.M., RANDALL, A.G., MARTIN, G.B., PRINCE, R.L., Phytoestrogens reduce bone loss and bone resorption in oophorectomized rats, *J. Nutr.*, 1997, **127**, 1795-1799.
8. MESSINA, M., Soy, soy phytoestrogens (Isoflavones), and breast cancer, *Amer. J. Clin. Nutr.*, 1999, **70**, 574-575.
9. SETCHELL, K.D.R., CASSIDY, A., Dietary isoflavones: Biological effects and relevance to human health, *J. Nutr.*, 1999, **129**, 758S-767S.
10. JUNG, H.G., VOGEL, K.P., Influence of lignin on digestibility of forage cell wall material, *J. Animal Sci.*, 1986, **62**, 1703-1712.
11. JUNG, H.G., MERTENS, D.R., PAYNE, A.J., Correlation of acid detergent lignin and Klason lignin with digestibility of forage dry matter and neutral detergent fiber, *J. Diary Sci.*, 1997, **80**, 1622-1628.
12. SEWALT, V.J.H., NI, W., JUNG, H., DIXON, R.A., Lignin impact on fiber degradation: Increased enzymatic digestibility of genetically engineered tobacco(*Nicotiana tabacum*) stems reduced in lignin content, *J. Agric. Food Chem.*, 1997, **45**, 1977-1983.
13. CHAPPLE, C., Molecular-genetic analysis of plant cytochrome P450-dependent monooxygenases, *Annu. Rev. Plant Physiol. Plant Mol. Biol.*, 1998, **49**, 311-343.
14. BRITSCH, L., GRISEBACH, H., Purification and characterization of (2S)-flavanone 3-hydroxylase from *Petunia hybrida*, *Eur. J. Biochem.*, 1986, **156**, 569-577.
15. IBRAHIM, R.K., BRUNEAU, A., BANTIGNIES, B., Plant *O*-methyltransferases: molecular analysis, common signature and classification, *Plant Mol. Biol.*, 1998, **36**, 1-10.
16. JOSHI, C.P., CHIANG, V.L., Conserved sequence motifs in plant S-adenosyl-L-methionine-dependent methyltransferases, *Plant Mol. Biol.*, 1998, **37**, 663-674.
17. GOWRI, G., BUGOS, R.C., CAMPBELL, W.H., MAXWELL, C.A., DIXON, R.A., Stress responses in alfalfa (*Medicago sativa* L.) X. Molecular cloning and expression of S-adenosyl-L-methionine: caffeic acid 3-O-methyltransferase, a key enzyme of lignin biosynthesis, *Plant Physiol.*, 1991, **97**, 7-14.
18. OSAKABE, K., TSAO, C.C., LI, L., POPKO, J.L., UMEZAWA, T., CARRAWAY, D.T., SMELTZER, R.H., JOSHI, C.P., CHIANG, V.L., Coniferyl aldehyde 5-hydroxylation and methylation direct syringyl lignin biosynthesis in angiosperms, *Proc. Natl. Acad. Sci. USA*, 1999, **96**, 8955-8960.
19. SCHMITT, D., PAKUSCH, A.-E., MATERN, U., Molecular cloning, induction, and taxonomic distribution of caffeoyl-CoA 3-*O*-methyltransferase, an enzyme involved in disease resistance, *J. Biol. Chem.*, 1991, **266**, 17416-17423.
20. YE, Z.H., KNEUSEL, R.E., MATERN, U., VARNER, J.E., An alternative methylation pathway in lignin biosynthesis in *Zinnia*, *Plant Cell*, 1994, **6**, 1427-1439.

21. SCHRÖDER, J., A family of plant-specific polyketide synthases: facts and predictions, *Trends Plant Sci.*, 1997, **2**, 373-378.
22. WELLE, R., GRISEBACH, H., Isolation of a novel NADPH-dependent reductase which coacts with chalcone synthase in the biosynthesis of 6'-deoxychalcone, *FEBS Lett.*, 1988, **236**, 221-225.
23. KOCHS, G., GRISEBACH, H. 1986. Enzymic synthesis of isoflavones, *Eur. J. Biochem.*, 1986, **155**, 311-318.
24. DIXON, R.A., Isoflavonoids: biochemistry, molecular biology and biological functions, *in* Comprehensive Natural Products Chemistry (Sankawa, U., ed.), Elsevier, Oxford, 1999, pp. 773-823.
25. HAKAMATSUKA, T., MORI, K., ISHIDA, S., EBIZUKA, Y., SANKAWA, U., Purification of 2-hydroxyisoflavanone dehydratase from the cell cultures of *Pueraria lobata*, *Phytochemistry*, 1998, **49**, 497-505.
26. STEELE, C.L., GIJZEN, M., QUTOB, D., DIXON, R.A., Molecular characterization of the enzyme catalyzing the aryl migration reaction of isoflavonoid biosynthesis in soybean, *Arch. Biochem. Biophys.*, 1999, **367**, 147-150.
27. AKASHI, T., AOKI, T., AYABE, S., Cloning and functional expression of a cytochrome P450 cDNA encoding 2-hydroxyisoflavanone synthase involved in biosynthesis of the isoflavonoid skeleton in licorice, *Plant Physiol.*, 1999, **121**, 821-828.
28. JUNG, W., YU, O., LAU, S.-M.C., O'KEEFE, D.P., ODELL, J., FADER, G., MCGONIGLE, B., Identification and expression of isoflavone synthase, the key enzyme for biosynthesis of isoflavones in legumes, *Nature Biotech.*, 2000, **18**, 208-212.
29. EDWARDS, R., DIXON, R.A., Isoflavone O-methyltransferase activities in elicitor-treated cell suspension cultures of *Medicago sativa*, *Phytochemistry*, 1991, **30**, 2597-2606.
30. HE, X.-Z., DIXON, R.A., Affinity chromatography, substrate/product specificity and amino acid sequence analysis of an isoflavone O-methyltransferase from alfalfa (*Medicago sativa* L.), *Arch. Biochem. Biophys.*, 1996, **336**, 121-129.
31. HE, X.-Z., REDDY, J.T., DIXON, R.A., Stress responses in alfalfa (*Medicago sativa* L.) XXII. cDNA cloning and characterization of an elicitor-inducible isoflavone 7-*O*-methyltransferase, *Plant Mol. Biol.*, 1998, **36**, 43-54.
32. HE, X.-Z., DIXON, R.A., Genetic manipulation of isoflavone 7-*O*-methyltransferase enhances the biosynthesis of 4'-*O*-methylated isoflavonoid phytoalexins and disease resistance in alfalfa, *Plant Cell*, 2000, **12**, 1689-1702.
33. ZUBIETA, C., DIXON, R.A., NOEL, J.P., Crystal structures of chalcone O-methyltransferase and isoflavone O-methyltransferase reveal the structural basis for substrate specificity in plant O-methyltransferases, *Nature Struct. Biol.*, 2000, in press.
34. FRANKE, A.A., HANKIN, J.H., YU, M.C., MASKARINEC, G., LOW, S.H., CUSTER, L.J., Isoflavone levels in soy foods consumed by multiethnic populations in Singapore and Hawaii, *J. Agric. Food Chem.*, 1999, **47**, 977-986.

35. FISCHER, D., EBENAU-JEHLE, C., GRISEBACH, H., Phytoalexin syntheis in soybean: purification and characterization of NADPH:2'-hydroxydaidzein oxidoreductase from elicitor-challenged soybean cell cultures, *Arch. Biochem. Biophys.*, 1990, **276**, 390-395.
36. WU, Q., PRESIG, C.L., VANETTEN, H.D., Isolation of the cDNAs encoding (+)6a-hydroxymaackiain 3-*O*-methyltransferase, the terminal step for the synthesis of the phytoalexin pisatin in *Pisum satium*, *Plant Mol. Biol.*, 1997, **35**, 551-560.
37. KHOURI, H.E., DE LUCA, V., IBRAHIM, R.K., Enzymatic synthesis of polymethylated flavonols in *Chrysosplenium americanum*, *Arch. Biochem. Biophys.*, 1988, **265**, 1-7.
38. GAUTHIER, A., GULICK, P., IBRAHIM, R.K., cDNA cloning and characterization of a 3'/5'-*O*-methyltransferase for partially methylated flavonols from *Chrysosplenium americanum*, *Plant Mol. Biol.*, 1996, **32**, 1163-1169.
39. SEGUIN, J., MUZAC, I., IBRAHIM, R.K., Purification and immunological characterization of a recombinant trimethylflavonol 3'-*O*-methyltransferase, *Phytochemistry*, 1998, **49**, 319-325.
40. LEWIS, N.G., YAMAMOTO, E., Lignin: occurrence, biogenesis and biodegradation, *Annu. Rev. Plant Physiol. Plant Mol. Biol.*, 1990, **41**, 455-496.
41. TERASHIMA, N., FUKUSHIMA, K., HE, L.-F., TAKABE, K., Comprehensive model of the lignified plant cell wall, *in* Forage Cell Wall Structure and Digestibility (Jung, H. G., Buxton, D. R., Hatfield, R. D. and Ralph, J., eds.), ASA-CSSA-SSSA, Madison, 1993, pp. 247-270
42. HE, L., TERASHIMA, N., Formation and structure of lignin in monocotyledons IV. Deposition process and structural diversity of the lignin in the cell walls of sugarcane and rice plant studied by ultraviolet microscopic spectroscopy, *Holzforschung*, 1991, **45**, 191-198.
43. INOUE, K., SEWALT, V.J.H., BALLANCE, G.M., NI, W., STÜRZER, C., DIXON, R.A., Developmental expression and substrate specificities of alfalfa caffeic acid 3-*O*-methyltransferase and caffeoyl CoA 3-*O*-methyltransferase in relation to lignification, *Plant Physiol.*, 1998, **117**, 761-770.
44. CASLER, M.D., In vitro digestibility of dry matter and cell wall constituents of smooth bromegrass forage, *Crop Sci.*, 1987, **27**, 931-934.
45. GRABBER, J.H., RALPH, J., HATFIELD, R.D., QUIDEAU, S., p-hydroxyphenyl, guaiacyl, and syringyl lignins have similar inhibitory effects on wall degradability, *J. Agric. Food Chem.*, 1997, **45**, 2530-2532.
46. DIXON, R.A., CHEN, F., GUO, D., PARVATHI, K., The biosynthesis of monolignols: a "metabolic grid", or independent pathways to guaiacyl and syringyl units?, *Phytochemistry*, 2001, in press.
47. KERSEY, R., INOUE, K., SCHUBERT, K.R., DIXON, R.A., Immunolocalization of two lignin *O*-methyltransferases in stems of alfalfa (*Medicago sativa* L.), *Protoplasma*, 1999, **209**, 46-57.

48. PARVATHI, K., CHEN, F., GUO, D., BLOUNT, J.W., DIXON, R.A., Substrate preferences of *O*-methyltransferases in alfalfa suggest new pathways for 3-*O*-methylation of monolignols, *Plant J.*, 2001, **25**, 193-202.
49. HUMPHREYS, J.M., HEMM, M.R., CHAPPLE, C., New routes for lignin biosynthesis defined by biochemical characterization of recombinant ferulate 5-hydroxylase, a multifunctional cytochrome P450-dependent monooxygenase, *Proc. Natl. Acad. Sci. USA*, 1999, **96**, 10045-10050.
50. INOUE, K., PARVATHI, K., DIXON, R.A., Substrate preferences of caffeic acid/5-hydroxyferulic acid 3-*O*-methyltransferases in developing stems of alfalfa (*Medicago sativa* L.), *Arch. Biochem. Biophys.*, 2000, **375**, 175-182.
51. GUO, D., CHEN, F., INOUE, K., DIXON, R.A., Down-regulation of caffeic acid 3-*O*-methyltransferase and caffeoyl CoA 3-*O*-methyltransferase in transgenic alfalfa (*Medicago sativa* L.): impacts on lignin structure and evidence for independent pathways for the biosynthesis of G and S lignin, *Plant Cell*, 2001, **13**, 73-88.
52. SEWALT, V.J.H., NI, W., BLOUNT, J.W., JUNG, H.G., HOWLES, P.A., MASOUD, S.A., LAMB, C., DIXON, R.A., Reduced lignin content and altered lignin composition in transgenic tobacco down-regulated in expression of phenylalanine ammonia-lyase or cinnamate 4-hydroxylas, *Plant Physiol.*, 1997, **115**, 41-50.
53. FISHER, R.F., LONG, S.R., *Rhizobium*-plant signal exchange, *Nature*, 1992, **387**, 655-660.
54. MAXWELL, C.A., HARTWIG, U.A., JOSEPH, C.M., PHILLIPS, D.A., A chalcone and two related flavonoids released from alfalfa roots induce nod genes of *Rhizobium meliloti*, *Plant Physiol.*, 1989, **91**, 842-847.
55. MAXWELL, C.A., EDWARDS, R., DIXON, R.A., Identification, purification and characterization of S-adenosyl-L-methionine: isoliquiritigenin 2'-O-methyltransferase from alfalfa (*Medicago sativa* L.), *Arch. Biochem. Biophys.*, 1992, **293**, 158-166.
56. MAXWELL, C.A., HARRISON, M.J., DIXON, R.A., Molecular characterization and expression of alfalfa isoliquiritigenin 2'-O-methyltransferase, an enzyme specifically involved in the biosynthesis of an inducer of *Rhizobium meliloti* nodulation genes, *Plant J.*, 1992, **4**, 971-981.
57. DIXON, R.A., LAMB, C.J., MASOUD, S., SEWALT, V.J.H., PAIVA, N.L., Metabolic engineering: prospects for crop improvement through the genetic manipulation of phenylpropanoid biosynthesis and defense responses- a review, *Gene*, 1996, **179**, 61-71.
58. BURBULIS, I.E., WINKEL-SHIRLEY, B., Interactions among enzymes of the Arabidopsis flavonoid biosynthetic pathway, *Proc. Natl. Acad. Sci. USA*, 1999, **96**, 12929-12934.
59. JEZ, J.M., BOWMAN, M.E., DIXON, R.A., NOEL, J.P., Structure and mechanism of the evolutionarily unique plant enzyme chalcone isomerase, *Nature Struct. Biol.*, 2000, **7**, 786-791.
60. SCHLUCKEBIER, G., O'GARA, M., SAENGER, W., CHENG, X., Universal catalytic domain structure of AdoMet-dependent methyltransferases, *J. Mol. Biol.*, 1995, **247**, 16-20.

61. FERRER, J.-L., JEZ, J.M., BOWMAN, M.E., DIXON, R.A., NOEL, J.P., Structure of chalcone synthase and the molecular basis of plant polyketide biosynthesis, *Nature Struct. Biol.*, 1999, **6**, 775-784.
62. KAPULNIK, Y., JOSEPH, C.M., PHILLIPS, D.A., Flavone limitations to root nodulation and symbiotic nitrogen fixation in alfalfa, *Plant Physiol.*, 1987, **84**, 1193-1196.

Chapter Eight

MECHANISMS AND APPLICATIONS OF TRANSCRIPTIONAL CONTROL OF PHENYLPROPANOID METABOLISM

Cathie Martin, Hailing Jin, and Kathy Schwinn

Department of Genetics,
John Innes Centre,
Colney, Norwich,
NR4 7UH, UK

INTRODUCTION

Plants are voracious and versatile biosynthesizers. Many of the specialized secondary metabolites unique to particular species are beyond the range of chemical synthesis. Plant secondary metabolites play significant roles as medical products; anti-cancer agents, sedatives, analgesics, and anti-inflammatory agents, and are increasingly being recognized as providing health promoting components in the diet, including vitamins and antioxidants. Plant secondary metabolites also play important roles in plants themselves as sunscreens, strengthening and waterproofing compounds, signalling compounds, defense-related agents, pigments, and antioxidants.

Plants, therefore, represent a source of valuable phytochemical products. Some of their potential is already mined; much more is available for bioprospecting for new applications and uses. However, many of the more important secondary metabolites made by plants are made in relatively small amounts. Metabolic engineering of plant productivity, which through the application of molecular techniques is rapidly becoming a reality, is as much dependent upon manipulation of flux to alter production as it is on creating novel products. Whether the goal is to reduce lignin content of wood to facilitate paper and pulping industries or the digestibility of forage crops, or to produce tropane alkaloids in plants amenable to mass production, considerations of the amount of product in addition to the type of product are central to engineering strategies.

A central task in pathway engineering is, therefore, the manipulation of flux. However, even in secondary metabolism it is now appreciated that control is usually vested in many steps of a pathway,[1] and that the important, rate-limiting catalytic conversions may change depending on the tissue, the developmental stage, the environmental conditions, and the prevailing balance between primary and secondary metabolism. This means that classical strategies of manipulation of 'key regulatory steps' to achieve changes in synthesis of particular products are often limited in their impact. One example of an early attempt to increase anthocyanin production in *Petunia hybrida* involved increasing the activity of the first enzyme committed to flavonoid biosynthesis - chalcone synthase (CHS). Although this enzyme has some features typical of a 'key rate-limiting step' (being the first committed enzyme and catalyzing a non-readily reversible condensation of *p*-coumaroyl CoA and malonyl CoA), experiments failed to achieve any increase in anthocyanin biosynthesis as a result of expressing the CHS gene under the control of the strong CaMV35S promoter. In fact, the most significant outcome of this work was the discovery of gene silencing by sense-suppression in plants.[2] Analysis of *CHS* mutants suggests that expression levels can be reduced to less than 15% before a significant impact on anthocyanin production is observed. Therefore, CHS is not a suitable target for manipulation of rates of flavonoid biosynthesis. However, it is unlikely that large increases in flux can be achieved by increasing the activity of any single bioysnthetic step. As the flux increases as a result of lifting the limitation placed on it by one step, another step becomes limiting, because control is distributed across pathways.[1]

The solution to the problem of limitations on flux is to modulate the activity of all the steps in a biosynthetic pathway co-ordinately. This is how plants turn on or turn off biosynthesis in response to environmental or developmental triggers. Manipulation of the genes encoding enzymes of desirable pathways is possible provided all the genes are cloned, but this approach is technically demanding and open to the potential of enormous instability in the manipulated plants. A simpler strategy is to identify the regulatory genes that control induction or repression of the pathways and to manipulate the activity of these genes. This strategy is also the course most commonly followed in evolution of metabolic capacity in plants.[3,4] Most of the major differences in capability among species are associated with modifications of regulatory gene activity.

Manipulation of regulatory genes requires their identification and an understanding of how they work. This field is relatively well-explored for the regulation of phenylpropanoid biosynthesis, but is only just being opened up for other areas of plant secondary metabolism.[5-7] By understanding how regulatory genes control phenylpropanoid biosynthesis, how different branches are selected under different conditions, and how flux can be manipulated by engineering of regulatory genes, models for the control of other metabolic pathways can be formulated, which should greatly facilitate regulatory gene discovery and the resultant pathway manipulation.

IDENTIFICATION OF GENES CONTROLLING FLAVONOID METABOLISM

Our understanding of genes regulating flavonoid biosynthesis (Fig. 8.1) in plants is good because of pioneering biochemical analysis on the induction of enzymes synthesizing flavonoids in response to light and fungal elicitors,[8] and because mutants affecting the production of colored flavonoid pigments are easy to identify.[9] Biochemical analysis has established the principle of co-ordinate induction of genes encoding enzymes in related sectors of a pathway, implying that induction is achieved through the activity of a common regulatory system. Mutational analysis led to the identification of loci required for the co-ordinate activity of biosynthetic genes[9] and to the subsequent isolation of the regulatory genes involved.[10-12] In maize, anthocyanin biosynthesis requires induction of at least nine genes encoding biosynthetic enzymes. Two regulatory genes, *C1* and *R* (encoding MYB-related and

Phenylpropanoid metabolism

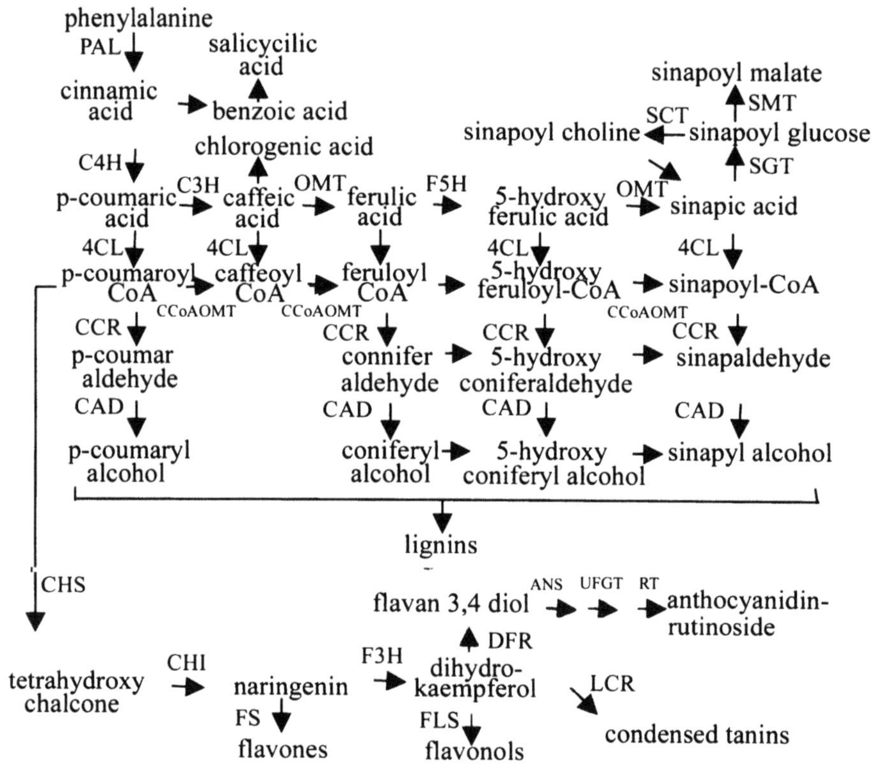

Figure 8.1: Pathways of phenylpropanoid metabolism including flavonoid synthesis, hydroxycinnamic acid metabolism, and monolignol synthesis: PAL, phenylalanine ammonia lyase; C4H, cinnamate 4-hydroxylase; 4CL, p-coumaroyl 4-CoA ligase; CCR, caffeoyl CoA reductase; CAD, cinnamoyl alcohol dehydrogenase; C3H, p-coumarate 3-hydroxylase; F5H, ferulate 5-hydroxylase; OMT, caffeic acid O-methyl transferase; CCoAOMT, caffeoyl CoA o-methyl transferase; SGT, sinapic acid UDPG:sinapoyl transferase; SMT, sinapoyl glucose malate: sinapoyl transferase; SCT, sinapoyl glucose choline: sinapoyl transferase; CHS, chalcone synthase; CHI, chalcone flavone isomerase; F3H, flavanone 3-hydroxylase; DFR, dihyroflavonol 4-reductase; LCR, leucoanthocyanidin reductase; FS, flavone synthase; FLS, flavonol synthase; ANS, anthocyanidin synthase; UFGT, UDPglucose; flavonol 3-O-glucosyl transferase; RT, rhamnosyl transferase.

bHLH related transcription factors, respectively), are required for induction of anthocyanin production in aleurone. The products of these genes have been shown to activate transcription of most of the structural genes of the pathway (the chalcone isomerase gene is not induced by C1/R, but its activity may not limit anthocyanin production as the isomerization of chalconone narigenin can occur spontaneously), and introduction of these genes into other species has been demonstrated to induce ectopic anthocyanin production.[13,14]

This control system is a simplified version of that operating in other species. Flavonoid production in flowers is important for insect pollinated species. Not only anthocyanins, but also other flavonoids affect the appearance of flowers to insect pollinators, through absorption of light in the UV-range and through co-pigmentation. In flowers of several dicotyledonous species, the genes regulating flavonoid biosynthesis are equivalent to those in maize (MYB- and bHLH-related), but the target genes they regulate represent a smaller subset of genes encoding enzymes of the pathway than in maize.[12,15-17] In *Antirrhinum*, for example, the *Delila* gene encodes a bHLH transcription factor (equivalent to R in maize) that is required for expression of genes late in the anthocyanin biosynthetic pathway.[18] The early steps (chalcone synthase and chalcone isomerase) are under independent regulation. In *Antirrhinum,* there are also genes encoding MYB-related transcription factors required for anthocyanin production (*Rosea1, Rosea2* and *Venosa*).[18] The products of these genes interact with the bHLH factors to activate the late steps of the pathway. This separate regulation of the early steps from the late steps of flavonoid metabolism allows for metabolic flexibility. Independent regulation of CHS and CHI can allow for the synthesis of the precursors of flavones independently of anthocyanin production, and flavones are important in identifying zones on the petal for insect attraction in *Antirrhinum.*[18]

DEPLOYMENT OF REGULATORY GENES IN THE CONTROL OF FLORAL PIGMENTATION

The regulatory genes in *Antirrhinum* are used to control anthocyanin biosynthesis in a rather similar way to those in maize. In maize, paralogous genes (duplicate genes with identical functions in a single species) of the *R* family (*B, Sn, r-P, r-*S) with equivalent regulatory functions control anthocyanin biosynthesis in different tissues.[19] Similarly the *C1* and the *Pl* genes are paralogous, controlling pigmentation in seed and vegetative tissues, respectively.[20] In the control of floral pigmentation in *Antirrhinum*, there are also duplicated functions regulating pigmentation in different parts of the flower (Fig. 8.2). *Delila* is required for anthocynain biosynthesis only in the corolla tube; it is expressed in the corolla lobes, but in lobes another gene can complement its function. This is likely to be the *Mutabilis* gene; loss of function of this gene causes loss of pigmentation in the petal lobes, but only when in double mutant combination with *Delila* (Fig. 8.2).[21] The *Rosea* gene family also provides specificity to floral pigmentation. The *Rosea2* gene gives weak activation of late anthocyanin biosynthetic genes specifically in flowers. The *Rosea1* gene gives strong activation of late anthocyanin biosynthetic genes to give strong pigmentation, typical of that observed in wild type *Antirrhinum majus*. The *Venosa* gene product activates the same target genes, but this time in the epidermal tissue immediately above the veins, giving rise to striking pigmentation patterns. This specificity in function is achieved through specificity in expression of

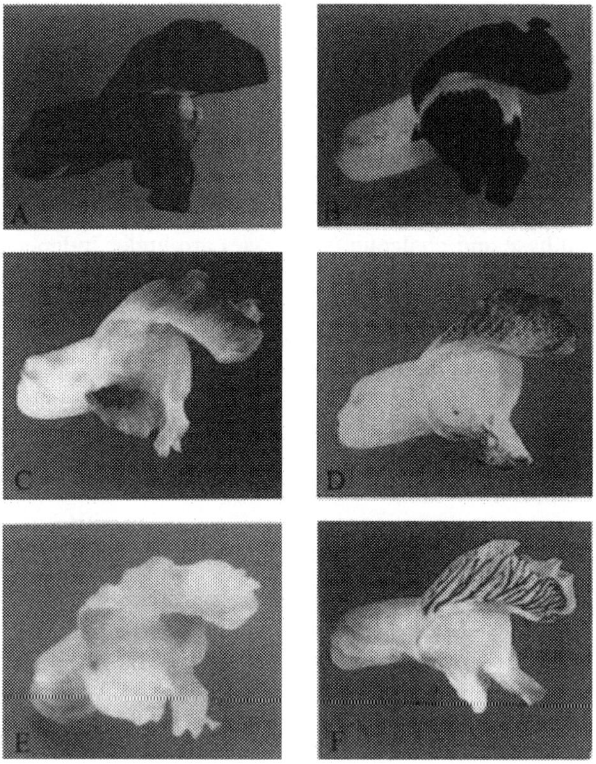

Figure 8.2: Phenotypes of mutants of anthocyanin regulatory genes in *Antirrhinum majus*.
A. *wild type*, B. *delila*. C. *rosea1*, D. *mutabilis/delila*, E. *rosea1/rosea2*, F. *Venosa/rosea1/delila*. See also cover.

the *Venosa* gene. In this way, the regulation of anthocyanin biosynthesis is used to build up elaborate patterns of pigmentation of importance in attracting insect pollinators. There is also emerging evidence that in the evolution of different *Antirrhinum* species with different floral pigmentation patterns and intensities, differences have been achieved through selection for changes in the activity of the *Rosea* gene family members.[21]

GENES REGULATING THE EARLY STEPS OF FLAVONOID BIOSYNTHESIS AND GENERAL PHENYLPROPANOID METABOLISM

If the genes encoding late steps in anthocyanin biosynthesis are regulated by genes equivalent to *R* and *C1* in maize, what genes are likely to be regulating the early steps of flavonoid metabolism in dicotyledonous plants? Genetic analysis is less helpful here, due to a current lack of mutants, although biochemical analysis of CHS expression suggests that bZIP and MYB factors may be involved.[22,23] Biochemical analysis can not prove a regulatory role *in vivo*, but has provided compelling data to support the idea that MYB-related transcription factors also control the early steps in general phenylpropanoid metabolism and in anthocyanin biosynthesis.[23] A motif within the promoter of the bean phenylalanine ammonia lyase (*PAL*) gene was shown to be necessary and sufficient for expression in tobacco petals. This motif resembles a MYB-recognition site, and by screening with specific antibodies the product of a MYB gene from *Antirrhinum* (*AmMYB305*) and the equivalent protein from tobacco were shown to be able to bind this motif and activate from the *PAL* promoter in yeast and in transfection assays. This established that *MYB* genes were involved in the regulation of steps in general phenylpropanoid metabolism as well as flavonoid metabolism. Interestingly, in this case the MYB protein does not seem to require an interacting bHLH partner to induce expression. Although encoding an R2R3 MYB protein, the *AmMYB305* gene is not particularly closely related structurally to *C1* or *Rosea*, which may reflect the ability of its protein product to activate transcription alone. Regulation of different steps in anthocyanin biosynthesis may depend on distinctions in target site binding preferences of these different MYB proteins.

AmMYB305 can activate transcription of genes encoding CHI and F3H as well as PAL. This, and its expression pattern, suggest a primary role in flavonol biosynthesis.[24] There is a closely related protein from *Antirrhinum*, AmMYB340, which has the same binding site specificity as AmMYB305 and a similar expression pattern. Binding and transfection assays that use the CHI gene promoter suggest that there are subtle differences in the activity of the two transcription factors, which imply that they serve distinct roles in the activation of target gene expression; AmMYB340 ensures the rapid activation of transcription during petal expansion, and

AmMYB305 ensures stable maintenance transcription in fully expanded petals. These differences in regulatory protein function are probably achieved by differential phosphorylation of the proteins, and illustrate the intricacy of regulatory mechanisms involved in controlling metabolism.

MANIPULATION OF FLAVONOID METABOLISM BY USING REGULATORY GENES

Considerable success has been achieved in upregulating anthocyanin biosynthesis through overexpression of regulatory genes.[12,13] This has worked best for overexpression of the maize *R* gene family members that enhance pigmentation levels in a range of species. In combination with overexpression of the *C1* gene, ectopic anthocyanin production has been claimed although anthocyanin production does not become constitutive. The results from overexpression of the genes from dicots are more subtle, although pigmentation is enhanced in tobacco and tomato by overexpression of *Delila.*[13] Unlike the results for *C1* on its own,[12] overexpression of *Rosea* genes can also increase anthocyanin production, although the effect, observed in both petunia and lisianthus is restricted to particular tissues under certain environmental conditions. Limitations on the efficacy of *Rosea* may be that it needs high levels of its bHLH partner to activate transcription effectively, or that it is the factor limiting the rate of anthocyanin biosynthesis only in certain cells under certain conditions. Alternatively, it may be because, unlike *C1*, it only activates the late biosynthetic genes, in which case, elevated anthocyanin levels can only be achieved if the supply of precursors from the early biosynthetic enzymes is ensured. Nevertheless, such manipulations show that overexpression of regulatory genes can achieve modified floral pigmentation and patterning. Such changes may be of interest in providing new garden varieties in the future.

IDENTIFICATION OF GENES REGULATING HYDROXYCINNAMIC ACID METABOLISM

Many other genes involved in phenylpropanoid metabolism contain P or L-box motifs in their promoters, indicative of sites of regulation by R2R3 plant MYB proteins. In trying to assign function to novel members of the R2R3 MYB gene class, we ectopically expressed several different MYB genes in transgenic tobacco and looked for marked phenotypes that might indicate the function of the genes *in vivo*. The most dramatic results came from overexpression of *AmMYB308* and *AmMYB330*, where plants showed dramatic reductions in the levels of soluble and wall-bound hydroxycinnamic acid derivatives and in levels of lignin in vascular tissue.[25] These results indicate that distinct MYB family members might regulate another branch of phenypropanoid metabolism, but this time as negative regulators.

We have investigated the *in vivo* function of these transcriptional regulators by identifying a knockout mutant in *Arabidopsis* of the orthologous gene to *AmMYB308* (the gene from *Arabidopsis* structurally most similar to *AmMYB308*; the two genes share a common ancestor), *AtMYB4*.[26] Analysis of this mutant in comparison to wild-type plants has confirmed the role for AtMYB4 in negative regulation of hydroxycinnamic acid metabolism; the mutant has elevated levels of sinapate esters in its leaves. However, the analysis of target genes in wild-type and mutant has shown that the expression of only one structural gene required for sinapate ester formation is upregulated in the atmyb4$^-$ mutant, that encoding cinnamate 4-hydroxylase. This means that the increase in flux achieved by mutation of *AtMYB4* results from increased expression of a single key target gene. With this insight, it might be possible to manipulate flux to sinapate esters effectively by overexpressing just the gene encoding cinnamate 4-hydroxylase.

The product of the *AtMYB4* gene operates as a transcriptional repressor, and under some conditions the flux along the general phenylpropanoid pathway may be modulated primarily by derepression. The way that the transcription factor represses transcription appears to involve two biochemical actions; first, a competition with activators for a common DNA binding site within the target promoter and second, a repression operating through its C-terminal domain that probably involves direct interaction with the basal transcriptional machinery. This means that activation of the pathway (and *C4H* gene expression) also depends on activators (which are likely to be MYB factors due to their shared binding site preferences with AtMYB4). These remain to be identified, but when they have been cloned they will provide powerful tools in combination with the AtMYB4 repressor (or its orthologues in other species) for modulating flux through hydroxycinnamic acid metabolism. Engineering flux along this pathway can have important consequences for the plant. The increased sinapate ester content of the atmyb4$^-$ mutant effectively provided enhanced tolerance to UV-B.[26] Enhanced levels of cell wall bound phenolics may have beneficial effects on the texture and mechanical properties of food. Diferulic acid cross-linking within cell walls is associated with improved textural properties of food, such as the crunchiness of Chinese water chestnut.[27] We are currently investigating means to upregulate the amount of ferulic acid in plant cell walls and to discover whether this will enhance cross-linking, wall firmness, and texture. In addition, there has been considerable publicity recently on the beneficial effects of antioxidants in the diet. Attention has focussed on the flavonoid content of red wine, which is responsible for the 'French paradox', the fact that, for equivalent fat intake, people living in the South of France have significantly lower incidence of arterial sclerosis and coronary thrombosis. The active ingredient is particular to red wine and thought to be the enhanced flavonoid content that provides additional antioxidants in the diet.[28,29] Many phenylpropanoid derivatives have antioxidant properties; some like the flavonoids serve as free radical scavengers. Hydroxycinnamic acid derivatives such as chlorogenic acid serve to protect membranes against free radical damage.[29] Therefore, strategies to increase the content of flavonoids and soluble phenolics in plants would improve foods. Although many plants synthesize these phenylpropanoids, increased production would improve their dietary value. We are investigating strategies to modulate flavonoid and phenolic productivity in plants and to test the effects of their inclusion in the diet on the incidence of arteriosclerosis and coronary problems.

Reductions in hydroxycinnamic acid metabolism may also provide beneficial traits in plants. In tobacco plants overexpressing *AmMYB308,* the down regulation in hydroxycinnamic acid metabolism is sufficient to impact on the supply of monolignols for lignin biosynthesis.[22,30] This results in plants with up to 50% less lignin, causing the stem tissue to have modified physical properties. Such modifications in lignin production are of significance to the paper industry, where it has been calculated that a reduction in lignin content of just 1% would make a significant improvement in the properties of wood as a raw material. This is because the cellulose of wood has to have its lignin removed before it can be used to produce paper. The inert nature of lignin means that chemical extraction is costly and creates a heavy chemical load on the environment. Small reductions in lignin content would provide significantly improved material for this purpose.

Similar arguments exist for reducing the lignin content of forage legumes, where the inert material reduces the digestibility and hence the nutritional value of the crop for cattle. Increasing the activity of these MYB repressors of hydroxycinnamic acid metabolism and monolignol synthesis provides an effective strategy to modulate flux appropriately, and one that appears to work in different plant species.[25,26]

IDENTIFICATION OF GENES CONTROLLING OTHER SECONDARY METABOLIC PATHWAYS

The examples given above provide some ideas about how phenylpropanoid metabolism is regulated in plants, and the strategies that could be adopted for manipulating flux to improve the yield of valuable plant products. However, there are many important secondary metabolites synthesized by other metabolic routes. How can the production of these pathways be modified? A general principle that emerges is that other pathways, (for example carotenoid biosynthesis, alkaloid biosynthesis, terpenoid biosynthesis, etc.) are induced through transcriptional control of the genes encoding the biosynthetic enzymes. Therefore, similar strategies to modify flux through control of regulatory genes should be effective. To what extent are the same or related transcription factors involved in controlling different pathways of secondary metabolism? There is as yet no clear answer to this question, although there is one MYB gene involved in regulation of tryptophan biosynthesis[7] in addition to the prominent role this gene family plays in the control of phenylpropanoid metabolism.

In an elegant approach to identification of genes regulating tropane indole alkaloid biosynthesis in *Catharanthus roseus*, activators of tryptophan decarboxylase and strictosidine synthase have been cloned. [5, 6] These regulatory genes encode products belonging to the AP2 family. The AP2 gene family includes members involved in transcriptional activation in response to stress, and it is thought that the synthesis of the important secondary metabolites of *Catharanthus* may represent a metabolic defense response. Having identified the first genes involved in regulation of tropane indole alkaloid biosynthesis, the next task will be to assign further regulatory gene function such that manipulation of flux becomes possible. If the phenylpropanoid example holds true, it is likely that other AP2-related genes play roles in the detailed control of tropane indole alkaloid metabolism.

Although there has been considerable recent success in identifying genes controlling secondary metabolism in plants, the examples represent results from enormous and lengthy research efforts. For many aspects of secondary metabolism, identification of regulatory genes seems remote because biosynthetic pathways have not yet been established to completion. A functional genomics approach to the control of secondary metabolism represents an efficient strategy for overcoming such technical limitations for developing understanding rapidly and for providing the tools for pathway manipulation. This type of approach is possible in *Arabidopsis*, and identification of the genes controlling secondary metabolism in this species, should provide a way into the mechanisms of control in species with more elaborate and more intractable pathways, in much the same way that *C1* and the *MYB* gene family have for phenylpropanoid metabolism. A functional genomics approach to the control of secondary metabolism is part of a current European program to identify the function of all transcription factors in plants. Genes are being identified from the annotated sequence through homology to known transcription factor families. cDNA clones for each gene are being isolated, and a portion of each cDNA encoding a unique identifying region of the transcription factor will be arrayed on chips or filters so that the expression patterns for each gene under a range of different treatments, environmental conditions, and in different tissues can be determined. Expression profiling will provide the first clues to the involvement of a particular factor in metabolic regulation.

Knock-out mutants will be sought for each gene through reverse genetic screening. Once a homozygous mutant line has been identified it will be characterized phenotypically. In the case of regulators of secondary metabolism, the most important component of phenotypic analysis will be metabolic profiling. This should identify genes responsible for modulating flux to specific metabolites. In some cases, it may be necessary to produce double mutants to observe significant changes. This would be the case for the regulation of flavonol metabolism by

AmMYB305 and *AmMYB340*, or their *Arabidopsis* orthologues, for example.

To supplement the analysis of the action of *Arabidopsis* transcription factors each gene will enter a systematic screen for interacting transcription factors using the yeast two-hybrid system. In this way, we hope to establish the regulatory networks involved in metabolic control and, by association, to link in transcription factors involved in metabolic regulation, for which no mutational proof of function is available.

CONCLUSION

One of the major challenges to metabolic engineering is the manipulation of flux. A solution to this problem can be provided through the use of regulatory genes controlling specific biosynthetic pathways. Use of regulatory genes may not solve all problems associated with engineering flux; the supply of carbon and nitrogen from primary metabolism may impose final, insurmountable limitations. However, transgenic experiments have already demonstrated that significant changes in productivity can be achieved.

The phenylpropanoid pathway represents a relatively well-understood metabolic route for which a number of regulatory genes have been identified and cloned. The use of regulatory genes to define patterns of metabolism and the kinetics of response of anthocyanin biosynthesis illustrates the versatility of this means of control. Proteins of the MYB family feature prominently in the control of phenylpropanoid metabolism; working in combination with other factors or alone, they control early and late parts of flavonoid biosynthesis. They also control other branches of phenylpropanoid metabolism and may serve as transcriptional activators or as repressors.

Manipulation of these regulatory genes can lead to manipulation of phenylpropanoid metabolism and the engineering of improved traits for both man and plants. By using this system as a model for the control of secondary metabolism generally, functional genomic approaches to the identification of transcription factors controlling secondary metabolism should represent the most direct means of generating the tools for the manipulation of flux of all secondary metabolic pathways in plants.

ACKNOWLEDGEMENTS

Many thanks to all members of the MYB consortium (1995-1999) and to all those participating in REGIA, the new European transcription factor consortium. Thanks also to Mandy Coventry for manuscript preparation. Parts of this work were financed by the EU, QLG2-CT-1999-00876 and BIO4-CT95-0129, and parts by the New Zealand Institute for Crop and Food Research.

REFERENCES

1. KACSER, H., BURNS, J.A., FELL, D.A., The control of flux, *Biochem. Soc. Trans.*, 1995, **23**, 341-366.
2. NAPOLI, C., LEMIEUX, C., JORGENSEN, R., Introduction of a chimeric chalcone synthase gene into petunia results in reversible co-suppression of homologous genes *in trans*, *Plant Cell*, 1990, **2**, 279-289.
3. DOEBLEY, J., Genetic development and plant evolution, *Curr. Op. Genet. Dev.*, 1993, **3**, 865-872.
4. DOEBLEY, J., LUKENS, L., Transcriptional regulators and the evolution of plant form, *Plant Cell*, 1998, **10**, 1075-1082.
5. VAN DER FITS, L., MEMELINK, J., ORCA3, a jasmonate-responsive transcriptional regulator of plant primary and secondary metabolism, *Science*, 2000, **289**, 295-297.
6. MENKE, F.L.H., CHAMPION, A., KIJNE, J.W., MEMELINK, J., A novel asmonate- and elicitor-responsive element in the periwinkle secondary metabolite biosynthetic gene Str interacts with a jasmonate- and elicitor-inducible AP2-domain transcription factor, ORCA2, *EMBO J.*, 1999, **18**, 4455-4463.
7. BENDER, J., FINK, G.R., A Myb homologue, ATR1, activates tryptophan gene expression in *Arabidopsis*, *Proc. Natl. Acad. Sci. USA*, 1998, **95**, 5655-5600.
8. RAGG, H., KUHN, D.N., HAHLBROCK, K., Coordinated regulation of 4-coumarate-CoA ligase and phenylalanine ammonia-lyase messenger-RNAs in cultured plant-cells, *J. Biol. Chem.*, 1981, **256**, 61-651.
9. DOONER, H.K., Coordinate genetic-regulation of flavonoid biosynthetic-enzymes in maize, *Mol. Gen. Genet.*, 1983, **189**, 136-141.
10. PAZARES, J., WIENAND, U., PETERSON, P.A., SAEDLER, H., Molecular-cloning of the C-locus of *Zea mays* - a locus regulating the anthocyanin pathway, *EMBO J.*, 1986, **5**, 829-833.
11. LUDWIG, S.R., HABERA, L.F., DELLAPORTA, S.L., WESSLER, S.R., *Lc*, a member of the maize *R*-gene family responsible for tissue-specific anthocyanin production, encodes a protein similar to transcriptional activators and contains the myc-homology region, *Proc. Natl. Acad. Sci. USA*, 1989, **86**, 7092-7096.
12. GOODRICH, J., CARPENTER, R., COEN, E.S., A common gene regulates pigmentation pattern in diverse plant-species, *Cell*, 1992, **68**, 955-964.
13. LLOYD, A.M., WALBOT, V., DAVIS, R.W., *Arabidopsis* and *Nicotiana* anthocyanin production activated by maize regulator-R and regulator-C1, *Science*, 1992, **258**, 1773-1775.
14. MOONEY, M., DESNOS, T., HARRISON, K., JONES, J., CARPENTER, R., COEN, E., Altered regulation of tomato and tobacco pigmentation genes caused by the Delila gene of *Antirrhinum*, *Plant J.*, 1995, **7**, 333-339

15. QUATTROCCHIO, F., WING, J.F., VAN DER WOUDE, K., MOL, J.N.M., KOES, R., Analysis of bHLH and MYB domain proteins: species-specific regulatory differences are caused by divergent evolution of target anthocyanin genes, *Plant J.,* 1998, **13**, 475-488.
16. DE VETTEN, N., QUATTROCCHIO, F., MOL, J., KOES, R., The *an11* locus controlling flower pigmentation in petunia encodes a novel WD-repeat protein conserved in yeast, plants, and animals, *Genes Dev.*, 1997, **11**, 1422-1434.
17. QUATTROCCHIO, F., WING, J.F., LEPPEN, H.T.C., MOL, J.N.M., KOES, R.E., Regulatory genes-controlling anthocyanin pigmentation are functionally conserved among plant-species and have distinct sets of target genes, *Plant Cell*, 1993, **5**, 1497-1512.
18. MARTIN, C., PRESCOTT, A., MACKAY, S., BARTLETT, J., VRIJLANDT, E., Control of anthocyanin biosynthesis in flowers of *Antirrhinum majus, Plant J.*, 1991, **1**, 37-49.
19. CHANDLER, V.L., RADICELLA, J.P., ROBBINS, T.P., CHEN, J.C., TURKS, D., Two regulatory genes of the maize anthocyanin pathway are homologous – isolation of *B* utilizing *R* genomic sequences, *Plant Cell*, 1989, **1**, 1175-1183.
20. CONE, K.C., COCCIOLONE, S.M., BURR, F.A., BURR, B., Maize anthocyanin regulatory gene *Pl* is a duplicate of *C1* that functions in the plant, *Plant Cell*, 1993, **5**, 1795-1805.
21. SCHWINN, K., The Control of Anthocyanin Bisynthesis in *Antirrhinum majus*, PhD Thesis University of East Anglia, 2000.
22. WEISSHAAR, B., ARMSTRONG, G.A., BLOCK, A., DE SILVA, O., HAHLBROCK, K., Light-inducible and constitutively expressed DNA-binding proteins recognizing a plant promoter element with functional relevance in light responsiveness, *EMBO J.*, 1991, **10**, 1777-1786.
23. SABLOWSKI, R.W.M., MOYANO, E., CULIANEZ-MACIA, F.A., SCHUCH, W., MARTIN, C., BEVAN, M., A flower-specific MYB protein activates transcription of phenylpropanoid biosynthetic genes, *EMBO J.*, 1994, **13**, 128-137.
24. MOYANO, E., MARTINEZGARCIA, J.F., MARTIN, C., Apparent redundancy in *Myb* gene function provides gearing for the control of flavonoid biosynthesis in Antirrhinum flowers, *Plant Cell*, 1996, **8**, 1519-1532
25. TAMAGNONE, L., MERID, A.A., PARR, A., MACKAY, S., CULIANEZ-MACIA, F.A., ROBERTS, K., MARTIN, C., The AmMYB308 and AmMYB330 transcription factors from antirrhinum regulate phenylpropanoid and lignin biosynthesis in transgenic tobacco, *Plant Cell*, 1998, **10**, 135-154.
26. JIN, H.L., COMINELLI, E., BAILEY, P., PARR, A., MEHRTENS, F., JONES, J., TONELLI, C., WEISSHAAR, B., MARTIN, C., Transcriptional repression by AtMYB4 controls production of UV-protecting sunscreens in Arabidopsis, *EMBO J.*, 2000, **19**, 6150-6161.
27. PARR, A.J., WALDRON, K.W., NG, A., PARKER, M.L., The wall-bound phenolics of Chinese water chestnut (*Eleocharis dulcis*), *J. Sci. Food Agric.*, 1996, **71**, 501-507.

28. DE GAULEJAC, N.S., GLORIES, Y., VIVAS, N., Free radical scavenging effect of anthocyanins in red wines, *Food Res. Int.*, 1999, **32**, 327-333.
29. RICE-EVANS, C.A., MILLER, N.J., PAGANGA, G., Antioxidant properties of phenolic compounds, *Trends Plant Sci.*, 1997, **2**, 152-159.
30. TAMAGNONE, L., MERIDA, A., STACEY, N., PLASKITT, K., PARR, A., CHANG, C.F., LYNN, D., DOW, J.M., ROBERTS, K., MARTIN, C., Inhibition of phenolic acid metabolism results in precocious cell death and altered cell morphology in leaves of transgenic tobacco plants, *Plant Cell*, 1998, **10**, 1801-1816.

Chapter Nine

ENGINEERING ISOPRENOID METABOLISM AND BIOCHEMISTRY IN PLANTS

Mark A. Schoenbeck and Joseph Chappell

Plant Physiology/Biochemistry/Molecular Biology Program
Agronomy Department
University of Kentucky
Lexington, Kentucky 40546

INTRODUCTION

Isoprenoids are a family of highly diverse, though structurally related compounds, performing a wide variety of functions in the life processes of plants and nearly every biological entity. Stated simply, an isoprenoid is any compound that is derived wholly or in part from the basic five-carbon building unit, isopentenyl diphosphate (IPP).[1] In their biological contexts, isoprenoids may be primary metabolites, essential for the completion of an organism's life cycle. Sterols, carotenoids, and quinones are examples of such primary metabolites common to many organisms, while others, such as the growth regulators cytokinin and gibberellin, are isoprenoids more specific to plants.[2] Isoprenoids also constitute a long list of secondary metabolites such as capsidiol, an antimicrobial compound associated with plant defense responses,[3] and linolool, a component of floral scent.[4] In addition to their roles in biological systems, isoprenoids are also important for their utility to humans: For example, natural rubber used in manufacturing,[5] carotenoids (as dietary precursors to vitamin A and antioxidants[6]), essential oils used in food and flavorings,[7] and a variety of chemical species with potential pharmacological applications (*i.e.*, artemisinin[8,9] and taxol[10]).

Though isoprenoid compounds and their derivatives share as their origin one or more units of IPP, their diversity of form and function comes from the many ways these units can be assembled and modified, and the distinctive chemical properties associated with these compounds. IPP units may be joined together in different numbers and conformations, and the resulting structures further elaborated by a host of enzymatic activities, including cyclases, desaturases, methylases, and hydroxylases. Because so many distinct and useful chemical species are derived from this simple origin, much effort has been put towards understanding the underlying biochemistry and physiology of isoprenoid metabolism. Ultimately, this knowledge may be employed in the rational design of systems for the enhanced production of known products, and exploited for the biosynthesis of novel isoprenoids.

The objective of this review is to summarize recent advances in the understanding of plant isoprenoid metabolism and molecular biology. In keeping with the spirit of the volume, special emphasis is given to those investigations that employ molecular techniques to modify plant isoprenoid metabolism. Studies designed to elucidate the *in vivo* biology of isoprenoid metabolism, as well as applied investigations aimed towards enhanced nutritional quality of foods, and those exploring the possibilities of new compounds produced through the rational design of isoprenoid synthetic enzymes, will be discussed.

ISOPRENOID METABOLISM IN PLANTS: A CELLULAR OVERVIEW

Isoprenoid metabolism is compartmentalized in plant cells with each site of synthesis producing a distinct array of compounds (Fig. 9.1). In both the cytoplasm and the plastid, isoprenoid synthesis begins with the generation of the basic building block of all terpenes, the "active isoprene" IPP. However, the synthetic mechanisms employed in these two compartments are strikingly distinct. Cytosolic IPP is synthesized via the well-characterized acetate-mevalonate pathway, with acetyl-CoA providing the starting material.[11] By contrast, the plastid uses the more recently recognized 2-*C*-methyl-D-erythritol 4-phosphate (MEP) pathway, starting from pyruvate and glyceraldehyde 3-phosphate.[12] Isoprenoid metabolism continues in both the cytosol and the plastid with the isomerization of IPP to form dimethylallyl diphosphate (DMAPP), an allylic donor substrate of prenyltransferases, that catalyzes the polymerization of isoprene units to form 10 carbon mono-, 15 carbon sesqui-, or 20 carbon diterpene diphosphate intermediates. The cytosol is typically viewed as the site of sesquiterpene and triterpene (30 carbon) synthesis, both derived from the sesquiterpenoid diphosphate (FPP) intermediate.[2] In contrast, monoterpenes derived from geranyl diphosphate (GPP), and diterpenes and tetraterpenes (40 carbon) derived from geranylgeranyl diphosphate (GGPP), are usually associated with the plastid.[2,13] In both the cytosol and the plastid, nascent prenyl diphosphates are direct substrates of terpene synthases resulting in an intramolecular rearrangement, often a cyclization, or head-to head condensations catalyzed by squalene synthase and phytoene synthase. In this way, intermediates are essentially committed to a particular metabolic fate, though these products are generally modified further by a host of enzymes before assuming their terminal form.

Subsequent sections of this review will address specific portions of isoprenoid metabolism in greater detail, describing briefly the biochemistry of synthesis and the status of efforts to identify and characterize the genes encoding isoprenoid biosynthetic enzymes. Recent comprehensive reviews of the biochemistry and molecular biology of specific aspects of isoprenoid metabolism are available for cytosolic[11] and plastidic[12] IPP biosynthesis and for carotenoid biosynthesis.[13] Though there is clear evidence of isoprenoid metabolism occurring in other cellular compartments, for example mitochondria[14] and microbodies,[15] this review will focus primarily on cytoplasmic and plastidic biosynthetic pathways.

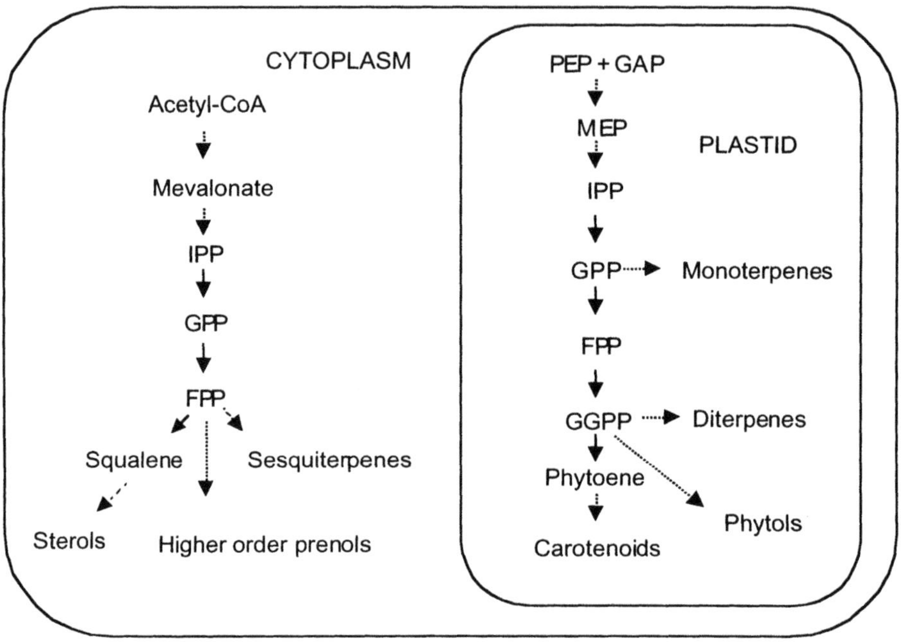

Figure 9.1: Distribution of isoprenoid biosynthesis in the plant cell. Biosynthesis in other compartments such as mitochondria and microbodies might occur, but have not been documented in plants. Dashed lines indicate conversions requiring multiple enzymatic steps. Abbreviations: Isopentenyl diphosphate (IPP), farnesyl diphosphate (FPP), phosphoenol pyruvate (PEP), glyceraldehyde 3-phosphate (GAP), 2-*C*-methyl-D-erythritol 4-phosphate (MEP), geranyl diphosphate (GPP), geranylgeranyl diphosphate (GGPP).

THE BUILIDING BLOCK OF ISOPRENOID METABOLISM: IPP BIOSYNTHESIS

Cytosolic IPP Synthesis: The Mevalonate-Dependent Pathway

In the cytoplasm, acetyl-CoA provides the raw material for isoprenoid synthesis. IPP synthesis in the cytoplasm begins with the condensation of two acetyl-CoA to form acetoacetyl-CoA (Figs. 9.1 and 9.2). Condensation with a third acetyl-CoA molecule yields 3-hydroxy-3-methylglutaryl-CoA (HMG-CoA). In fungi and mammals, the first and second condensations are catalyzed by distinct enzymatic activities, acetoacetyl-CoA thiolase (AACT) and HMG-CoA synthase (HMGS).[11,16] The situation in plants is less well understood. Characterization of AACT and HMGS activities in *Catharanthus roseus* showed these activities to have similar chromatographic behaviors.[17] Despite being unable to purify AACT and HMGS activities into separate fractions, van der Heijden *et al.* concluded that these were distinct reactions, based on the free reversibility of the AACT reaction and the consumption of acetoacetyl-CoA as a substrate.[17] cDNA clones encoding enzymes with AACT or HMGS activities have been characterized in plants.[11,18,19] In striking contrast, an enzyme with apparent dual functionality, catalyzing the direct synthesis of HMG-CoA from three acetyl-CoAs, has been purified to apparent homogeneity in radish.[20,21] Purification of AACT and HMGS activities corresponded with the purification of a single 55 kDa polypeptide.[20] While these observations suggest the possibility of a single polypeptide capable of performing both condensations, the corresponding gene has not yet been cloned.[11] Identification of such a gene and comparison to known AACT and HMGS genes could be informative in defining unique catalytic motifs participating in these condensations.

The next step in cytoplasmic IPP synthesis is the NADPH-dependent reduction of HMG-CoA to produce mevalonic acid (MVA, Fig. 9.2). This is an irreversible step and has long been viewed as the first committed step in isoprenoid metabolism, at least with regard to the mevalonate-dependent pathway. For this same reason it is regarded as a key regulatory point in the pathway.[22] Results from mammalian systems show that the enzyme mediating this reaction, HMG-CoA reductase (HMGR), is under stringent transcriptional and post-translational controls, supporting the hypothesis of its regulatory role.[23]

IPP synthesis continues with the phosphorylation of MVA twice in succession. The first phosphorylation generates mevalonic acid 5-phosphate (MVAP), a reaction catalyzed by MVA kinase. The second phosphorylation, catalyzed by MVAP kinase, adds a second phosphate to the first, generating mevalonic acid 5-diphosphate (MVAPP). Each of these reactions consumes one ATP. MVA kinase and MVAP kinase activities have been characterized in *Catharanthus* cell cultures.[24,25] The *Arabidopsis* MVA kinase cDNA has been cloned by functional complementation of a yeast mutant,[26] and the corresponding gene has also been cloned and characterized.[27] At the time of this writing, no plant MVAP kinases cDNAs or genes have been published.

The final step in the cytoplasmic synthesis of IPP is the ATP-dependent decarboxylation of MVAPP. This reaction releases the carboxyl function of MVAPP as CO_2, and an ATP is consumed along with the loss of the hydroxyl group of carbon 3. This reaction is catalyzed by MVAPP decarboxylase (MVD). An *Arabidopsis* cDNA clone with significant similarity to yeast and mammalian MVD has been identified.[28] Though the activity of the encoded MVD has not been demonstrated *in vitro*, it was sufficient to complement an MVD-deficient yeast strain.

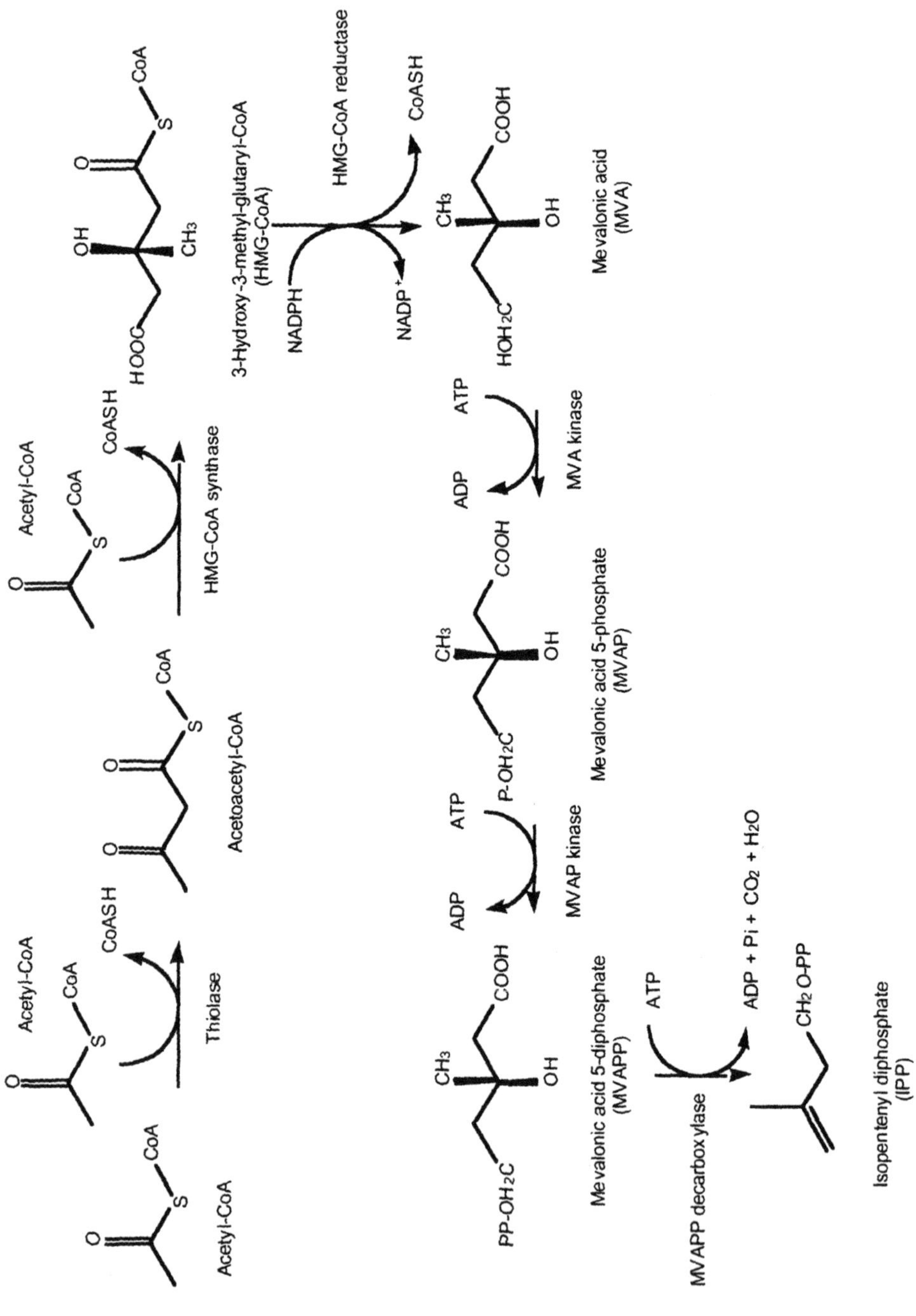
Acetyl-CoA
Acetyl-CoA
Thiolase
CoASH
Acetoacetyl-CoA
Acetyl-CoA
HMG-CoA synthase
CoASH
3-Hydroxy-3-methyl-glutaryl-CoA
(HMG-CoA)
NADPH
NADP+
HMG-CoA reductase
CoASH
Mevalonic acid
(MVA)
ATP
ADP
MVA kinase
Mevalonic acid 5-phosphate
(MVAP)
ATP
ADP
MVAP kinase
Mevalonic acid 5-diphosphate
(MVAPP)
ATP
ADP + Pi + CO2 + H2O
MVAPP decarboxylase
Isopentenyl diphosphate
(IPP)

Figure 9.2: Cytoplasmic IPP synthesis via the mevalonate pathway.

Plastidic IPP Synthesis: The 2-C-Methyl-D-Erythritol 4-Phosphate Pathway

One of the most exciting advances in the study of plant isoprenoid metabolism, and in plant physiology and biochemistry in general, is the recent recognition that plastids synthesize IPP by utilizing a distinct biosynthetic pathway independent of that operating in the cytoplasm.[12] The conservation of the enzymatic steps and the sequence identity among cloned eukaryotic and prokaryotic genes suggests that the plastidic pathway shares an origin with the biochemical processes employed by eubacteria in the synthesis of the sterol-like hopanoids,[12,29] Labeling studies with metabolic precursors suggest that this pathway produces a majority of the IPP used in the synthesis of plastidic monoterpenes and diterpenes.[30,31] Other isoprenoid products of plastidic origin include the carotenoids and xanthophylls, tetraterpene pigments derived from head-to-head condensations of GGPP active in light harvesting, coloration, and photoprotection,[13] and phytols, long-chain prenols derived from GGPP.[30]

IPP synthesis in the plastid proceeds from a thiamine pyrophosphate- (TPP) dependent transketolation that combines pyruvate and glyceraldehyde-3-phosphate (GAP, Fig. 9.3). Synthesis begins with the reaction of pyruvate with TPP, forming a hydroxyethyl-TPP-enzyme complex, accompanied by the release of the carbonyl function as CO_2. A transketolase activity then transfers the hydroxyethyl group to GAP, releasing TPP and forming 1-deoxy-D-xylulose-5-phosphate (DXP). The gene for this enzyme activity, DXP synthase, has recently been cloned and characterized from both bacterial,[32,33] and plant sources.[34,35] DXP undergoes a reduction and rearrangement to produce 2-*C*-methyl-D-erythritol 4-phosphate (MEP) in an NADPH-dependent reaction. A single plant cDNA for this enzyme, DXP reductase-isomerase (DXRI), has been described.[36]

Metabolic tracer studies with [^{13}C]DXP clearly showed that it is a precursor to isoprenoids such as phytol and β-carotene, and thus is an intermediate in IPP synthesis.[31] However, the biosynthetic steps between MEP and IPP have not yet been unequivocally determined. It is presumed that the last step or steps entail the addition of one or more phosphates to the nascent isoprene skeleton via a metabolite kinase. A candidate for the final kinase step has been reported in a plant system. A cDNA from *Mentha* x *piperita* was identified as having significant sequence similarity to known metabolite kinases.[37] *In vitro* assays of the recombinant enzyme expressed in *E. coli* demonstrated its ability to catalyze the phosphorylation

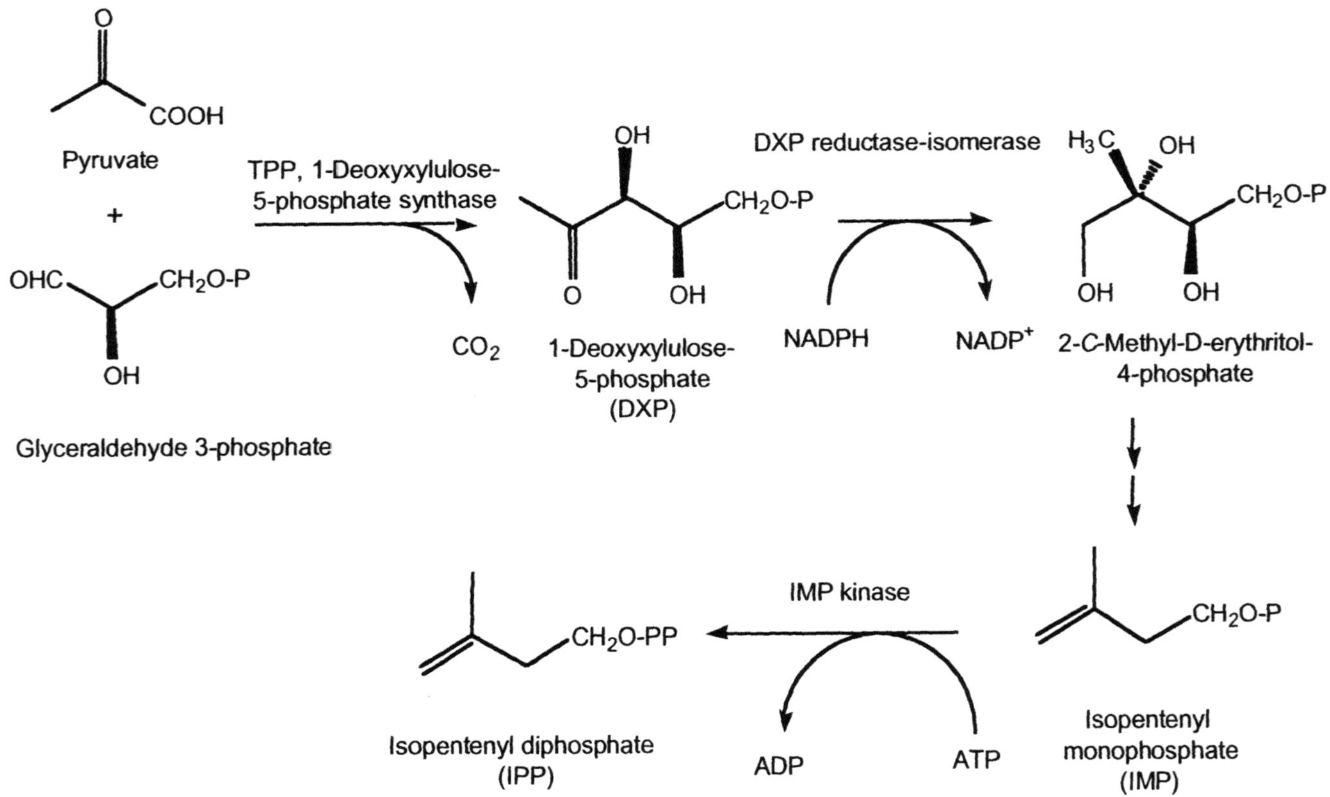

Figure 9.3: Plastidic IPP biosynthesis via the 2-*C*-methyl-D-erythritol 4-phosphate (MEP) pathway. Multiple arrows between MEP and IMP indicate reactions that have not yet been characterized. Abbreviation: thiamine pyrophosphate (TPP).

of isopentenyl 5-phosphate, producing IPP, and completing the pathway. It is too early, however, to rule out the possibility of alternative routes between MEP and IPP in plastids, as may exist in *E. coli.*[38]

The plastidic IPP biosynthetic pathway has been named the 2-*C*-methyl-D-erythritol 4-phosphate or MEP pathway by agreement of the 1999 European Symposium on Plant Isoprenoids, the basis for this being that MEP is the first committed intermediate solely to isoprenoid metabolism. Earlier intermediates in the pathway have alternative fates, isoprenoids being only one of these. The plastidic pathway is also variously referred to as the non-mevalonate pathway, the pyruvate-glyceraldehyde 3-phosphate pathway, or the deoxyxylulose pathway.

Though isoprenoid synthesis via the MEP pathway has been investigated largely in green tissues[31] and developing fruits,[39] where isoprenoid pigments of plastid origin are abundant, the operation of this pathway has also been demonstrated in non-green tissues of higher plants. During the formation of arbuscular mycorrhizal fungi associations with roots of grass species, the host plant produces apocarotenoids in the tissues surrounding the site of association.[40] Apocarotenoids are xanthophyll cleavage products, and are not typically produced in roots. Apocarotenoid production correlates with the induction of DXS and DXR mRNA transcripts in root tissues forming a mycorrhizal association, suggesting a specific induction of these genes by the microsymbiont.[40] As plastidic IPP synthesis in green tissues is thought to be sustained by the products of photosynthesis, the source of root plastid IPP carbon becomes an interesting question. In this respect, the root cell may have similarities to non-photosynthetic glandular trichomes that can readily utilize imported sucrose as a carbon source for monoterpene synthesis.[41]

Isomerization of IPP to DMAPP

Many plant primary and secondary metabolites are derived from carbon skeletons composed of two, three, or four IPP units joined together to form monoterpenes, sesquiterpenes, or diterpenes, respectively. The monoterpenoid GPP is produced by joining two IPP units in a head-to-tail fashion with the loss of an inorganic pyrophosphate. The addition of a third IPP to the chain generates the sesquiterpenoid FPP, and a fourth, diterpenoid GGPP. Addition of IPP units to a growing chain is catalyzed by a family of enzymes termed prenyl transferases. Because the reaction mechanism of a prenyl transferase uses a resonance-stabilized carbocation intermediate[42] (Fig. 9.4), generated by the withdrawal of the pyrophosphate moiety, IPP is not a suitable substrate for chain initiation. Instead, the first substrate in prenyl transferase-mediated polymerizations is the allylic isomer of IPP, dimethylallyl diphosphate (DMAPP). The freely reversible isomerization between IPP and DMAPP is catalyzed by IPP isomerase (IPI). As prenyl diphosphate chain elongations are observed to occur in both the cytosolic and plastidic cell compartments, it is not surprising to find multiple plant genes for IPI,[43,44] possibly encoding isozymes[45] that may be targeted to each of these distinct subcellular sites.

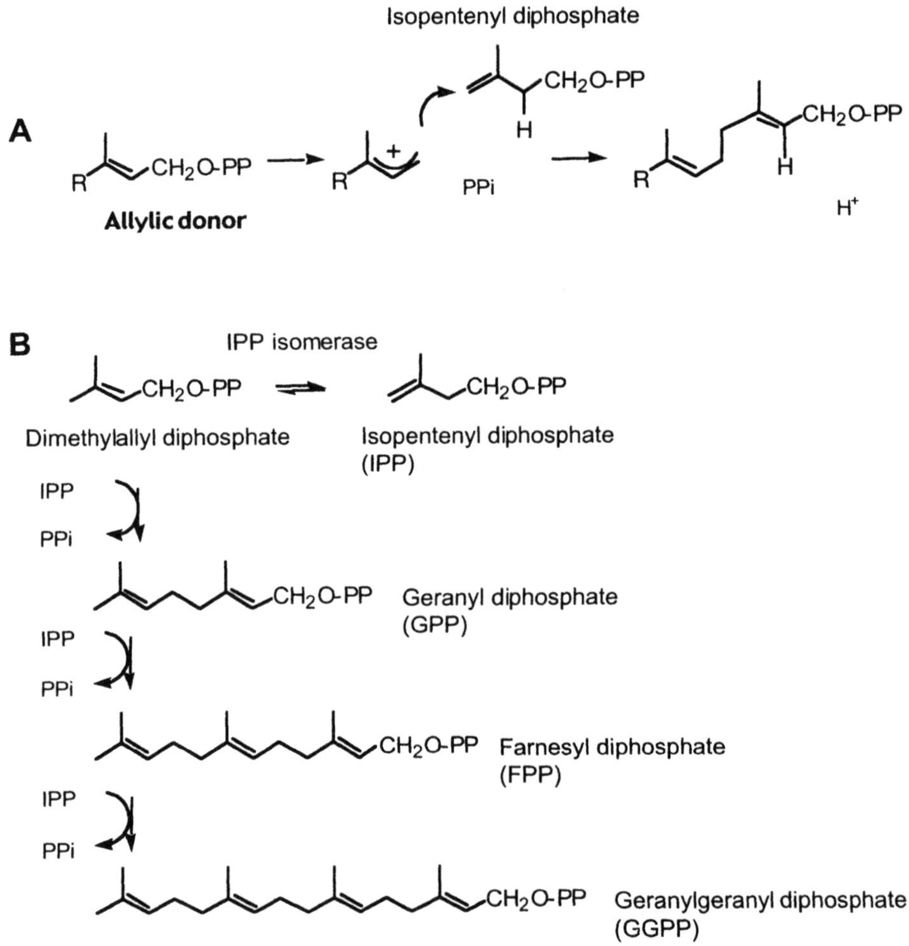

Figure 9.4: Polymerization of prenyl units to form short-chain prenyl diphosphates. A. Addition of IPP to an allylic donor by prenyl transferase proceeds via electrophilic attack of a resonance-stabilized carbocation. B. Short chain prenyl diphosphates are synthesized by the sequential addition of IPP to a growing polymer.

Genetic Engineering Towards Enhanced IPP and DMAPP Synthesis

Clearly any attempt to increase the overall content of isoprenoids *in planta* will be limited by the supply of the IPP, the building block of all isoprenoids. Investigations with transgenic plants overexpressing different IPP biosynthetic enzymes[46-48] have provided insights into the degree of control at different steps, suggesting means for manipulating the levels of IPP *in planta*.

HMG-CoA reductase (HMGR) has been the focus of much interest in plants as a potential key to increasing isoprenoid production. No doubt, much of this interest has been rationalized on the basis that this enzyme is rate-limiting for sterol synthesis in mammals. Evidence of its regulatory role in plants comes largely from observations of its inducibility in response to various stresses and pathogen challenge.[18,49-51] To determine whether HMGR activity represented a limiting step in isoprenoid synthesis, Chappell *et al.* introduced a recombinant hamster HMGR cDNA to tobacco under the control of the constitutive CaMV 35S promoter.[46] The N-terminal end of the hamster HMGR was truncated to remove the membrane-spanning domains responsible for localization to the ER membrane. Both mammalian and plant HMGRs have membrane-spanning domains at their N-terminal ends, though these domains differ in number of membrane spans and in amino acid sequence. In mammals, the membrane-spanning regions have been implicated in post-translational regulation, perhaps comprising a mechanism for sensing the sterol content of the ER membrane.[52] In plants, the role of these domains has not been well characterized, though they have been hypothesized to have a role in metabolite channeling into different isoprenoid pathways at the ER surface.[53]

Expression of recombinant HMGR in tobacco resulted in several-fold increases in both the soluble and, unexpectedly, microsomal HMGR activity.[46] The authors suggested that the overall amount of activity associated with the membrane preparations might arise from trapping during centrifugation. Alternatively, recent studies demonstrating that the catalytic domain of HMGR, the soluble C-terminal end, can form multimeric complexes that stabilize the enzyme against regulatory degradation[54] suggest that the recombinant and membrane-bound endogenous HMGR proteins might associate in this manner.

Total sterol content in the HMGR-overexpressing plants was increased three to eight-fold over control tissues.[46] In contrast, carotenoid levels were not consistently increased.[46] In retrospect, this makes sense from the cell biology of two isoprenoid pathways. Carotenoid biosynthesis occurs via the mevalonate-independent IPP route in plastids, whereas sterol biosynthesis occurs in the cytosol. A similar study by Schaller *et al.* examined the consequences of constitutively expressing a non-truncated HMGR gene from *Hevea brasiliensis* in tobacco.[47] Several transgenic lines with elevated microsomal HMGR activities were recovered. These plants accumulated higher levels of sterols and the sterol intermediates, cycloartenol and obtusifoliol, relative to control plants. Though no gross morphological differences were observed, a distinct phenotype was noted at the cellular level with the accumulation of cytoplasmic lipid droplets not observed in the control plants. The authors concluded that HMGR was likely to be the limiting activity in sterol synthesis.

Re *et al.* examined the effects of constitutive ectopic overexpression of the *Arabidopsis* HMGR cDNA driven by the CaMV 35S promoter in transgenic *Arabidopsis*.[48] In some transgenic lines, introduction of the transgene resulted in both the expression of the recombinant HMGR transcript and, remarkably, an increase in mRNA levels for the endogenous gene. Two- to three-fold increases in HMGR activity were measured in transgene-expressing lines. This increase was far less than the fold-increase in HMGR transcript levels, leading the authors to posit that regulation at the post-transcriptional level limited HMGR activity. Analysis of carotenoid, chlorophyll, and sterol content of control and HMGR-overexpressing plants showed no significant changes in levels of these isoprenoids. Based on these observations, the authors concluded that HMGR activity was not limiting to the synthesis of these compounds in *Arabidopsis*. However, the analysis of sterol content in these plants was limited to the non-esterified fraction, and may not have been an accurate reflection of total sterol content. In none of these three cited studies[46-48] were the activities of other metabolic steps measured, so the possibility of coordinate regulation of isoprenoid biosynthetic activities remains unresolved. Observations as to whether a parallel regulation of subsequent activities is a consequence of HMGR overexpression would be helpful in the design of further work.

Studies of metabolic regulation often focus on irreversible synthetic steps, as they commit metabolites to a particular pathway. There is evidence, however, that a degree of control may be exerted even by enzymes that catalyze freely reversible reactions. IPP isomerase is one such example. Experimental observations from microbial systems have demonstrated that increased IPP activity correlates with an increased rate of carotenoid synthesis. *E. coli* engineered with the carotenoid biosynthesis gene cluster from *Erwinia* synthesizes β-carotene, giving colonies a yellow color.[55] Color synthesis by *E. coli* can, thus, be used as a visible marker in functional cDNA library screening for the discovery of genes enhancing or suppressing the synthesis of carotenoid precursors. This approach has been used to screen cDNA libraries from yeast, algae, and several plant species. It is striking that carotenoid synthesis-enhancing cDNAs from each of these sources were found to encode IPP isomerase.[44,55] Subsequent experiments confirmed that *E. coli* expressing higher levels of IPP isomerase activity also had enhanced carotenoid synthesis rates.[55] Thus, it appears that although IPP isomerase catalyzes a freely reversible interconversion, it can still act as a limiting activity in isoprenoid metabolism. It is important to note that these experiments have only been observed in *E. coli*, which relies solely on the MEP pathway. The significance of IPP isomerase regulation in the plants, in either the plastids or cytosol, remains to be determined. However, at least one report suggests the importance of IPP isomerase for overall isoprenoid metabolism. Albrecht and Sandmann observed that among several carotenoid biosynthetic enzyme activities measured in the course of maize de-etiolation, IPP isomerase was the most strongly induced.[56] This, the authors speculate, is an indication of the limiting nature of isomerase for plastidic isoprenoid metabolism.

POLYMERIZATION OF PRENYL DIPHOSPHATES

Synthesis of *trans*-prenyl diphosphates, as previously mentioned, occurs by the sequential addition of IPP units to the allylic donor, DMAPP (Fig. 9.4). X-ray crystallographic studies of FPP synthase reveal that the prenyl transferase catalytic site resides in a hydrophobic pocket in which the allylic donor substrate is bound.[57] With each round of IPP addition, the growing prenyl chain elongates into the hydrophobic pocket of the enzyme[57] until it can proceed no further. As a consequence, the depth of the hydrophobic pocket appears to be a determining factor of product chain length.[57,58]

Distinct enzymes catalyze the generation of determined chain lengths in different cell compartments. Plastids are typically the site of mono- and diterpene synthesis. Accordingly, GPP and GGPP synthases are localized to the plastid. In contrast, the cytoplasm is the site of FPP-dependent isoprenoid biosynthesis, and FPP synthase is most generally considered as a cytosolic activity. Exceptions to these generalizations have been reported. For example, one of two cloned Arabidopsis FPP synthases encodes an enzyme that is targeted to the mitochondria,[59] and the Arabidopsis GGPP synthase gene family includes two plastidic, one mitochondrial, and two ER lumen-targeted enzymes.[14,15]

Some plant cells are specialized for the synthesis of very long isoprenoid chains, such as rubber.[5] The biosynthesis of rubber, however, requires a unique *cis*-prenyl transferase activity. While *trans*-prenyl transferases bear amino acid sequence and structural similarities to one another (typically homodimers, heterodimeric GPP synthase being the exception[60]), *cis*-prenyl transferases bear no recognizable similarity to the *trans* prenyl transferases.[61]

Daudonnet *et al.* have generated transgenic tobacco plants to examine the physiological consequences of increased cytoplasmic FPP synthase activity.[62] A 12-fold increase in FPP synthase activity was achieved by introducing the *Saccharomyces cerevisiae ERG20* gene, encoding FPP synthase, fused to the constitutive CaMV 35S promoter. Transgenic plants with increased FPP synthase activity showed a 2- to 4-fold increase in [^{14}C]acetate incorporation into isoprenoid compounds such as sterols and carotenoids.[62] In addition, these plants were more tolerant of higher levels of FPP synthase inhibitor, which induces bleaching, than were control plants. The authors interpret these results as evidence that FPP synthase is limiting for isoprenoid synthesis.[62] They suggest that the increased carotenoid synthesis indicates that FPP from the cytosol can reach the plastidic carotenoid synthetic pathway, presumably incorporated through the action of GGPP synthase. Because increased FPP synthase activity resulted in an increase of [^{14}C]acetate incorporation into both sterols and carotenoids, they conclude that neither squalene synthase nor GGPP synthase represent a limiting step in their experimental system. It could be argued, however, that, as neither squalene synthase nor GGPP synthase activities were measured, it remains to be determined whether increased metabolite availability results in parallel up-regulation of subsequent biosynthetic activities that could also be limiting. As will be illustrated in the subsequent section on enhanced carotenoid synthesis in plants, there may be precedence for regulation of gene expression by metabolites of the corresponding pathway.[63] It is difficult to reconcile the observations on increased FPP synthase activity with those of HMGR overexpression, because HMGR overexpression resulted in an increase of sterols specifically, and not carotenoids. Clearly, there remains much work in understanding how cytoplasmic isoprenoid metabolism is regulated, not the least of which entails how prenyl diphosphate pools are shared among cellular compartments.

TERPENE SYNTHASES AND CYCLASES

Biochemistry and Molecular Biology

Prenyl diphosphates synthesized in the cytoplasm (FPP) or plastid (GPP and GGPP) can be further modified through the formation of inter- or intramolecular carbon bonds. The enzymes that catalyze these different reactions can be categorized into groups according to the type of reaction they mediate (*i.e.*, intramolecular cyclization vs. intermolecular bond formation), and by the enzymatic mechanism employed. One group includes monoterpene, sesquiterpene, and certain diterpene synthases, many of which catalyze the stereospecific cyclization of GPP, FPP, or GGPP, respectively. These enzymes employ a carbocation, generated by withdrawal of the pyrophosphate group, to initiate an electrophilic attack on electron-rich double bonds within the same chain. The presence of two double bonds in FPP, each representing two possible sites of new C-C bond formation, and the potential for electrophilic attacks, either above or below the plane of the molecule, suggest a large number of potential products (Fig. 9.5A). Despite this large number of possibilities, cyclases such as the tobacco 5-*epi*-aristolochene synthase or the *Hyoscyamus muticus* vetispiradiene synthase each produce only a single major product,[64,65] indicating that the process of internal C bond formation is stereospecific. In contrast, Steele *et al.* have described mono- and sesquiterpene synthases that generate more than 30 reaction products.[66] Individually, the products of these enzymes comprise from less than 1% to over 25% of the total mixture of products.

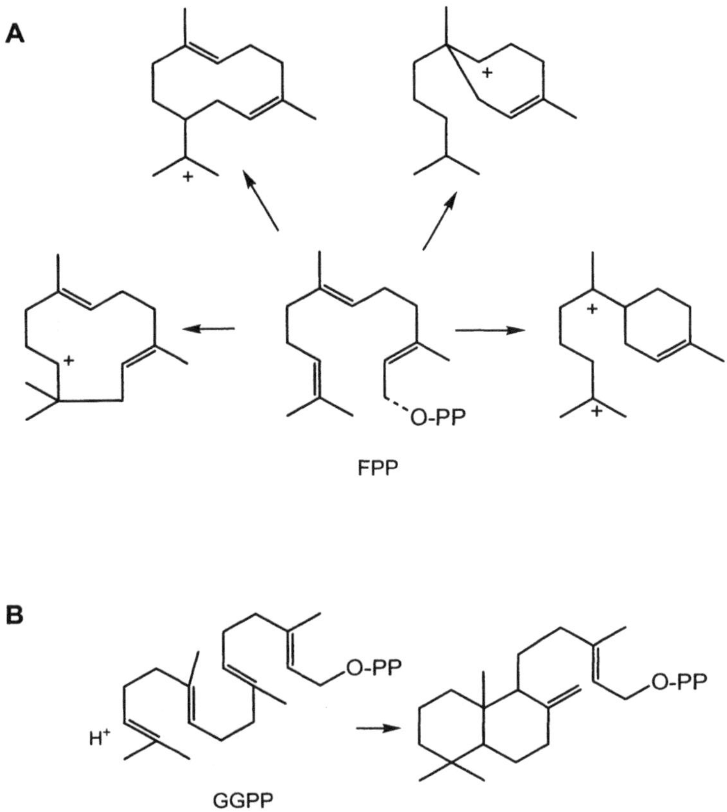

Figure 9.5: Cylization reactions catalyzed by sesquiterpene and diterpene synthases. A. Cyclization of FPP by sesquiterpene cyclase begins with carbocation generation by pyrophosphate group withdrawal. Several different products can result, depending on the site of electrophilic carbocation attack. B. Cyclization of GGPP by diterpene synthase can be initiated by proton addition to create a carbocation. Carbocation attack on electron-rich double bonds results in a cascade of C bond formations.

Squalene synthase and phytoene synthase, a second group of terpene synthases, use a carbocation-generating mechanism similar to that described for the terpene cyclases and the prenyl transferases to initiate catalysis. However, instead of cyclizing a single allylic molecule or joining prenyl groups head-to-tail, these two enzymes catalyze the head-to-head condensation of two molecules of FPP or GGPP, respectively.[2] The condensation catalyzed by these synthases proceeds via the initial formation of an unusual C1'-C2-C3 cyclopropyl intermediate, generating either presqualene diphosphate or prephytoene diphosphate (Fig. 9.7). A second reaction step breaks the cyclopropyl ring and, in the case of squalene synthase, produces a saturated C1-C1' bond while consuming one NADPH.[2]

A third group of terpene synthases is exemplified by select diterpene cyclases and most of the triterpene cyclases. While the prenyl transferases, most terpene cyclases, and squalene and phytoene synthases all rely on the initial withdrawal of a pyrophosphate group to generate a reactive carbocation, these other di- and triterpene synthases use proton addition at a double bond to generate the first electrophilic center that will initiate a cascade of stereospecific bond formations within GGPP or oxidosqualene (Fig. 9.5B). Examples of these types of synthases include abietadiene synthase,[67] β-amyrin synthase,[68] and cycloartenol synthase.[69]

Progress Towards Rational Design of Stereospecificity

Cyclized terpenes and their derivatives have many interesting and useful properties. For example, plant-derived isoprenoids are known to serve as chemical attractants or repellants to insects, and as allelochemicals limiting the growth of neighbor plants in resource-limited environments.[70] Taxol, an anti-tumor drug derived from the Pacific yew (*Taxus brevifolia*), is derived from GGPP via cyclization by taxadiene synthase and subsequent modifications.[10] Artemisinin, an antimalarial compound found in *Artemisia annua* (sweet wormwood), is derived from cyclization of FPP by amorpha-4,11-diene synthase.[8,9] Because of these useful activities and the likelihood of other uncharacterized cyclic terpene compounds having unique biological activities, it has become desirable to engineer the synthase enzymes for the stereospecific synthesis of novel products.

Sesquiterpene cyclases, as previously described, catalyze the intramolecular C bond formation of FPP to produce cyclic sesquiterpenes. Many of these activities are inducible,[71,72] and they are associated with pathogen defense or environmental stress responses in angiosperms, particularly in solanaceous genera.[3] Many sesquiterpene cyclases catalyze the synthesis of phytoalexins or phytoalexin precursors. Examples of sesquiterpene cyclases include tobacco (*Nicotiana tabacum*) 5-*epi*-areistolochene synthase (EAS),[64] potato[73] and *Hyoscyamus muticus* vetispiradiene synthase[65] (HVS), and the tomato germacrene synthase.[74] Alignments of cloned cyclase genes or their predicted amino acid sequences show that many features are conserved among sesquiterpene cyclases, including gene intron-exon structure.[2,75]

Amino acid sequence comparisons among the terpene cyclases provide useful information for elucidation of the molecular determinants of product specificity. Tobacco EAS and *H. muticus* vetispiradiene synthase (HVS) catalyze the syntheses of products that differ in ring structure and methyl group position. The rationalized enzymatic mechanisms for both EAS and HVS proceed via multiple intermediate

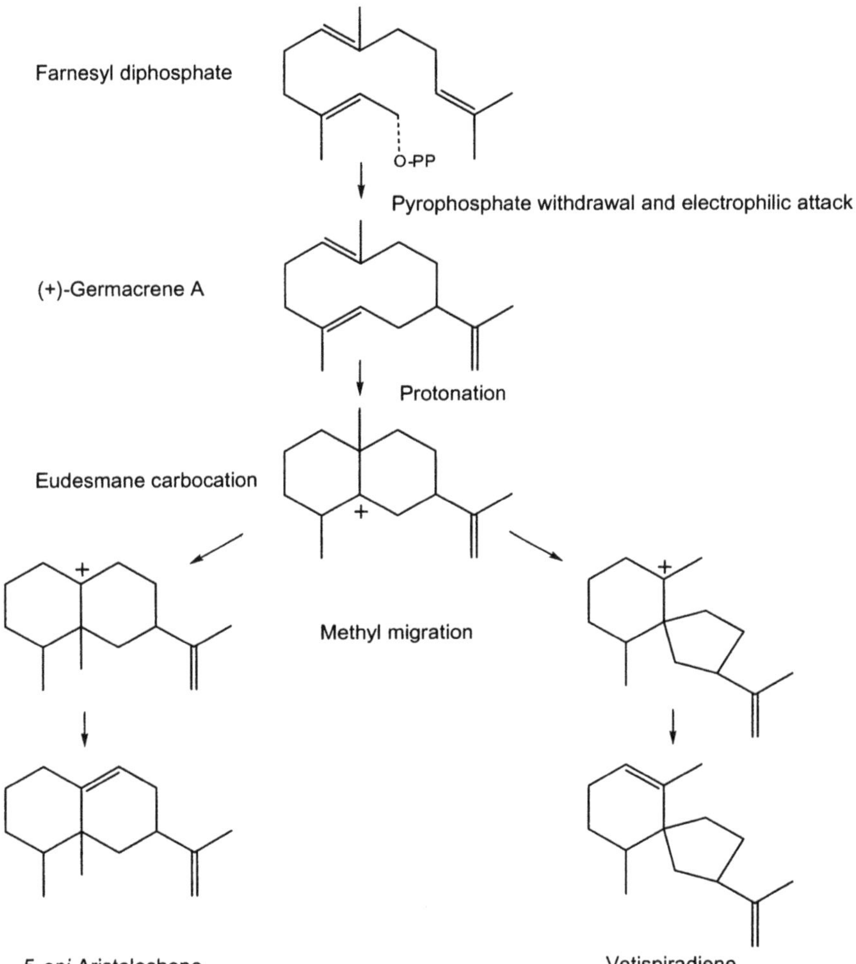

Figure 9.6: Comparison of rationalized FPP cyclization reactions catalyzed by tobacco 5-*epi*-aristolochene synthase and *Hyoscyamus* vetispiradiene synthase.

steps,[76] including the generation of a reactive carbocation upon withdrawal of the pyrophosphate group (Fig. 9.6). An electrophilic attack of the carbocation on the distal electron-rich double bond results in the generation of (+)-germacrene A, a reaction intermediate. Reprotonation of the macrocyclic germacrene A produces a second carbocation intermediate that can react to ultimately form the spirane (bicyclic, five- and six-membered rings) vetispiradiene, or alternatively, the eremophilane (bicyclic, two six-membered rings) 5-*epi*-aristolochene. It was proposed earlier that steps in common between the synthesis of 5-*epi*-aristolochene and vetispiradiene could be attributed to conserved portions of respective synthetic enzymes, which share over 75% amino acid identity,[65] while the differences might be manifest as simple differences in their amino acid sequences.

The identification of amino acids crucial to product specificity was approached through a series of domain-swapping experiments between EAS and HVS.[77] In these experiments, defined sequences of the enzymes, delimited by the exon boundaries, were exchanged to produce novel, chimeric enzymes. Among fourteen chimeric enzymes containing different portions of EAS or HVS, twelve had measurable sesquiterpene cyclase activity. Interestingly, chimeric enzymes containing exon 4 of EAS retained the capacity for synthesizing 5-*epi*-aristolochene, while enzymes containing exon 6 of HVS produced vetispiradiene. Chimeric enzymes having both exon 4 of EAS and exon 6 of HVS produced both 5-*epi*-aristolochene and vetispiradiene, implying that the chimera possessed the full complement of functional groups for both synthetic mechanisms. The exchange of other exons between EAS or HVS had the effect of varying the ratio of products in the bifunctional enzymes.

Though the influence of non-conservative amino acid substitutions elsewhere in the enzyme cannot be ruled out, it is presumed that amino acid functional groups in or near the substrate binding pocket and catalytic active site are the best candidates for affecting reaction specificity. Determination of the tobacco three-dimentional structure[78] has made it possible to focus further on amino acids residing in or near the catalytic site.

Recent site-directed mutagenesis studies have been directed to demonstrating the existence of proposed chemical intermediates in the cyclization reaction.[79] For example, three-dimentional analysis of the EAS enzyme with substrate analogs bound suggested that proton donation by a tyrosine residue at position 520 (Y520) resulted in activation of germacrene A, leading to the second series of cyclization reactions. Introduction of a Y520F mutation replaces the tyrosine with phenylalanine, an amino acid of comparable size but unable to donate a proton.[79] Although the mutant Y520F EAS has a catalytic rate only 3% of the wild-type enzyme, it produced a single reaction product identified as germacrene A. These results confirm that Y520 is essential to a particular partial reaction within the cyclase catalytic mechanism, and that the reaction progresses via a germacrene A intermediate.

While current efforts towards understanding terpene cyclase catalytic mechanisms are proceeding via a series of defined mutational studies, future investigations will employ the rapidly accumulating information about terpene cyclases from many plant sources. As previously mentioned, plant sesquiterpene cyclases share significant sequence identity. A phylogenetic analysis of mono-, sesqui-, and diterpene cyclases indicates a common origin.[75] This is true even for the diterpene synthases that employ proton addition as the initial mechanism for starting catalysis. Alignments of these different classes of cyclases shows conservation or conservative substitution at several sites that appear to participate in catalysis, suggesting conservation of some aspects of mechanism and product specificity.[2,75] Furthermore, x-ray crystallographic data from plant sesquiterpene cyclase,[78] fungal sesquiterpene cyclase,[80] and avian prenyl transferase[81] suggest that these enzymes have retained common three-dimensional structural motifs, despite limited sequence conservation.[82] Superimposition of conserved sequences over a three-dimensional cyclase framework makes it possible to predict which amino acid side groups participate in catalysis in enzymes for which the three-dimensional structures have not been solved. With such information in hand, it is possible to consider the generation of a large number of site-specific mutants that generate a large array of functional activities, some known and some unknown.

Indeed, this process appears to have occurred already in nature. Cloning of linolool synthase from *Clarkia brewerii* revealed that the gene coding for this enzyme appears to have the N-terminal half of a diterpene cyclase-type enzyme and the C-terminal half of a monoterpene synthase.[83] Whether this was the result of a naturally occurring domain swap, possibly a mechanism by which the diversity among terpene synthase enzymes could be created, remains to be determined.

CAROTENOID BIOSYNTHESIS

Carotenoids are a class of isoprenoid compounds essential to the plants that produce them and the animals that consume them. In plants, carotenoids are found in the plastid, where they facilitate light harvesting and photoprotection.[13] Carotenoids are equally important in animal systems, as precursors to vitamin A and retinoic acid, and may perform similar antioxidant functions in animals as in plants. Animals do not synthesize carotenoids; they are acquired exclusively through the diet. The identification and characterization of carotenoid biosynthesis genes from plant and microbial sources has provided the knowledge and material basis for the metabolic engineering of plants for enhanced or specialized carotenoid synthesis.[13] Molecular techniques have been employed to introduce recombinant genes directing tissue and organ-specific carotenoid synthesis into rice,[63] a carotenoid-poor staple for much of the developing world, and canola,[84] an oil crop, with the goal of increasing the nutritional value of these foods. Furthermore, techniques permitting rapid *in vitro* recombination among carotenoid biosynthetic genes and functional screening methods have led to the discovery of enzymes that synthesize novel carotenoids. The biochemistry of and molecular biology of carotenoid biosynthesis in plants,[13] and the roles of carotenoids in human nutrition[6] have recently been reviewed.

Biochemistry of Carotenoid Synthesis

The biosynthetic steps in the formation of common plant carotenoids from phytoene are illustrated in Figure 9.7. Carotenoids, and their oxygenated derivatives xanthophylls, are derived from the tetraterpene phytoene. The head-to-head condensation of two GGPPs to form phytoene is the first committed step toward carotenoid biosynthesis, and it proceeds via two intermediate steps, both catalyzed by the enzyme phytoene synthase. In the first step, a withdrawal of the pyrophosphate group from one GGPP generates a carbocation. Electrophilic attack upon the terminal double bond of the second GGPP results in the formation of the cyclopropyl intermediate, prephytoene diphosphate (PPPP). In the second reaction step, the cyclopropyl ring is broken and an unsaturated C1-C1' bond is formed at the middle of the symmetrical phytoene molecule. This reaction is similar to that catalyzed by squalene synthase, except that phytoene synthase does not utilize NADPH and the final C1-C1' bond is unsaturated.

As a substrate for carotenoid biosynthesis, phytoene undergoes a series of four desaturation reactions (Fig. 9.7). These desaturation reactions produce, in order, phytofluene, ζ-carotene, neurosporene, and finally, lycopene. In plants, this series of reactions is catalyzed by two enzymes, phytoene desaturase and ζ-carotene desaturase.[13] Phytoene desaturase catalyzes the desaturation of the C11-12 bond of phytoene and the C11'-12' of phytofluene, while ζ-carotene desaturase mediates the desaturation of the C7-8 bond of ζ-carotene and the C7'-8' bond of neurosporene. In contrast to plants, bacteria and fungi employ a single enzyme for all four of these desaturations.[13]

Further steps in the synthesis of carotenoids entail the cyclization of one or both ends of lycopene (Fig. 9.7). α-Carotene and β-carotene each have two ring structures, one at each end. These rings are described as either ε-rings or β-rings, differing in the location of the single desaturation. α-Carotene is asymmetric, having one ε-ring and one β-ring, while the symmetric β-carotene has β-rings at both ends. The synthesis of β-carotene from lycopene proceeds via two successive cyclizations, with the one-β-ring γ-carotene as an intermediate. The synthesis of α-carotene

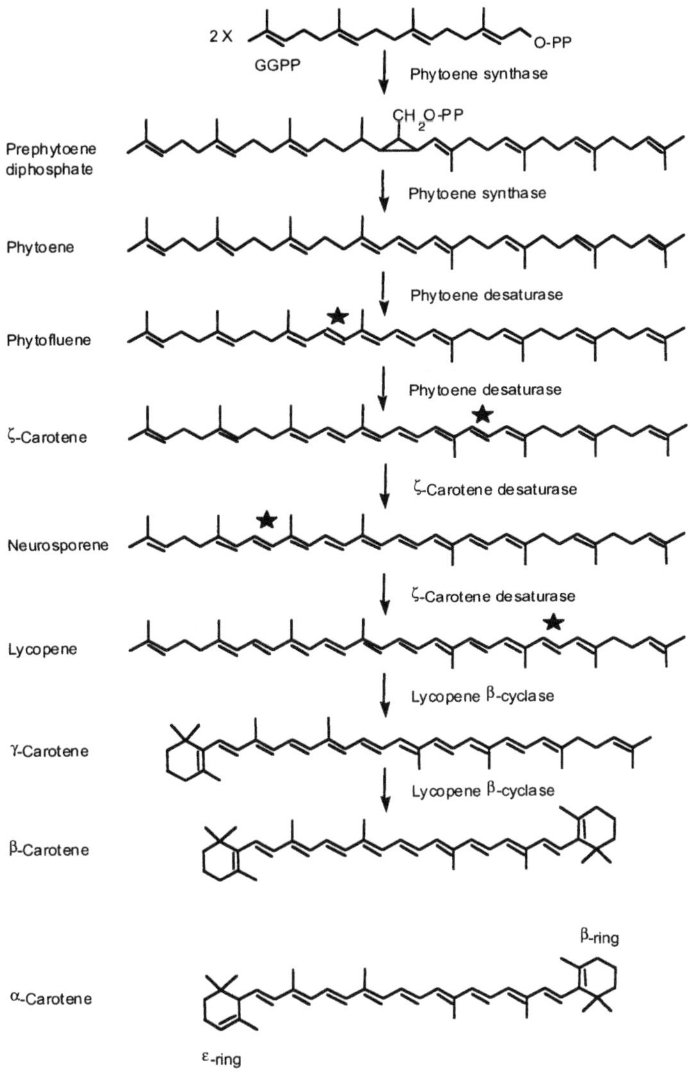

Figure 9.7: Biosynthesis of carotenoids from GGPP. Dedication of GGPP to carotenoid biosynthesis occurs with the head-to-head condensation of two GGPP to form phytoene. Four desaturations, indicated by stars, are catalyzed by phytoene desaturase and ζ-carotene desaturase. Formation of two β-rings by lycopene β-cyclase produces β-carotene. Alternatively, formation of one ε-ring, by lycopene ε-cyclase and one β-ring by lycopene β-cyclase produces α-carotene.

progresses similarly, but with the initial synthesis of the one-ε-ring δ-carotene as the intermediate. Synthesis of the ε-ring member of δ-carotene and α-carotene is catalyzed by lycopene ε-cyclase. The β-ring cyclizations of lycopene, γ-carotene, and δ-carotene are catalyzed by a single enzyme, lycopene β-cyclase.[85]

Enhanced Carotenoid Biosynthesis in Rice, Canola, and Tomato

One objective of genetically altering food crop plants has been to increase their nutritional value. Recent successes in the engineering of rice (*Oryza sativa*)[63,86] and canola (*Brassica napus*)[84] for the seed specific production of β-carotene have provided a vivid demonstration of this strategy. Rice endosperm is devoid of β-carotene and its precursor phytoene, though the precursor to phytoene, GGPP is present.[86] In a first attempt to introduce carotenoid biosynthesis to rice endosperm, the phytoene synthase cDNA of daffodil, *psy*, (*Narcissus pseudonarcissus*) was transformed into rice via particle bombardment.[86] Two versions of the transgene were tested, the first driven by the constitutive CaMV 35S promoter, and the second by the endosperm-specific rice glutein gene promoter. Transgene expression in regenerated plants was confirmed by RT-PCR, and relative phytoene synthase levels were determined by immunodetection. Immunoreactive protein was detected only in the endosperm of plants containing the endosperm-specific transgene. The highest phytoene content among the transformed lines was 0.74 μg per gram of seed dry weight, whereas phytoene in control seed was undetectable or present only in trace amounts.[86] Activities of the subsequent enzyme activities in carotenoid synthesis, phytoene desaturase and ζ-carotene desaturase, were not detected, nor were any of their corresponding products.[86]

A second effort by the same research group employed transgenes for phytoene desaturase and lycopene β-cyclase, in addition to the endosperm-specific phytoene synthase, to complete the biosynthetic route to β-carotene.[63] Rather than introduce two separate plant genes for the series of four desaturations required for the synthesis of lycopene from phytoene, a single gene from *Erwinia uredovora*, *crtI*, encoding an enzyme catalyzing all four desaturations, was chosen. As this enzyme lacks a plastid targeting sequence, it was joined to the plastid transit peptide sequence of the pea Rubisco small subunit, and placed under the regulation of the constitutive CaMV 35S promoter. The daffodil lycopene β-cyclase *lcy*, which contains its own transit peptide sequence, was placed under the control of the endosperm-specific glutein gene promoter. Transgenic rice lines were examined for carotenoid content. In many cases, the accumulation of carotenoid pigments was evident from the yellow color of the rice endosperm. β-Carotene levels in transformed rice were found as high as 1.6 μg per gram dry weight.[63] Remarkably, the endosperm coloration and pigment accumulation depended only upon the presence of phytoene synthase and phytoene desaturase transgenes. The authors conclude that lycopene cyclase activity is either constitutive in the rice endosperm, or that it is induced by lycopene synthesis.

While seed-specific production of carotenoids in rice required the introduction of more than one gene, the introduction of phytoene synthase alone has been sufficient for the overexpression of carotenoids in the oil crop canola.[84] In this study, the *E. uredovora* phytoene synthase gene, *crtB*, was joined to the Rubisco small subunit transit peptide for targeting to the plastid. This recombinant gene was placed next to the seed-specific *Brassica* napin gene promoter. Transgenic canola containing this construct produced orange seeds as a consequence of enhanced carotenoid synthesis.[84] Chemical analysis of control and transgenic canola showed an altered pattern of carotenoid accumulation. While control seed contained the carotenoid lutein and trace amounts of β-carotene, transgenic seed had a 50-fold increase in carotenoids, predominantly α- and β-carotene, with trace levels of lycopene. The increased carotenoid phenotype was also correlated with a relative decrease of fatty acid unsaturation, and the appearance of inclusion bodies at the cellular level. Though it is clear that desaturase and lycopene cyclase activities were present in the seed, the influence of enhanced phytoene synthesis on these activities was not reported.

While strategies for engineering carotenoid synthesis in rice and canola rely on the introduction of biosynthetic activities to tissues or organs that typically produce none, improvement of food quality can also be approached through the modification of plants already producing high levels of carotenoids. Tomato fruit has been the subject of efforts to alter the relative abundance of β-carotene among other carotenoid species. Early experiments in which tomato was transformed with constitutively-expressed phytoene synthase resulted in ectopic pigmentation.[87] Analysis of unripe fruit showed an accumulation of phytoene, ζ-carotene, and lycopene, several-fold over that in untransformed plants. Furthermore, overexpression of phytoene synthase was associated with stunted plant growth, apparently as a result of depleted gibberellin supply.

Because phytoene synthase is at the head of the carotenoid biosynthetic pathway, its constitutive and ectopic expression may have undesirable effects on plant growth, as evidenced by stunted growth. A subsequent study used instead the *E. uredovora crtI* phytoene desaturase, constitutively expressed, to affect higher β-carotene synthesis.[88] Fruits of tomato expressing phytoene desaturase did not have greater quantities of total carotenoids, but the fraction occurring as β-carotene was increased from 10% in the control fruit to up to 45% in the transgenic fruit. Morphologically, transformed tomato plants and their fruits were unaffected by increased phytoene desaturase expression. However, tomato fruit with high desaturase activity had an orange color owing to increased proportions of β-carotene. RT-PCR quantitation of the endogenous phytoene synthase, phytoene desaturase, ζ-carotene desaturase, and lycopene β-cyclase gene expression levels in control and transgenic plants showed the desaturases and cyclase to be induced by 50% or more, while phytoene synthase transcripts were reduced by half.[88] Collectively, these results show that it is practicable to modify carotenoid metabolism by altering the expression of genes subsequent to phytoene synthase. Perhaps future studies will assess the limits of coordinate regulation of the steps leading up to IPP synthesis during tomato and pepper fruit ripening.[39,89] Theoretically, increasing these early steps would increase all the plastid isoprenoids, including the carotenoids.

Introduction of Novel Carotenoid Biosynthetic Pathways to Plants

Xanthophylls are oxygenated carotenoids. The ε- and β-rings of α- and β-carotene are substrates for the addition of hydroxyl, keto, and epoxy functional groups by distinct enzyme activities. As an example, hydroxylation of the C3 ring member of β-carotene produces the asymmetric β-cryptoxanthin; a second hydroxylation at the 3'C ring member produces the symmetric zeaxanthin, a common plant xanthophyll. Xanthophylls are economically and agronomically important for their use as food colorants and animal feeds. The engineering of plants for the production of useful xanthophylls could, therefore, increase crop values.

Astaxanthin is an important food additive in aquaculture. It is a symmetrical xanthophyll derived from β-carotene (Fig. 9.8). Each of the β-rings is hydroxylated at the C3 and C3' ring members, and the adjacent C4 and C4' each have a keto function. While tobacco possesses the enzyme necessary for C3 and C3' hydroxylation (β-carotene hydroxylase), as is evident from its synthesis of zeaxanthin, it does not produce astaxanthin. To examine the possibility of producing a novel xanthophyll in a crop plant, tobacco was transformed with the *CrtO* gene of *Haematococcus pluvialis*,[90] encoding β-carotene ketolase. The transgene construct was fused to the promoter of the phytoene synthase gene of tomato to provide fruit-specific gene expression. Additionally, though the algae-derived *CrtO* gene possesses a putative plastid transit peptide sequence, it was exchanged for the phytoene synthase transit peptide to insure correct processing. Regenerated transgenic tobacco showed no change in growth or development in comparison to control plants. However, while nectaries of control tobacco flowers had typical yellow coloration, plants expressing the *CrtO* transgene were red, indicating a change in carotenoid accumulation. Control plants had no detectable ketocarotenoid in leaf tissue, and *CrtO* plants had only trace levels. Comparison of flower tissue carotenoids showed that, while control flowers primarily accumulated β-carotene (57% of total carotenoids) and violaxanthin (an epoxy-derivative of zeaxanthin, 40%), transgenic plants accumulated astaxanthin (23%) and other ketocarotenoid derivatives. Furthermore, the total carotenoid content of the transgenic flowers was 170% of the control flowers.

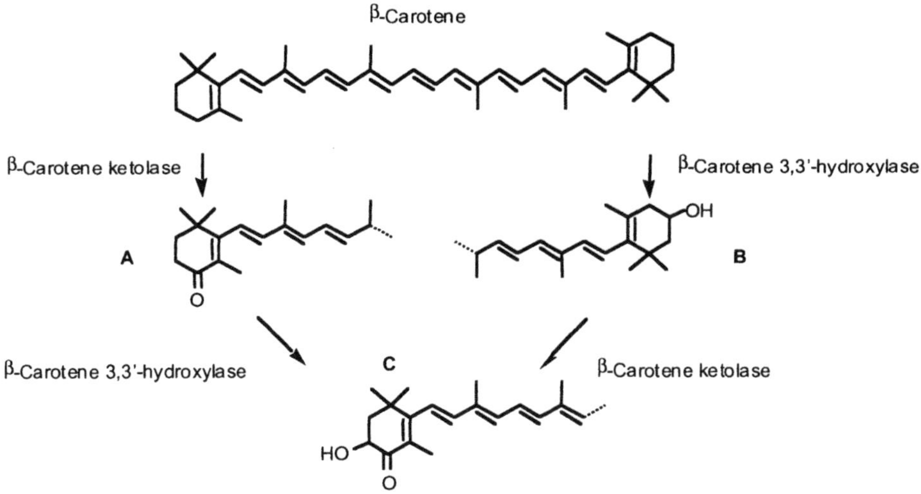

Figure 9.8: Biosynthesis of astaxanthin from β-carotene. The symmetrical carotenoid astaxanthin (structure C) results from the 3,3' hydroxylation and 4,4' ketolation of β-carotene. Biosynthesis of astaxanthin in transgenic tobacco engineered with β-carotene ketolase may occur via several intermediates (combinations of structures A, B, and C).

These encouraging results demonstrate that novel products can be produced by plants through the engineering of select biosynthetic genes. It is not necessarily the case that the possibilities are limited to those pathways of isoprenoid metabolism found in nature, however. Two studies, employing different methods, report success in generating novel carotenoids through combinatorial approaches. One study reported the co-expression of three different carotenoid desaturases in combination with a carotenoid hydratase, a carotenoid cyclase, and a carotenoid hydroxylase in *E. coli*.[91] Four new compounds were identified using this approach that had not been detected in biological systems previously. Further analysis suggested that these compounds may have useful antioxidant and other novel properties.

The second study reported the use of rapid recombinatorial methods, or "molecular breeding" to generate novel carotenoid desaturases.[92] This method, which involved the random breaking and reconstruction of recombinant, functional genes, was originally described for generating enhanced antibiotic resistance.[93] This approach was applied to the generation of recombinant phytoene desaturases by using the *crtI* genes of *Erwinia uredovora* and *E. herbicola* as starting material. Ten thousand recombinant desaturases were screened for novel catalytic properties in an *E. coli* background that had been engineered for the synthesis of colorless phytoene, but not for subsequent steps in carotenoid synthesis. Roughly a third of the recombinant enzymes lost catalytic activity as a result of recombination, as evidenced by colorless colonies, and a majority produced red-orange colonies, indicating unaltered functionality. A significant number produced yellow colonies indicating a reduced number of desaturations by the enzyme. One pink colony harbored a recombinant desaturase that increased the number of desaturations introduced into phytoene from four to six. Through DNA sequence analysis and further directed sequence exchanges, it was possible to identify the mutation sites responsible for altered catalysis. This same method was applied successfully in the generation of novel lycopene β-cyclases.[92] In principle, this method should be applicable to the modification of any enzyme involved in the synthesis of many possible products in a conserved family of genes, such as terpene cyclases.

CONCLUSIONS

At the time of this writing, research publications addressing the metabolic engineering of plant isoprenoid metabolism and biochemistry are appearing at a rate of several per month, an indication of how vigorous this area of investigation has become. While much has been learned, much also remains to be discovered. We can expect that with each new discovery, more applications will become evident. Lange and Croteau[7] have proposed a strategy for improved essential oil production in *Mentha*. This strategy suggests several points in terpene biosynthesis where efficiency might be improved, whether by increasing metabolic flow to useful products, or by decreasing the amount of resources spent in the production of unwanted compounds. They list several areas in which fundamental knowledge is lacking, knowledge that would augment rational approaches to metabolic engineering. These areas include an understanding of developmental regulation of biosynthetic activities, of metabolic branch points in the control of flux, of possible feedback regulatory properties of metabolic intermediates, and of mechanisms for the transport of products from the site of biosynthesis. To this list we might suggest the utility of a clearer understanding of the number and kinds of potential interactions between cytosolic and plastidic isoprenoid pools, and contributions from other cellular compartments.

REFERENCES

1. SPURGEON, S., PORTER, J., Introduction, *in* Biosynthesis of Isoprenoid Compounds (J. Porter and S. Spurgeon, eds.), Wiley-Interscience, New York, 1981, pp. 1-46.
2. CIIAPPELL, J., Biochemistry and molecular biology of the isoprenoid biosynthetic pathway in plants, *Annu. Rev. Plant Physiol. Plant Mol. Biol.*, 1995, **46**, 521-547.
3. STOESSL, A., STOTHERS, S., WARD, E., Sesquiterpenoid stress compounds of the *Solanaceae, Phytochemistry*, 1976, **15**, 855-872.
4. DUDAREVA, N., CSEKE, L., BLANC, V., PICHERSKY, E., Evolution of floral scent in *Clarkia*: Novel patterns of S-linoolol synthase gene expression in the *C. breweri* flower, *Plant Cell*, 1996, **8**, 1137-1148.
5. KUSH, A., Isoprenoid biosynthesis: The *Hevea* factory!, *Plant Physiol. Biochem.*, 1994, **32**, 761-767.
6. VAN DEN BERG, H., FAULKS, R., GRANADO, H.F., HIRSCHENBERG, J., OLMEDILLA, B., SANDMANN, G., SOUTHON, S., STAHL, W., The potential for the improvement of carotenoid levels in foods and the likely systemic effects, *J. Sci. Food Agric.*, 2000, **80**, 880-912.
7. LANGE, B., CROTEAU, R., Genetic engineering of essential oil production in mint, *Curr. Opin. Plant Biol.*, 1999, **2**, 139-144.
8. BOUWMEESTER, H., WALLAART, T., JANSSEN, M., VAN LOO, B., JANSEN, B., POSTHUMUS, M., SCHMIDT, C., DE KRAKER, J., KONIG, W., FRANSSEN, M., Amorpha-4,11-diene synthase catalyses the probable first step in artemisinin biosynthesis, *Phytochemistry*, 1999, **52**, 843-854.
9. MERCKE, P., BENGTSSON, M., BOUWMEESTER, H., POSTHUMUS, M., BRODELIUS, P., Molecular cloning, expression, and characterization of amorpha-4,11-diene synthase, a key enzyme of artemisinin biosynthesis in *Artemisia annua, Arch. Biochem. Biophys.*, 2000, **381**, 173-180.
10. LIN, X., HEZARI, M., KOEPP, A., FLOSS, H., CROTEAU, R., Mechanism of taxadiene synthase, a diterpene cyclase that catalyzes the first step of taxol biosynthesis in Pacific yew, *Biochemistry*, 1996, **35**.
11. BACH, T., BORONAT, A., CAMPOS, N., FERRER, A.,VOLLACK, K., Mevalonate biosynthesis in plants, *Crit. Rev. Biochem. Mol. Biol.*, 1999, **34**, 107-122.
12. LICHTENTHALER, H., The 1-deoxy-D-xylulose-5-phosphate pathway of isoprenoid biosynthesis in plants, *Annu. Rev. Plant Physiol. Plant Mol. Biol.*, 1999, **50**, 47-56.
13. CUNNINGHAM, F., GANTT, E., Genes and enzymes of carotenoid biosynthesis in plants, *Annu. Rev. Plant Physiol. Plant Mol. Biol.*, 1998, **49**, 557-583.
14. ZHU, X., SUZUKI, K., SAITO, T., OKADA, K., TANAKA, K., NAKAGAWA, T., MATSUDA, H., KAWAMUKAI, M., Geranylgeranyl pyrophosphate synthase encoded by the newly isolated gene *GGPS6* from *Arabidopsis thaliana* is localized in the mitochondria, *Plant Mol. Biol.*, 1997, **35**, 331-341.
15. OKADA, K., SAITO, T., NAKAGAWA, T., KAWAMUKAI, M., KAMIYA, Y., Five geranylgeranyl diphosphate synthases expressed in different organs are localized into three subcellular compartments in Arabidopsis. *Plant Physiol.*, 2000, **122**, 1045-1056.

16. QURESHI, N., PORTER, J., Conversion of acetyl-coenzyme A to isopentenyl pyrophosphate, *in* Biosynthesis of Isoprenoid Compounds (J. Porter and S. Spurgeon, eds.), Wiley-Interscience, New York, 1981, pp. 47-94.
17. VAN DER HEIJDEN, R., VERPOORTE, R., DUINE, J., Biosynthesis of 3S-hydroxy-3-methylglutaryl-coenzyme A in *Catharanthus roseus*: acetoacetyl-CoA thiolase and HMG-CoA synthase show similar chromatographic behaviour, *Plant Physiol. Biochem.*, 1994, **32**, 807-812.
18. ALEX, D., BACH, T., CHYE, M., Expression of *Brassica juncea* 3-hydroxy-3-methylglutaryl CoA synthase is developmentally regulated and stress-responsive, *Plant J.*, 2000, **22**, 415-426.
19. MONTAMAT, F., GUILLOTON, M., KARST, F., DELROT, S., Isolation and characterization of a cDNA encoding *Arabidopsis thaliana* 3-hydroxy-3-methylglutaryl-coenzyme A synthase, *Gene*, 1995, **167**, 197-201.
20. WEBER, T., BACH, T., Conversion of acetyl-coenzyme A into 3-hydroxy-3-methylglutaryl-coenzyme A in radish seedlings, Evidence of a single monomeric protein catalyzing a FeII/quinone-stimulated double condensation reaction, *Biochim. Biophys. Acta*, 1994, **1211**, 85-96.
21. BACH, T., RAUDOT, V., VOLLACK, K., WEBER, T., ZEILER, S., Further studies on the enzymatic conversion of acetyl-coenzyme A into 3-hydroxy-3-methylglutaryl-coenzyme A in radish, *Plant Physiol. Biochem.*, 1994, **32**, 775-783.
22. DUGAN, R., Regulation of HMG-CoA reductase, *in* Biosynthesis of Isoprenoid Compounds (J. Porter and S. Spurgeon, eds.), Wiley-Interscience, New York, 1981, pp. 95-160.
23. GOLDSTEIN, J., BROWN, M., Regulation of the mevalonate pathway, *Nature*, 1990, **343**, 425-430.
24. SCHULTE, A., VAN DER HEIJDEN, R., VERPOORTE, R., Microplate enzyme-coupled assays of mevalonate and phosphomevalonate kinase from *Catharanthus roseus* suspension cultured cells, *Anal. Biochem.*, 1999, **269**, 245-254.
25. SCHULTE, A., VAN DER HEIJDEN, R., VERPOORTE, R., Purification and characterization of mevalonate kinase from suspension-cultured cells of *Catharanthus roseus* (L.), *G. Don. Arch. Biochem. Biophys.*, 2000, **378**, 287-298.
26. RIOU, C., TOURTE, Y., LACROUTE, F., KARST, F., Isolation and characterization of a cDNA encoding *Arabidopsis thaliana* mevalonate kinase by genetic complimentation in yeast, *Gene*, 1994, **148**, 293-297.
27. LLUCH, M., MASFERRER, A., ARRÓ, M., BORONAT, A., FERRER, A., Molecular cloning and expression analysis of the mevalonate kinase gene from *Arabidopsis thaliana, Plant Mol. Biol.*, 2000, **42**, 365-376.
28. CORDIER, H., KARST, F., BERGÈS, T., Heterologous expression in *Saccharomyces cerevisiae* of an *Arabidopsis thaliana* cDNA encoding mevalonate diphosphate decarboxylase, *Plant Mol. Biol.*, 1999, **39**, 953-957.
29. SAHM, H., ROHMER, M., BRINGER-MEYER, S., SPRENGER, G., WELLE, R., Biochemistry and physiology of hopanoids in bacteria, *Adv. Microb. Physiol.*, 1993, **35**, 247-273.

30. SCHWENDER, J., ZEIDLER, J., MULLER, C., FOCKE, M., BRAUN, S., LICHTENTHALER, F., LICHTENTHALER, H., Incorporation of 1-deoxy-D-xylulose into isoprene and phytol by higher plants and algae, *FEBS Lett.*, 1997, **414**, 129-134.
31. ARIGONI, D., SANGER, S., LATZEL, C., EISENREICH, W., BACHER, A., ZENK, M., Terpenoid biosynthesis from 1-deoxy-D-xylulose in higher plants by intramolecular skeletal rearrangement, *Proc. Natl. Acad. Sci., USA*, 1997, **94**, 10600-10605.
32. SPRENGER, G., SCHÖRKEN, U., WIEGERT, T., GROLLE, S., DE GRAAF, A., TAYLOR, S., BEGLEY, T., BRINGER-MEYER, S., SAHM, H., Identification of a thiamine-dependent synthase in *Escherichia coli* required for the formation of the 1-deoxy-D-xylulose 5-phosphate precursor to isoprenoids, thiamin, and pyridoxol, *Proc. Natl. Acad. Sci., USA*, 1997, **94**, 12857-12862.
33. LOIS, L., CAMPOS, N., PUTRA, S., DANIELSEN, K., ROHMER, M., BORONAT, A., Cloning and characterization of a gene from *Escherichia coli* encoding a transketolase-like enzyme that catalyzes the synthesis of D-1-deoxyxylulose 5-phosphate, a common precursor for isoprenoid, thiamin, and pyridoxol biosynthesis, *Proc. Natl. Acad. Sci., USA*, 1998, **95**, 2105-2110.
34. MANDEL, M., FELDMANN, K., HERRERA-ESTRELLA, L., ROCHA-SOSA, M., LEÓN, P., *CLA1*, a novel gene required for chloroplast development, is highly conserved in evolution, *Plant J.*, 1996, **9**, 649-658.
35. LANGE, B., WILDUNG, M., MCCASKILL, D., CROTEAU, R., A family of transketolases that directs isoprenoid biosynthesis via a mevalonate-independent pathway, *Proc. Natl. Acad. Sci., USA*, 1998, **95**, 2100-2104.
36. LANGE, B., CROTEAU, R., Isoprenoid biosynthesis via a mevalonate-independent pathway in plants: Cloning and heterologous expression of 1-deoxy-D-xylulose-5-phosphate reductoisomerase from peppermint, *Arch. Biochem. Biophys.*, 1999, **365**, 170-174.
37. LANGE, B., CROTEAU, R., Isopentenyl diphosphate biosynthesis via a mevalonate-independent pathway: Isopentenyl monophosphate kinase catalyzes the terminal step, *Proc. Natl. Acad. Sci., USA*, 1999, **96**, 13714-13719.
38. CHARON, L., HOEFFLER, J., PALE-GROSDEMANGE, C., LOIS, L., CAMPOS, N., BORONAT, A., ROHMER, M., Deuterium-labelled isotopomers of 2-*C*-methyl-D-erythritol as tools for the elucidation of the 2-*C*-methyl-D-erythritol 4-phosphate pathway for isoprenoid biosynthesis, *Biochem. J.*, 2000, **346**, 737-742.
39. LOIS, L., RODRÍGUEZ-CONCEPCIÓN, M., GALLEGO, F., CAMPOS, N., BORONAT, A., Carotenoid biosynthesis during tomato fruit development: regulatory role of 1-deoxy-D-xylulose 5-phosphate synthase, *Plant J.*, 2000, **22**, 503-513.
40. WALTER, M., FESTER, T., STRACK, D., Arbuscular mycorrhizal fungi induce the non-mevalonate methylerythritol phosphate pathway of isoprenoid biosynthesis correlated with accumulation of the 'yellow pigment' and other apocarotenoids, *Plant J.*, 2000, **21**, 571-578.
41. MCCASKILL, D., GERSHENZON, J., CROTEAU, R., Morphology and monoterpene biosynthetic capabilities of secretory cell clusters isolated from trichomes of peppermint (*Mentha piperita* L.), *Planta*, 1992, **187**, 445-454.

42. POULTER, C., RILLING, H., Prenyl transferases and isomerases, *in* Biosynthesis of Isoprenoid Compounds (J. Porter and S. Spurgeon, eds.), Wiley-Interscience, New York., 1981, pp. 161-224.
43. CAMPBELL, M., HAHN, F., POULTER, C., LEUSTEK, T., Analysis of the isopentenyl diphosphate isomerase gene family from *Arabidopsis thaliana, Plant Mol. Biol.*, 1998, **36**, 323-328.
44. CUNNINGHAM, F., GANTT, E., Identification of multi-gene families encoding isopentenyl diphosphate isomerase in plants by heterologous complementation in *Escherichia coli, Plant Cell Physiol.*, 2000, **41**, 119-123.
45. RAMOS-VALDIVIA, A., VAN DER HEIJDEN, R., VERPOORTE, R., CAMARA, B., Purification and characterization of two isoforms of isopentenyl-diphosphate isomerase from elicitor-treated *Chinchona robusta* cells, *Eur. J. Biochem.*, 1997, **249**, 161-170.
46. CHAPPELL, J., WOLF, F., PROULX, J., CUELLAR, R., SAUNDERS, C., Is the reaction catalyzed by 3-hydroxy-3-methylglutaryl coenzyme A reductase a rate-limiting step for isoprenoid biosynthesis in plants?, *Plant Physiol.*, 1995, **109**, 1337-1343.
47. SCHALLER, H., GRAUSEM, B., BENVENISTE, P., CHYE, M., TAN, C., SONG, Y., CHUA, N., Expression of the *Hevea brasiliensis* (H.B.K.) Müll. Arg. 3-hydroxy-3-methylglutaryl-coenzyme A reductase 1 in tobacco results in sterol overproduction, *Plant Physiol.*, 1995, **109**, 761-770.
48. RE, E., JONES, D., LEARNED, R., Co-expression of native and introduced genes reveals cryptic regulation of HMG CoA reductase expression in *Arabidopsis*, *Plant J.*, 1995, **7**, 771-784.
49. CHOI, D., WARD, B., BOSTOCK, R., Differential induction and suppression of potato 3-hydroxy-3-methylglutaryl coenzyme A reductase genes in response to *Phytophthora infestans* and to its elicitor arachodonic acid, *Plant Cell*, 1992, **4**, 1333-1344.
50. YANG, Z., PARK, H., LACY, G., CRAMER, C., Differential activation of potato 3-hydroxy-3-methylglutaryl coenzyme A reductase genes by wounding and pathogen challenge, *Plant Cell*, 1991, **3**, 397-405.
51. NELSON, A., DOERNER, P., ZHU, Q., LAMB, C., Isolation of a monocot 3-hydroxy-3-methylglutaryl coenzyme A reductase gene that is elicitor-inducible, *Plant Mol. Biol.*, 1994, **25**, 401-412.
52. CHUN, K., SIMONI, R., The role of the membrane domain in the regulated degredation of 3-hydroxy-3-methylglutaryl coenzyme A reductase, *J. Biol. Chem.*, 1992, **267**, 4236-4246.
53. NEWMAN, J., CHAPPELL, J., Isoprenoid biosynthesis in plants: Carbon partitioning within the cytoplasmic pathway, *Crit. Rev. Biochem. Mol. Biol.*, 1999, **34**, 95-106.
54. CHENG, H., XU, L., KUMAGAI, H., SIMONI, R., Oligomerization state influences the degredation rate of 3-hydroxy-3-methylglutaryl-CoA reductase, *J. Biol. Chem.*, 1999, **274**, 17171-17178.
55. KAJIWARA, S., FRASER, P., KONDO, K., MISAWA, N., Expression of an exogenous isopentenyl diphosphate isomerase gene enhances isoprenoid biosynthesis in *Escherichia coli, Biochem. J.*, 1997, **324**, 421-426.

56. ALBRECHT, M., SANDMANN, G., Light-stimulated carotenoid biosynthesis during transformation of maize etioplasts is regulated by increased activity of isopentenyl pyrophosphate isomerase, *Plant Physiol.*, 1994, **105**, 529-534.
57. TARSHIS, L., PROTEAU, P., KELLOGG, B., SACCHETTINI, J., POULTER, C., Regulation of product chain length by isoprenyl diphosphate synthases, *Proc. Natl. Acad. Sci., USA*, 1996, **93**, 15018-15023.
58. OHNUMA, S., HIROOKA, K., OHTO, C., NISHINO, T., Conversion from archaeal geranylgeranyl diphosphate synthase to farnesyl diphosphate synthase. Two amino acids before the first aspartate-rich motif soley determine eukaryotic farnesyl diphosphate synthase activity, *J. Biol. Chem.*, 1997, **272**, 5192-5198.
59. CUNILLERA, N., BORONAT, A., FERRER, A., The *Arabidopsis thaliana FPS1* gene generates a novel mRNA that endcodes a mitochondrial farnesyl-diphosphate synthase isoform, *J. Biol. Chem.*, 1997, **272**, 15381-15388.
60. BURKE, C., WILDUNG, M., CROTEAU, R., Geranyl diphosphate synthase: Cloning, expression, and characterization of this prenyltransferase as a heterodimer, *Proc. Natl. Acad. Sci., USA*, 1999, **96**, 13062-13067.
61. OH, S., HAN, K., RYU, S., KANG, H., Molecular cloning, expression, and functional analysis of a *cis*-prenyltransferase from *Arabidopsis thaliana, J. Biol. Chem.*, 2000, **275**, 18482-18488.
62. DAUDONNET, S., KARST, F., TOURTE, Y., Expression of the farnesyldiphosphate synthase gene of *Saccharomyces cerevisiae* in tobacco, *Mol. Breed.*, 1997, **3**, 137-145.
63. YE, X., BABILI, S.A., KLOTI, A., ZHANG, J., LUCCA, P., BEYER, P., POTRYKUS, I., Engineering of the provitamin A (β-carotene) biosynthetic pathway into (carotenoid-free) rice endosperm, *Science*, 2000, **287**, 303-305.
64. BACK, K., YIN, S., CHAPPELL, J., Expression of a plant sesquiterpene cyclase gene in *Escherichia coli, Arch. Biochem. Biophys.*, 1994, **315**, 527-532.
65. BACK, K., CHAPPELL, J., Cloning and bacterial expression of a sesquiterpene cyclase from *Hyoscyamus muticus* and its molecular comparison to related sesquiterpene cyclases, *J. Biol. Chem.*, 1995, **270**, 7375-7381.
66. STEELE, C., CROCK, J., BOHLMANN, J., CROTEAU, R., Sesquiterpene synthases from grand fir (*Abies grandis*), Comparison of constitutive and wound-induced activities, and cDNA isolation, characterization and bacterial expression of δ-selinene synthase and γ-humulene synthase, *J. Biol. Chem.*, 1998, **273**, 2078-2089.
67. LAFEVER, R., VOGEL, B., CROTEAU, R., Diterpenoid resin acid biosynthesis in conifers: Enzymatic cyclization of geranylgeranyl pyrophosphate to abietadiene, the precursor of abietic acid, *Arch. Biochem. Biophys.*, 1994, **313**, 139-149.
68. KUSHIRO, T., SHIBUYA, M., EBIZUKA, Y., β-Amyrin synthase - Cloning of oxidosqualene cyclase that catalyzes the formation of the most popular triterpene among higher plants, *Eur. J. Biochem.*, 1998, **256**, 238-244.
69. STEELE, C., LEWINSOHN, E., CROTEAU, R., Induced oleoresin biosynthesis in grand fir as a defense against bark beetles, *Proc. Natl. Acad. Sci., USA*, 1995, **92**, 4164-4168.

70. ELAKOVICH, S., Sesquiterpenes as phytoalexins and allelopathic agents, *in* Ecology and Metabolism of Plant Lipids (G. Fuller and W. Nes, eds.), American Chemical Society, Washington, DC., 1987, pp. 93-108.
71. LOIS, A., WEST, C., Regulation of expression of the casbene synthetase gene during elicitation of castor bean seedlings with pectic fragments, *Arch. Biochem. Biophys.*, 1990, **276**, 270-277.
72. VÖGELI, U., CHAPPELL, J., Induction of sesquiterpene cyclase actvity and suppression of squalene synthetase activities in plant cell cultures treated with fungal elicitor, *Plant Physiol.*, 1988, **88**, 1291-1296.
73. YOSHIOKA, H., YAMADA, N., DOKE, N., cDNA cloning of sesquiterpene cyclase and squalene synthase, and expression of the genes in potato tuber infected with *Phytophthora infestans, Plant Cell Physiol.*, 1999, **40**, 993-998.
74. COLBY, S., CROCK, J., DOWDLE-RIZZO, B., LEMAUX, P., CROTEAU, R., Germacrene C synthase from *Lycopersicon esculentum* cv. VFNT cherry tomato: cDNA isolation, characterization, and bacterial expression of the multiple product sesquiterpene cyclase, *Proc. Natl. Acad. Sci., USA*, 1998, **95**, 2216-2221.
75. BOHLMANN, J., MEYER-GAUEN, G., CROTEAU, R., Plant terpenoid synthases: Molecular biology and phylogenetic analysis, *Proc. Natl. Acad. Sci., USA*, 1998, **95**, 4126-4133.
76. MATHIS, J., BACK, K., STARKS, C., NOEL, J., POULTER, C., CHAPPELL, J., Pre-steady-state study of recombinant sesquiterpene cyclases, *Biochemistry*, 1997, **36**, 8340-8348.
77. BACK, K., CHAPPELL, J., Identifying functional domains within terpene cyclases using a domain-swapping strategy, *Proc. Natl. Acad. Sci., USA*, 1996, **93**, 6941-6845.
78. STARKS, C., BACK, K., CHAPPELL, J., NOEL, J., Structural basis for cyclic terpene biosynthesis by tobacco 5-*epi*-aristolochene synthase, *Science*, 1997, **277**, 1815-1820.
79. RISING, K., STARKS, C., NOEL, J., CHAPPELL, J., Demonstration of germacrene A as an intermediate in 5-*epi*-aristolochene synthase catalysis, *J. Am. Chem. Soc.*, 2000, **122**, 1861-1866.
80. CARUTHERS, J., KANG, I., RYNKIEWICZ, M., CANE, D., CHRISTIANSON, D., Crystal structure determination of aristolochene synthase from the blue cheese mold *Pennecillium roqueforti, J. Biol.Chem.*, 2000, **275**, 25533-25539.
81. TARSHIS, L., YAN, M., POULTER, C., SACCHETTINI, J., Crystal structure of recombinant farnesyl diphosphate synthase at 2.6-Å resolution, *Biochemistry*, 1994, **33**, 10871-10877.
82. LESBURG, C., CARUTHERS, J., PASCHALL, C., CHRISTIANSON, D., Managing and manipulating carbocations in biology: Terpenoid cyclase structure and mechanism, *Curr. Opin. Struct. Biol.*, 1998, **8**, 695-703.
83. CSEKE, L., DUDAREVA, N., PICHERSKY, E., Structure and evolution of linolool synthase, *Mol. Biol. Evol.*, 1998, **15**, 1491-1498.
84. SHEWMAKER, C., SHEEHY, J., DALEY, M., COLBURN, S., KE, D., Seed-specific overexpression of phytoene synthase: Increase in carotenoids and other metabolic effects, *Plant J.*, 1999, **20**, 401-412.

85. CUNNINGHAM, F., POGSON, B., SUN, Z., MCDONALD, K., DELLAPENNA, D., GANTT, E., Functional analysis of the β and ε lycopene cyclase enzymes of *Arabidopsis* reveals a mechanism for control of cyclic carotenoid formation, *Plant Cell*, 1996, **8**, 1613-1626.
86. BURKHARDT, P., BEYER, P., WUNN, J., KLÖTI, A., ARMSTRONG, G., SCHLEDZ, M., VON LINTIG, J., POTRYKUS, I., Transgenic rice (*Oryza sativa*) endosperm expressing daffodil (*Narcissus pseudonarcissus*) phytoene synthase accumulates phytoene, a key intermediate in provitamin A biosynthesis, *Plant J.*, 1997, **11**, 1071-1078.
87. FRAY, R., WALLACE, A., FRASER, P., VALERO, D., HEDDEN, P., BRAMLEY, P., GRIERSON, D., Constitutive expression of a fruit phytoene synthase gene in transgenic tomatoes causes dwarfism by redirecting metabolites from the gibberellin pathway, *Plant J.*, 1995, **8**, 693-701.
88. RÖMER, S., FRASER, P., KIANO, J., SHIPTON, C., MISAWA, N., SCHUCH, W., BRAMLEY, P., Elevation of the provitamin A content of transgenic tomato plants, *Nature Biotech.*, 2000, **18**, 666-669.
89. BOUVIER, F., D'HARLINGUE, A., SUIRE, C., BACKHAUS, R., CAMARA, B., Dedicated roles of plastid transketolases during the early onset of isoprenoid biogenesis in pepper fruits, *Plant Physiol.*, 1998, **117**, 1423-1431.
90. MANN, V., HARKER, M., PECKER, I., HIRSCHENBERG, J., Metabolic engineering of astaxanthin production in tobacco flowers, *Nature Biotech.*, 2000, **18**, 888-892.
91. ALBRECHT, M., SHINICHI, T., STEIGER, S., WANG, Z., SANDMANN, G., Novel hydroxycarotenoids with improved antioxidative properties produced by gene combination in *Escherichia coli*, *Nature Biotech.*, 2000, **18**, 843-846.
92. SCHMIDT-DANNERT, C., UMENO, D., ARNOLD, F., Molecular breeding of carotenoid biosynthetic pathways, *Nature Biotech.*, 2000, **18**, 750-753.
93. STEMMER, W., Rapid evolutoin of a protein *in vitro* by DNA shuffling, *Nature*, 1994, **370**, 389-391.

Chapter Ten

PRODUCTION OF CELLULASES IN PLANTS FOR BIOMASS CONVERSION

Kathleen J. Danna

Department of Molecular, Cellular, and Developmental Biology
University of Colorado
Boulder, CO 80309-0347

INTRODUCTION

Consumption of fossil fuels in the U.S. contributes to many economic and environmental problems. These include a massive trade deficit; growing dependence on an unreliable and finite source of petroleum; air pollution; and global warming. Clean-burning fuels derived from renewable resources offer an attractive alternative to petroleum-based fuels.[1] For example, using ethanol as an additive to or a replacement for gasoline reduces emissions of CO, NOx, aromatic hydrocarbons, and other volatile organic compounds, thus also lowering the photochemical production of ozone. Moreover, because the carbon in ethanol comes from carbon dioxide recently fixed by plants through photosynthesis, ethanol-fueled vehicles operate on a closed CO_2 cycle and do not contribute to accumulation of CO_2 in the atmosphere.[2] Since 1990, when the U.S. Clean Air Act was amended, ethanol has been widely used as a gasoline additive. The recent ban on the petroleum-based additive MTBE will only increase demand for ethanol.

In Brazil, the U.S., and Canada, cane sugar and grain starch, primarily from corn, have been the predominant feedstocks for fuel ethanol production. Cane sugar is fermented directly by yeast, and the ethanol product is recovered by distillation. Grain processing requires the additional steps of starch extraction and enzymatic conversion of starch to glucose by α-amylase. The maximum amount of ethanol attainable from these sources is limited by the availability and expense of these feedstocks. Competing demand for sugar and grains as food causes frequent fluctuations in both availability and price.[2] In contrast, plant biomass, which includes agricultural waste, wood residues from timber and paper industries, municipal solid waste, and fast-growing bioenergy crops, offers a virtually unlimited, inexpensive feedstock for ethanol production. Through biomass conversion, lignocellulosic feedstocks can potentially provide enough ethanol to fulfill current U.S. demands for transportation fuel.[1] As succinctly stated by Wheals, *et al.*, "Fuel-ethanol production will only remain a significant industry and become a potentially self-sustainable agriculture system for the 21st century if the utilization of lignocellulose becomes a commercial reality."[2]

BIOMASS CONVERSION

With a synthesis rate of about 4 X 10^{13} kg/yr, cellulose is the most abundant organic polymer on earth and comprises up to 45% of the dry weight of plant biomass.[3] Consisting of long polymers of β-(1,4)-linked D-glucose (Fig. 10.1), cellulose molecules form overlapping parallel crystalline arrays through hydrogen bonding. In the plant cell wall, these microfibril arrays, which provide tensile strength and rigidity, are embedded in a matrix of hemicellulose and pectin networks that play a role in cross-linking.[4]

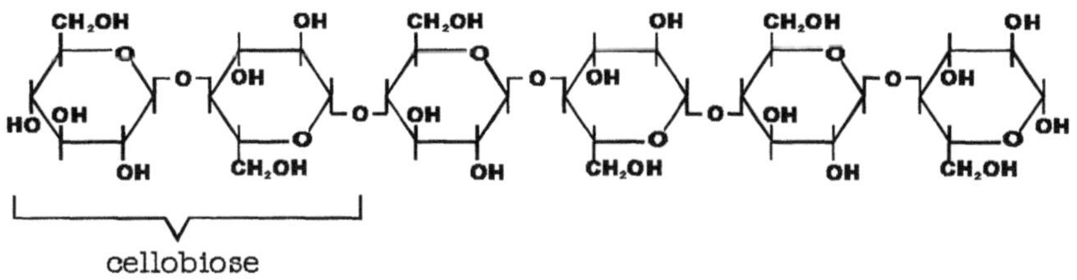

Figure 10.1: Cellulose structure. The repeating unit, cellobiose, is a β-(1,4)-linked glucose dimer.

Conversion of cellulosic biomass to ethanol involves pretreatment to reduce particle size and saccharification to release fermentable sugars (Fig. 10.2). Conventional saccharification, based on acid hydrolysis, discharges toxic pollutants and is expensive.[1] Enzymatic saccharification is a preferred alternative, but the cellulases needed for this process are expensive. At $0.30 to $0.80 per gram and an

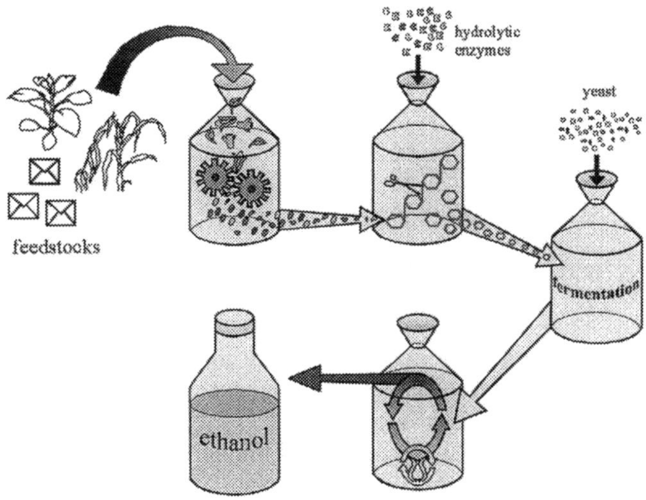

Figure 10.2: Outline of the biomass conversion process. In this simplified scheme, saccharification and fermentation are shown as separate steps. Simultaneous saccharification and fermentation (SSF) is a popular alternative.[7]

input ratio of 1 ton of enzyme per 50 tons of cellulose,[5] the enzymes currently account for up to $0.33/gal of ethanol production costs.[6] Reducing the cost of cellulases would greatly accelerate the development of biofuels.

Enzymatic Degradation of Cellulose

Many pathogenic and wood-rotting microorganisms convert cellulose to fermentable sugars through the synergistic action of three classes of secreted enzyme (Fig. 10.3), described in detail in several excellent reviews.[3,5,8-10] Genes for a variety of fungal and bacterial cellulases have been cloned and sequenced.[11] Typically, an endo-1,4-β-D-glucanase cleaves internal linkages within amorphous regions in crystalline cellulose to create free ends. An exoglucanase releases oligosaccharides (*e.g.*, cellobiose) from these ends. Finally, a β-D-glucosidase converts oligosaccharides to D-glucose. Although endo- and exoglucanases work synergistically to degrade crystalline cellulose, the exoglucanase component confers the ability to attack crystalline cellulose *efficiently*.[8,9]

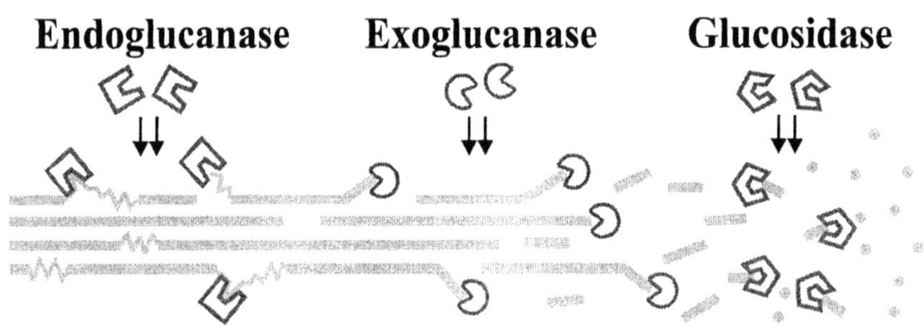

Figure 10.3: Enzymatic degradation of cellulose.

Tertiary structure of the active site of a glucanase determines its mode of action. As a rule, the active site of an endoglucanase is an open cleft capable of binding in the middle of a substrate chain.[5,9,12] In contrast, the active site of an exoglucanase characteristically resides in a tunnel formed by long polypeptide loops.[9,13] The tunnel can bind substrate only from a free end and can accommodate no more than a single glucan chain. Moreover, by preventing dissociation, longer tunnels that have more potential binding sites for substrate promote strong processivity.[13]

In practice, classification of an enzyme as an endoglucanase or exoglucanase is based largely on substrate specificity.[9,14] Endoglucanases readily cleave substituted soluble celluloses such as carboxymethylcellulose, a substrate resistant to attack by strict exoglucanases. In contrast, release of *soluble* reducing sugars from crystalline cellulose typifies exoglucanase activity. The possession of both endo- and exoglucanase activities by some enzymes probably reflects the ability of the active site to assume alternate conformations (cleft ←→ tunnel).

Most fungal and bacterial cellulases minimally have three distinct domains (Fig. 10.4): a large catalytic domain (cat), a short linker, and a small cellulose-binding domain (CBD).[5,9] The CBD is an important feature for *efficient* attack of crystalline cellulose by exoglucanases,[8,15] but the catalytic domain can retain activity even if the linker and CBD are removed.[16,17]

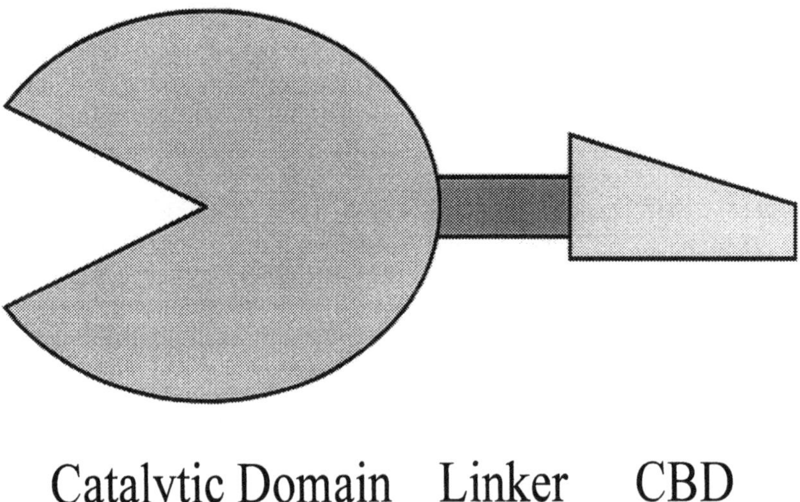

Figure 10.4: Typical structure of endo- and exo-glucanases. A large catalytic domain is linked to a smaller cellulose-binding domain (CBD) via a short linker peptide sequence.[5,9]

Enzymes Used for Biomass Conversion

Enzymes from thermophilic organisms[18,19] are particularly well suited for industrial applications, as exemplified by the thermostable α-amylase used for bioconversion of starch.[20] These enzymes not only have high temperature optima but also are generally thermostable, resistant to protease,[21] and tolerant toward other stresses such as pH extremes.[22] The most important structural features associated with thermostability are those that produce a rigid protein. These include dense packing of hydrophobic amino acids in the core of the molecule and ionic interactions along the protein's surface.[18,19] An added benefit of thermostable enzymes is that specific activity typically increases as the temperature optimum increases.[22,23]

The most popular enzyme source for biomass conversion is the cellulolytic fungus, *Trichoderma reesei*, which secretes both endo- and exoglucanases, the exoglucanase cellobiohydrolase I (CBHI) being the most abundant.[24-27] Although *T. reesei* productivity (units of enzyme/L/h) can be high, all cellulolytic fungi grow slowly, and specific activity of fungal enzymes is generally low.[5] There has been considerable interest in improving CBHI so that it will become an economically viable industrial enzyme.[6,28] Several features of CBHI are undesirable. The temperature optimum is only 50°C, and it rapidly denatures above 55°C.[29-31] In addition to being labile, CBHI has a specific activity about ten-fold lower than that of many bacterial enzymes.[5] Moreover, because the enzyme is strongly inhibited by its product (cellobiose), β-glucosidase must be used during saccharification, adding to process costs. Attempts to develop *Escherichia coli,*[32] *Saccharomyces cerevisiae,*[33] and *Pichia pastoris*[34] as alternative expression systems have not been successful. Recombinant CBHI from these heterologous hosts has a glycosylation pattern different from that of the native enzyme, an alteration that affects enzymatic activity. This limitation will seriously impede efforts to improve the thermostability and specific activity of CBHI. Rapid screening of variants produced through protein engineering or mutation will not be possible without an easily transformed organism that can produce active enzyme.

PRODUCTION OF MICROBIAL CELLULASES IN TRANSGENIC PLANTS

Rationale

Immense quantities of cellulase will be required for industrial-scale biomass conversion, and commercial cellulases are now prohibitively expensive for biofuels applications.[5,6,35] Hence, more highly active enzymes and/or a different source of abundant, low-cost enzymes must be developed. Transgenic plants provide an attractive alternative to microbial production systems for recombinant proteins, particularly for industrial use. Generating large amounts of plant biomass is inexpensive; the infrastructure for downstream processing of materials already exists; and production is readily scaled up by planting more acreage.[36-41] Moreover, significant progress has been made toward establishing parameters for extraction and purification of recombinant proteins from transgenic plants. Examples include chicken avidin and *E. coli* β-glucuronidase, which have now been produced in maize on a commercial scale.[42,43] Because of their inherent stability and high temperature optima (*i.e.*, low activity at plant growth temperatures), cellulases from thermophilic microbes are the most promising candidates for production in plants. An additional advantage of such industrial enzymes is that they could be used without extensive purification. A critical economic consideration for production of industrial enzymes in transgenic plants is the concentration of enzyme in extracts,[44] which depends on the extent of both transgene expression and accumulation of the protein product. Efforts to identify and characterize elements important for regulation of plant gene expression have produced a plethora of tools for achieving high expression of transgenes in plants and for controlling both the tissue specificity and timing of expression. These elements are described in several excellent publications.[36,45-50] Accumulation of a recombinant protein is governed by both the stability of the protein and tolerance of the host plant, which in turn might depend on the subcellular site in which the protein accumulates. Now that pathways for protein trafficking in plants are being elucidated,[51-55] targeting recombinant proteins to subcellular compartments, such as the apoplast, endoplasmic reticulum,[56-63] vacuoles,[64-74] and plastids,[75-78] is feasible.

Endo- and Exoglucanase Genes Expressed in Transgenic Plants

As described in the following sections, several research groups, including my own, have produced transgenic plants that express genes for thermostable microbial glucanases (summarized in Table 10.1). The reported yields of enzyme in leaf extracts have ranged from 0.001% to 26% of total soluble protein (TSP) and have been particularly low for non-targeted enzymes that accumulate in the cytosol. Because these studies involved different host plants, glucanase genes, and enzyme expression and targeting strategies, it is not yet possible to identify the best overall scheme for producing glucanases in plants. Nevertheless, we can conclude that transgenic plants hold considerable promise as a production system for these enzymes. With one exception (*T. reesei* EGI), the plants exhibited no alteration in growth or development, even when they accumulated large amounts of enzyme.

Trichoderma reesei Endo-1,4-β-D-glucanase I (EGI)

With the aim of producing an enzyme for reducing wort viscosity during beer production, Aspegren *et al.* expressed the gene for *T. reesei* endoglucanase I (EGI) in a rapidly growing, easily transformed barley suspension cell line, P1.[79] Though formally classified as an endo-1,4-β-D-glucanase, EGI has such low specificity that it would be more appropriately grouped with non-specific β-D-glucanases, as it has

Table 10.1: Microbial cell-wall hydrolyzing enzymes produced in transgenic plants

Enzyme	**Site**	**Plant**	**Maximum % TSP**
Endo-1,4-β-D-glucanases			
Trichoderma reesei EGI[79]	Apoplast	barley suspension cells	0.4% of secreted protein
Trichoderma reesei EGI[80]	apoplast expected	germinating barley seeds	0.025% in seed
Thermomonospora fusca E2[83]	Cytosol	alfalfa tobacco potato	0.01% 0.1% ND
Acidothermus cellulolyticus E1[87]	Chloroplast	tobacco	1.35%
Acidothermus cellulolyticus E1-cat[90]	Apoplast	tobacco BY-2 cells *A. thaliana*	ND 26%
Exo-1,4-β-D-glucanases			
Trichoderma reesei CBH I[84]	Cytosol	tobacco callus tobacco leaf	0.082% 0.11%
Thermomonospora fusca E3[85]	Cytosol	alfalfa tobacco potato	0.002% 0.02% ND
1,3;1,4-β-D-glucanases			
Bacillus amyloliquefaciens – B. macerans hybrid	Apoplast	barley protoplasts	ND
Bacillus amyloliquefaciens -- B. macerans hybrid	Apoplast	germinating barley seeds	ND
Clostridium thermocellum licB[108]	Cytosol Apoplast	tobacco	ND, developed as a reporter
Endo-1,4-β-xylanase			
Clostridium thermocellum XynZ8[9]	Apoplast	tobacco	4.1%

[a] TSP = total soluble protein

[b] ND = not determined

high activity on mixed (1,3-1,4)-β-glucans and can hydrolyze xylan.[24] In this study, expression of EGI cDNA in barley cells was driven by the CaMV 35S promoter.[48] The native fungal signal peptide included in the protein sequence was apparently functional in plant cells because EGI was efficiently secreted into the culture medium, where it accounted for 0.4% of extracellular protein. Recombinant EGI was glycosylated and retained both activity and thermostability.

An unusual feature of the P1 cell line used for these experiments is that the cell walls are resistant to protoplasting enzymes. Thus, the cells could readily produce and secrete a glucanase with no detrimental effects. Attempts to generate transgenic tobacco plants that expressed EG1 were unsuccessful. Furthermore, although EGI could be transiently expressed in tobacco protoplasts, the cells were unable to regenerate walls.[79] It is unclear whether enzymatic activity of EGI on tobacco cell walls or some other factor accounts for this observation.

This work was extended to generation of transgenic barley plants that express EGI under the control of the high-pI α-amylase promoter to restrict expression of the gene to germinating seeds.[80] Although EGI accumulated to only 0.025% of soluble seed protein during germination, the activity was sufficient to reduce wort viscosity during mashing.

Thermomonospora fusca E2 and E3

The endoglucanase E2 and cellobiohydrolase E3, two of the six glucanases produced by the actinomycete *T. fusca*,[81,82] act synergistically on crystalline cellulose. These enzymes were produced in the cytosol of transgenic alfalfa, potato, and tobacco, under the control of a strong, constitutive hybrid promoter (Mac).[83] As estimated from the signal in immunoblots, recombinant E2 represented up to 0.1% of total soluble leaf protein in tobacco and 0.01% in alfalfa. E3 accumulation was lower (0.02% and 0.001-0.002% in tobacco and alfalfa, respectively). In spite of the lower percentage of enzyme in alfalfa extracts compared with tobacco, the yield of recombinant protein per fresh weight would be comparable, since alfalfa extracts contain about ten times the concentration of soluble protein. Recombinant E2 was enzymatically active *in vitro* and exhibited the expected thermostability when assayed on carboxymethylcellulose. Activity of E3, albeit low, was detectable in an ensilement assay.

Trichoderma reesei cellobiohydrolase I (CBHI)

A recent report describes the expression of a CBHI cDNA, under the control of the CaMV 35S promoter, in tobacco leaves and calli.[84] The enzyme accumulated in the cytosol. As roughly estimated from band intensity in immunoblots, recombinant CBHI accounted for up to 0.11% and 0.082% of total soluble protein in leaves and calli, respectively. Enzymatic activity in leaf extracts could be detected on the soluble substrate, 4-methylumbelliferylcellobioside (MUC), when assayed at 55°C.

Acidothermus cellulolyticus Endoglucanase 1 Holoenzyme (E1)

A. cellulolyticus, an actinomycete collected in 1986 from acidic hot springs in Yellowstone Park, grows optimally at 55°C and pH 5.5 and can use cellulose as a sole carbon source.[85] E1, the major cellulase secreted by this organism, has a temperature optimum of 83°C and a pH optimum of 6.0. It is somewhat resistant to inhibition by cellobiose and is stable for several days at 75°C.[86] This enzyme is typical of microbial cellulases in that it has a large catalytic domain (E1-cat) connected to a cellulose-binding domain (CBD) *via* a short linker sequence (see Fig. 10.4). When synthesized in *Streptomyces lividans* without the linker and CBD, the catalytic domain retains full activity on soluble substrates and has twice the half-life of the holoenzyme at 80°C.[17] In a study of synergism among 16 pairs of purified bacterial and fungal cellulases, E1 in combination with *T. reesei* CBHI exhibited the highest degree of synergy on crystalline cellulose,[29] performing significantly better than the coevolved *T. reesei* EGI.

Dai *et al.* expressed the gene for mature E1 holoenzyme in transgenic tobacco under the control of the leaf-specific promoter (RbcS-3C) of the tomato Rubisco small subunit.[87] The recombinant enzyme was targeted to the chloroplast by means of a Rubisco transit peptide (tomato RbcS-2A) engineered into the gene. Among the more than 30 primary transformants analyzed, E1 accumulated in leaves at concentrations up to 1.35% of TSP, as estimated from the signals on immunoblots. In leaf extracts, the enzyme exhibited a temperature optimum of 81°C and a pH optimum of 5.25, closely agreeing with values for the native enzyme. Analysis of the recombinant enzyme as a function of leaf age revealed that nearly all of the E1 was proteolytically cleaved to an enzymatically active form slightly larger than the catalytic domain. This cleavage was also observed during dehydration of young leaves, but to a lesser extent. Even the plant that accumulated the highest amount of E1 grew normally and maintained normal rates of photosynthesis.

Acidothermus cellulolyticus Endoglucanase 1 Catalytic Domain (E1-cat)

Given the successful production of xylanase (4% of total soluble leaf protein)[88] and phytase (14% TSP)[89] in the apoplast of transgenic tobacco, my laboratory has developed an apoplast-targeting cassette for the catalytic domain of *A. cellulolyticus* E1 (E1-cat).[90] The cassette includes several features to promote efficient expression of the E1-cat gene in plants. To ensure high rates of mRNA synthesis, we incorporated the strong constitutive 35S promoter of cauliflower mosaic virus (CaMV 35S)[48,49] upstream of the E1-cat gene and an efficient polyadenylation signal (nos 3')[45] downstream of the gene. We included the tobacco mosaic virus translational enhancer Ω[91] to provide an optimal context in the mRNA for ribosome binding and initiation of translation. To direct recombinant E1-cat to the endoplasmic reticulum, which serves as the entry point into the default secretory pathway, we cloned the E1-cat gene in frame with the coding sequence of the tobacco PR1a amino terminal signal peptide.[50]

Analysis of transgenic tobacco BY-2 calli[92,93] expressing the gene in this cassette provided evidence that the plant cells produced enzymatically active E1-cat (Fig. 10.5).[90] A protein of the expected size (~42 kDa) was detected in Western blots by using monoclonal antibody specific for E1. In contrast, we detected no cross-reacting protein in extracts of non-transformed BY-2 cells.[94] Moreover, suspension cell lines established from transgenic calli exhibited normal morphology and growth rate and secreted the enzyme into the culture medium.

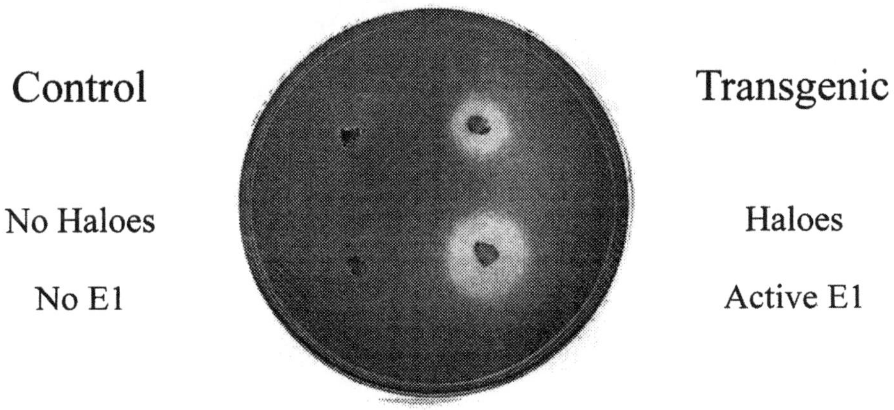

Figure 10.5: Congo Red endoglucanase assay on transgenic tobacco BY-2 calli. Control (non-transformed) calli and two independently transformed calli expressing apoplast-targeted E1-cat were grown on maintenance medium for 24 hr. Calli were overlaid with agarose containing the substrate carboxymethylcellulose (CMC). After the plate was heated at 65°C for 45 min to activate the enzyme, the CMC was stained with Congo Red.[93] Clear haloes indicate endoglucanase activity.

We then determined that transgenic *Arabidopsis thaliana* plants expressing the cassette accumulated large amounts of E1-cat in the leaves.[90] We estimated the amount of E1-cat protein in extracts of primary transformants by ELISA by using a monoclonal antibody specific for E1.[94] We also compared enzymatic activity in extracts with that of a known amount of E1-cat purified from *S. lividans* by using the substrate MUC (Fig. 10.6).[95] The results from the two assays agreed quite closely (Fig. 10.7). Of the 15 plants randomly screened, expression of E1-cat ranged from 0.01 to 25.7% of total soluble leaf protein, with a mean of 7.3%. Five of the 15 plants expressed E1-cat at >10% TSP. The enzyme could be recovered in intercellular fluid by vacuum infiltration and centrifugation of leaves. Despite having a large amount of E1-cat in the apoplast, plants exhibited normal growth and development and were fertile. These results suggest that apoplast targeting might be a preferred strategy for producing endoglucanases in transgenic plants.

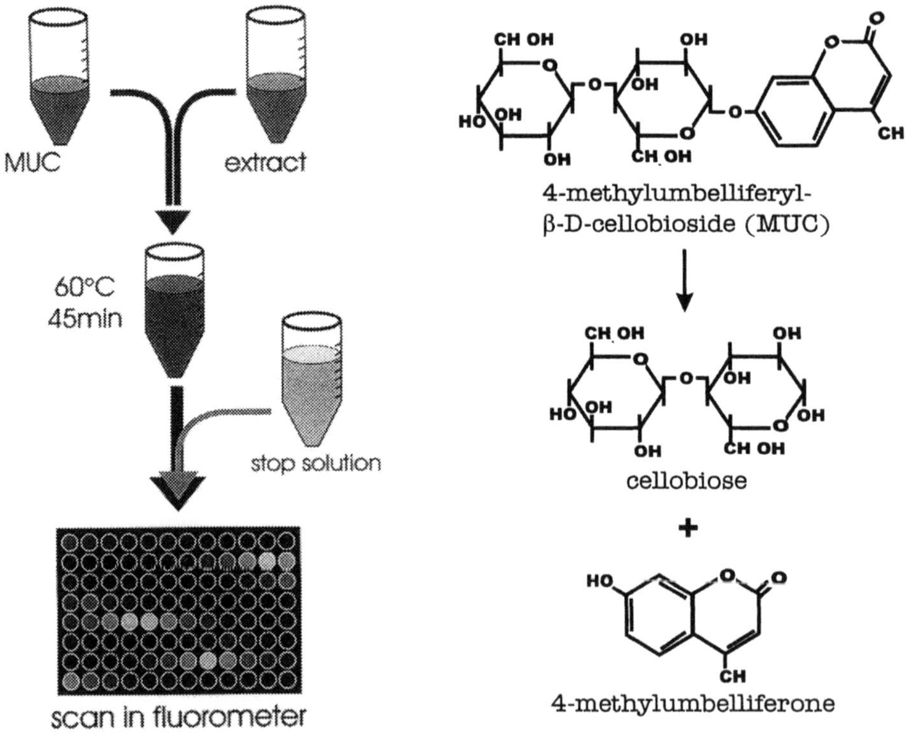

Figure 10.6: Cellulase assay on the fluorigenic substrate 4-methylumbelliferyl-β-D-cellobioside (MUC).

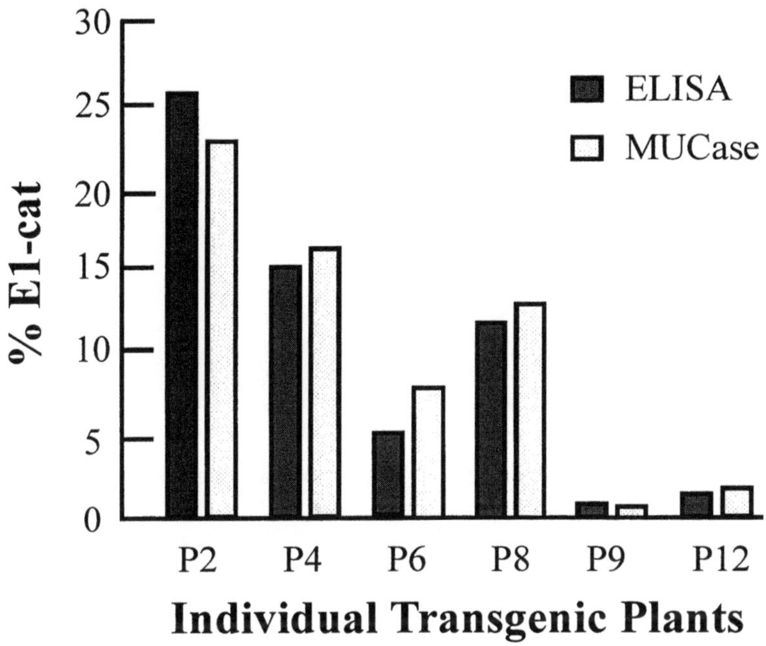

Figure 10.7: Amount of E1-cat in primary *Arabidopsis thaliana* transformants. Estimates were based on both ELISA and enzymatic activity (MUCase) relative to known amounts of purified E1-cat produced in *Streptomyces lividans*.[90]

E1-cat can be partially purified from *A. thaliana* leaf extracts by heat treatment (Fig. 10.8).[90] Leaves of untransformed *A. thaliana* contain several abundant proteins detected in Coomassie-stained protein gels. Because E1-cat is thermotolerant, it is stable at temperatures that would denature most plant proteins and should, therefore, be detectable after heat denaturation that would precipitate most endogenous proteins. In heated extracts of both untransformed and transformed plants, we detected no endogenous proteins after SDS/PAGE and Coomassie staining. In heated extracts of transgenic plants, E1-cat was not denatured and was the only protein detected.

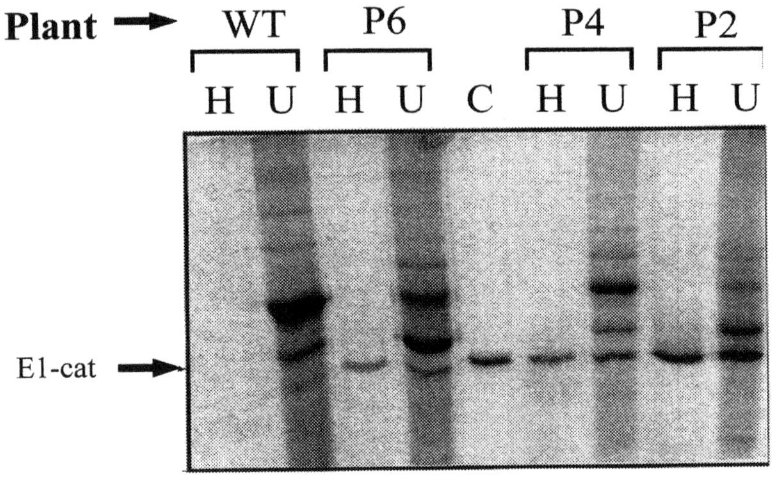

Figure 10.8: Enrichment of E1-cat in *Arabidopsis thaliana* leaf extracts by heat denaturation.[90] Proteins in leaf extracts from wild-type (wt) and transgenic *A. thaliana* (plants P2, P4, and P6) were subjected to SDS/PAGE and stained with Coomassie Blue R250. An aliquot of each sample was heated to 65° for 45 min and centrifuged at 10,000 x g for 10 min to precipitate denatured proteins. H = heated samples. U = unheated samples. Lane C: 1.5 μg purified E1-cat. All lanes containing unheated samples were loaded with ~15 μg of total leaf protein. The same volume was loaded from heated and unheated samples.

Although we have demonstrated that apoplast targeting has been highly effective for E1-cat, it is important for several reasons to develop additional repositories for cellulases. First, it is unknown whether some classes of CBD (with or without the catalytic domain) will be deleterious if they are secreted or even if they simply enter the Golgi, where (1-4)-β-D-glucans are synthesized.[96] Since holoenzymes will be most effective for degradation of crystalline cellulose, the ER, vacuoles, and chloroplast offer reasonable alternatives. Second, it is not clear whether some cellulases will be active if secreted (*e.g.*, if they are glycosylated improperly in plants). The difficulty in producing CBHI in heterologous systems is but one example of the effect of glycosylation on activity.[32-34] Third, it may be desirable to produce combinations of enzymes in the same plant. Even if each enzyme alone can accumulate in the apoplast to high level with no deleterious effects, the combination of two might be devastating. Targeting one enzyme to the apoplast and another to the storage vacuole, for example, has a much greater chance of success. Fourth, apoplast targeting might not be successful for some species of plant. Lastly, an alternate targeting strategy or combination of strategies might result in even higher yields of enzyme than is achieved with a single strategy.

As an extension of our earlier work, we have developed a set of expression cassettes for targeting cellulases to not only the apoplast but also the ER, neutral vacuole, and acidic vacuole (Fig. 10.9; Prikryl, Erdmann, Applewhite, and Danna, unpublished). We designed these pUC-based plasmids so that a cloned cellulase gene inserted in frame with the Pr1a signal peptide can be easily shuttled to each of the other cassettes after excision with the rare-cutting restriction enzyme *Not*I. Because the regulatory elements and cloning site in these cassettes are identical, the plasmids will be useful for *systematically* determining the best targeting strategy for a given enzyme.

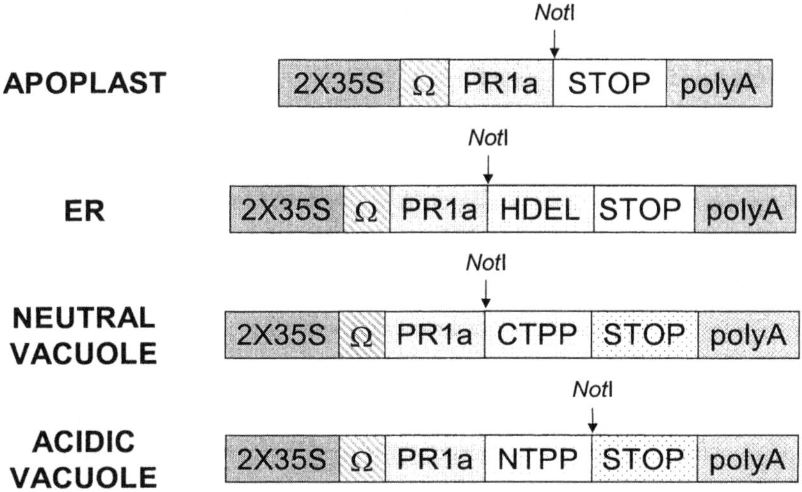

Figure 10.9: Plant expression cassettes for targeting glucanases to cellular compartments. A *Not*I cloning site is used because the eight-base-pair recognition site is unlikely to occur in a protein-coding region. The gene to be inserted is amplified by PCR with *Not*I sites engineered at either end. The cloned gene sequence is in frame with the Pr1a signal peptide.[50] 2X35S is the CaMV double 35S promoter,[48,49] Ω is the TMV translational enhancer,[90] and poly A is the translational terminator of CaMV. STOP encodes the four-base-pair translational terminator UAAA.[45] The targeting elements (HDEL, CTPP, and NTPP) are explained in the text.

The PR1a amino terminal signal peptide in all of the cassettes mediates translocation of the recombinant protein into the endoplasmic reticulum (ER) lumen, where the signal peptide is removed.[50] After transit through the ER and Golgi, the protein is secreted into the apoplast by default unless it contains a secondary targeting signal for subsequent sorting.[51-55] Hence, the ER cassette encodes a carboxyl-terminal retention signal, His-Asp-Glu-Leu (HDEL), which will prevent secretion of the recombinant protein by the default secretory pathway.[56] Similarly, the cassette for targeting proteins to neutral vacuoles contains a carboxyl-terminal propeptide sequence (CTPP) that contains an acidic amino acid preceded by several hydrophobic residues.[52] This peptide is removed when the protein reaches the vacuole. The cassette for targeting proteins to the acidic vacuole contains an amino terminal propeptide (NTPP) as a sorting signal.[70] The sequence Asn-Pro-Ile-Arg (NPIR) is a necessary motif of the NTPP.

Targeting E1-cat to the ER is a promising approach (Prikryl and Danna, unpublished). As a control for ER-targeting, we used a line of tobacco BY-2 cells expressing the gene for ER-localized green fluorescent protein (GFP).[62,63] GFP, from the jellyfish *Aequoria victoria*, is a useful marker protein because it can be observed in living cells.[97,98] Laser confocal microscopy of these cells illustrates the high degree of GFP localization obtained with an ER-targeting cassette (Fig. 10.10A). We used centrifugation to prepare microsomes from this cell line (Fig. 10.10B) [59] as well as from a BY-2 line expressing E1-cat from the ER-targeting cassette and another that secretes E1-cat into the apoplast. Fluorescence in gradient fractions from the GFP-positive cells was measured directly (Fig. 10.10C). Fractions from the E1-cat transformants were assayed for enzymatic activity on the fluorigenic substrate MUC (see Figure 10.6). We detected recombinant protein (GFP or E1-cat) in the microsomal peak of all three samples (Fig. 10.10C). As expected, only apoplast-targeted E1-cat was also detected in the upper fractions that contain soluble proteins. We are now evaluating the E1-cat ER-targeting cassette in *A. thaliana* and are testing cassettes for vacuolar targeting.

PRODUCTION OF OTHER CELL-WALL HYDROLYZING ENZYMES

Even though cellulose is the major source of fermentable sugars in plant biomass, the hemicellulose component of plant cell walls is significant. Hemicellulose polymers, which include mixed-linked glucans as well as xylans and xyloglucans, contain not only glucose but also arabinose, galactose, and the pentose sugar xylose. Now that both bacteria and yeast have been engineered to ferment a variety of hemicellulose-derived sugars,[99-104] enzymatic release of these saccharides is desirable.

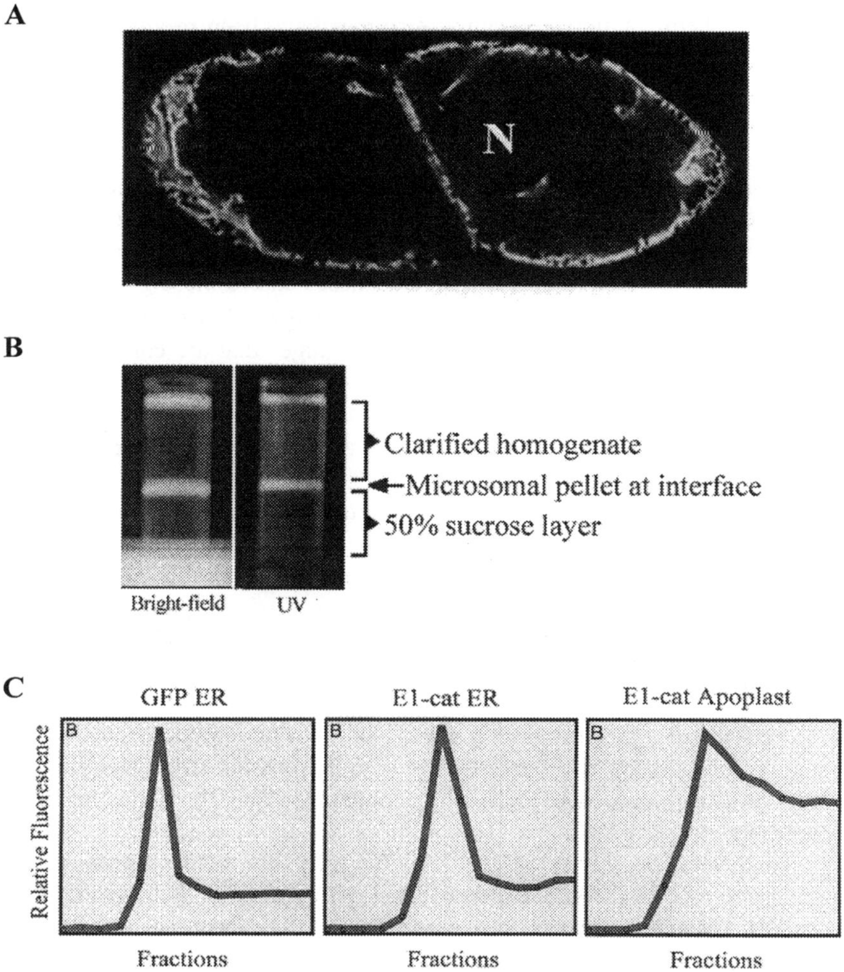

Figure 10.10: Targeting E1-cat to the endoplasmic reticulum. A. Laser confocal image of control tobacco BY-2 cells expressing GFP in the ER. N denotes the nucleus. B. Isolation of microsomes by centrifugation. Clarified cell homogenates were layered over 50% sucrose, and microsomes were pelleted onto the interface. The gradient for GFP-expressing cells is shown under both visible and UV illumination. Microsomes were also isolated from cells with E1-cat targeted to the ER or to the apoplast. C. Gradient fractions of the GFP sample (GFP ER) were assayed directly for fluorescence. Gradient fractions of the other two samples were assayed for E1-cat activity on MUC. "B" indicates the bottom of each gradient.

Two classes of hemicellulose-degrading enzymes have been produced in plants, as described in the following sections.

Endo-(1,3-1,4)-β-D-glucanases

Unbranched glucans, which contain a mixture of β-1,4- and β-1,3-linked glucose residues, represent a significant hemicellulose component of cell walls in cereals and other grasses.[4] In barley endosperm, these glucans are the major cell wall component. Specific endo-(1,3-1,4)-β-D-glucanases cleave these glucans but do not cleave polymers containing pure β-1,4 or β-1,3 linkages. For example, *Clostridium thermocellum* lichenase hydrolyzes only β-1,4 linkages that are contiguous to β-1,3 linkages.[105]

Phillipson transiently expressed a hybrid gene for a heat-stable, *Bacillus*-derived endo-(1,3-1,4)-β-D-glucanase in barley aleurone protoplasts and was able to recover small amounts of active enzyme.[106] The 5' and 3' halves of β-glucanase genes from *B. amyloliquefaciens* and *B. macerans*, respectively, were combined to obtain a hybrid enzyme with both a low pH optimum and thermostability. The sequence encoding the amino terminal signal peptide of the low pI α-amylase was added to direct efficient secretion of the enzyme, and transcription was driven by a constitutive promoter, pEmu (derived from the OCS promoter and elements from the maize AdhI gene).

Jensen, *et al.* obtained transgenic barley plants that synthesized a similar enzyme specifically in germinating grains but not in leaves, stems, or roots.[107] Secretion of the enzyme was mediated by a signal peptide from barley high-pI α-amylase, and expression was limited to germinating seeds by using the high-pI α-amylase promoter.

To develop an easily assayed reporter gene for plant expression systems, Piruzian *et al.* modified the *lic*B gene of the thermophile, *C. thermocellum.*[108] This gene, which encodes lichenase, an endo-(1,3-1,4)-β-D-glucanase, was truncated at both the 5' and 3' ends to produce an enzyme that retained activity and thermostability. The enzyme without an amino terminal signal peptide was produced in transgenic tobacco plants under the control of either the constitutive octopine T_R2' promoter or the light-inducible *A. thaliana* rbcS promoter. A secreted version of the enzyme (containing the amino terminal signal peptide from carrot extensin) was also expressed by using the rbcS promoter. The enzyme was detectable in all cases, but recovered activity appeared to be highest when the enzyme was secreted into the apoplast (about two-fold higher than non-targeted protein expressed from the same promoter), though it is unclear how many individual transgenic plants were screened in this study. The transgenic plants exhibited normal growth and development.

Xylanase

Xylans are substituted polymers of the pentose sugar xylose (in 1,4-β linkages). Common substitutions include α-linked arabinose and 4-O-methylglucuronic acid. Xylans are an abundant constituent of primary and secondary cell walls of monocots as well as secondary walls in woody tissues of dicots. The cell walls of herbaceous dicots, however, contain little xylan.[4]

The *xynZ* gene from *C. thermocellum* encodes a thermostable xylanase with a temperature optimum of 70°C.[109] Herbers *et al.* produced transgenic tobacco plants that accumulated this protein in the apoplast at 4% TSP.[90] The coding region of the enzyme was fused to the sequence encoding the amino terminal signal peptide of the proteinase inhibitorII and expression was driven by the CaMV 35S promoter. The recombinant enzyme was readily purified up to 31-fold from leaf extracts by heat denaturation and centrifugation. Since transgenic plants had normal phenotypes, the apoplast of dicots could be used as a repository for xylanase.

FUTURE DIRECTIONS

Research described in this review has provided proof of the principle that plants can be used to produce large amounts of thermostable microbial glucanases and xylanase, but several issues must be resolved before this technology can be commercialized. The most effective enzymes for industrial application must be identified, and the best targeting strategy and host plant for their production must be determined. Details of downstream processing schemes must be worked out to ensure an adequate year-round supply of enzyme and to evaluate the performance of recombinant glucanases in biomass conversion applications. Lastly, assessment of environmental risks and compliance with guidelines for handling specific transgenic plants must be carried out before large-scale field trials are implemented.

SUMMARY

Plant biomass, the most abundant renewable resource on earth, is a potential source of fermentable sugars for production of clean-burning, alternative transportation fuels such as ethanol. Bioconversion of plant biomass to fermentable glucose involves enzymatic hydrolysis of cellulose, a major polysaccharide constituent of the plant cell wall. Enzymes from thermophilic organisms are particularly suited for this application because they are thermostable, resistant to protease, and tolerant toward other stresses. Commercially available microbial cellulases are prohibitively expensive for bioethanol processes. Manufacturing heterologous cellulases in crop plant bioreactors could significantly reduce costs associated with enzyme production and could offer a potentially high-volume alternative to traditional methods. Because transgenic plants that produce large amounts of bacterial glucanases (up to 26% of total soluble leaf protein) exhibit normal growth and development, this approach offers much promise.

ACKNOWLEDGMENTS

I thank Jana Prikryl for her excellent technical assistance and skill with computer graphics. Work in the author's laboratory has been supported by grants from the Colorado Institute for Biotechnology Research; the University of Colorado Council on Research and Creative Work; the U.S. Department of Energy Advanced Energy Projects (DE-FG03-96ER12194); and the NRI Competitive Grants Program/USDA (98-35504-6964 and 00-35504-9369).

REFERENCES

1. LYND, L.R., CUSHMAN, J.H., NICHOLS, R.J., WYMAN, C.E., Fuel ethanol from cellulosic biomass, *Science*, 1991, **251**, 1318-1323.
2. WHEALS, A.E., BASSO, L.C., ALVES, D.M.G., AMORIM, H.V., Fuel ethanol after 25 years, *TRENDS BIOTECH.*, 1999, **17**, 482-487.
3. BEGUIN, P., Molecular biology of cellulose degradation, *Annu. Rev. Microbiol.*, 1990, **44**, 219-248.
4. BRETT, C., WALDRON, K., Physiology and Biochemistry of Plant Cell Walls, 2nd Edition. Chapman & Hall, London, 1996, 255 p.
5. HIMMEL, M.E., ADNEY, W.S., BAKER, J.O., NIEVES, R.A., THOMAS S.R., Cellulases: Structure, function, and applications, *in* Handbook on Bioethanol: Production and Utilization (C. E. Wyman, ed.), Taylor and Francis, Washington, D.C., 1996, pp. 143-161.
6. HIMMEL, M.E., RUTH, M.F., WYMAN, C.E., Cellulase for commodity products from cellulosic biomass, *Curr. Opinion Biotechnol.*, 1999, **10**, 358-364.
7. PHILIPPIDIS, G.P., SMITH, T.K., WYMAN, C.E., Study of the enzymatic hydrolysis of cellulose for production of fuel ethanol by the simultaneous saccharification and fermentation process, *Biotech. Bioeng.*, 1993, **41**, 846-853.
8. BEGUIN, P., AUBERT, J.-P., The biological degradation of cellulose, *FEMS Microbiol. Rev.*, 1994, **13**, 25-58.
9. TEERI, T.T., Crystalline cellulose degradation: new insight into the function of cellobiohydrolases, *Trends Biotechnol.*, 1997, **15,** 160-167.
10. KLYOSOV, A.A., Trends in biochemistry and enzymology of cellulose degradation, *Biochemistry*, 1990, **29**, 10577-10585.
11. HENRISSAT, B., BAIROCH, A., New families in the classification of glycosyl hydrolases based on amino acid sequence similarities, *Biochem. J.*, 1993, **293**, 781-788.

12. SAKON, J., ADNEY, W. S., HIMMEL, M.E., THOMAS, S.R., KARPLUS, P.A., Crystal structure of thermostable family 5 endocellulase E1 from *Acidothermus cellulolyticus* in complex with cellotetraose, *Biochemistry*, 1996, **35**, 10648-10660.
13. DIVNE, C., STAHLBERG, J., REINIKAINEN, T., ROHONEN, L., PETTERSSON, G., KNOWLES, J.K.C., TEERI, T.T., JONES, A., The three-dimensional crystal structure of the catalytic core of cellobiohydrolase I from *Trichoderma reesei*, *Science*, 1994, **265**, 524-528.
14. WOOD, T.M., BHAT, K.M., Methods for measuring cellulase activities, *Meth. Enzymol.*, 1988, **160**, 87-113.
15. DONNER, R.R., EVANS, B.R., AFFHOLTER, K.A., WOODWARD, J., Role of cellulose-binding domain of cellobiohydrolase I in cellulose hydrolysis, *ACS Symposium Series*, 1994, **506**, 75-83.
16. SRISODSUK, M., REINIKAINEN, T., PENTTILA, M., TEERI, T.T., Role of the interdomain linker peptide of *Trichoderma reesei* cellobiohydrolase I in its interaction with crystalline cellulose, *J. Biol. Chem.*, 1993, **268**, 20756-20761.
17. ADNEY, W.S., THOMAS, S.R., BAKER, J.O., HIMMEL, M.E., CHOU, Y.-C., Method for increasing thermostability in cellulase enzymes, 1998, U.S. Patent 5712142.
18. ADAMS, M.W.W., KELLY, R.M., Finding and using hyperthermophilic enzymes, *Trends Biotechnol.*, 1998, **16**, 329-332.
19. MADIGAN, M.T., OREN, A., Thermophilic and halophilic extremophiles. *Curr. Opinion in Microbiol.*, 1999, **2**, 265-269.
20. ELANDER, R.T., PUTSCHE, V.L., Ethanol from corn: technology and economics, *in* Handbook on Bioethanol: Production and Utilization (C. E. Wyman, ed.), Taylor and Francis, Washington, D.C., 1996, pp. 329-349.
21. DANIEL, R.M., COWAN, D.A., MORGAN, H.W., CURRAN, M.P., A correlation between protein thermostability and resistance to proteolysis, *Biochem. J.*, 1982, **207**, 641-644.
22. ZEIKUS, J.G., VIEILLE, C., SAVCHENKO, A., Thermozymes: biotechnology and structure-function relationships, *Extremophiles*, 1998, **2**, 179-183.
23. ZEIKUS, J.G., Molecular determinants of thermozyme activity and stability: Analysis of xylose isomerase and amylopullulanase, *in* Enzymes for Carbohydrate Engineering (K.H. Park, J.F. Robyt, and Y-D. Choi, eds.), Elsevier Science, London, 1996, pp. 145-161.
24. KUBICEK, C.P., The cellulase proteins of *Trichoderma reesei*: structure, multiplicity, mode of action and regulation of formation, *Adv. Biochem. Eng./Biotechnol.*, 1992, **45**, 1-27.
25. SHOEMAKER, S., SCHWEICKART, V., LADNER, M., GELFAND, D., KWOK, S., MYAMBO, K., INNIS, M., Molecular cloning of exo-cellobiohydrolase I derived *from Trichoderma reesei* strain L27, *Bio/Technology*, 1983, **1**, 691-696.
26. SHOEMAKER, S., WATT, K., TSITOVSKY, G., COX, R., Characterization and properties of cellulases purified from *Trichoderma reesei* strain L27, *Bio/Technology*, 1983, **1**, 687-690.

27. TEERI, T., SALOVUORI, I., KNOWLES, J., The molecular cloning of the major cellulase gene from *Trichoderma reesei, Bio/Technology*, 1983, **1**, 696-699.
28. http://www.ceassist.com/executive.htm
29. BAKER, J.O., ADNEY, W.S., NIEVES, R.A., THOMAS, S.R., WILSON, D.B., HIMMEL, M.E., A new thermostable endoglucanase, *Acidothermus cellulolyticus* E1. Synergism with *Trichoderma reesei* CBH I and comparison to *Thermomonospora fusca* E_5, *Appl. Biochem. Biotech.*, 1994, **45/46**, 245-256.
30. BAKER, J.O., HIMMEL, M.E. 1993. Thermal and pH stress in thermal denaturation of *Trichoderma reesei* cellobiohydrolase I, *in* Biocatalyst Design for Stability and Specificity, ACS Symposium Series 516. (M.E. Himmel and G. Georgiou, eds.), American Chemical Society, Washington, D.C., 1993, pp. 83-101.
31. BAKER, J.O., TATSUMOTO, K., GROHMANN, K., WOODWARD, J., WICHERT, J.M., SHOEMAKER, S.P., Himmel, M.E., Thermal denaturation of *Trichoderma reesei* cellulases studied by differential scanning calorimetry and tryptophan fluorescence, *Appl. Biochem. Biotechnol.,* 1992, **34/35**, 217-231.
32. LAYMON, R.A., ADNEY, W.S., MOHAGHEGHI, A., HIMMEL, M.E., THOMAS, S.R., Cloning and expression of full-length *Trichoderma reesei* cellobiohydrolase I cDNAs in *Escherichia coli, Appl. Biochem. Biotechnol.*, 1996, **57/58**, 389-397.
33. PENTTILA, M.E., ANDRE, L., LEHTHOVAARA, P., BAILEY, M., TEERI, T.T., KNOWLES, J.K., Efficient secretion of two fungal cellobiohydrolases by *Saccharomyces cerevisiae*, *Gene*, 1988, **63**, 103-112.
34. GODBOLE, S., DECKER, S.R., NIEVES, R.A., ADNEY, W.S., VINZANT, T.B., BAKER, J.O., THOMAS, S.R., HIMMEL, M.E., Cloning and expression of *Trichoderma reesei* cellobiohydrolase I in *Pichia pastoris*, *Biotechnol. Prog.*, 1999, **15,** 828-833.
35. KADAM, K., Cellulase production., *in* Handbook on Bioethanol: Production and Utilization (C. E. Wyman, ed.), Taylor and Francis, Washington, D.C., 1996, pp. 213-252.
36. KUSNADI, A.R., NIKOLOV, Z.L., HOWARD, J.A., Production of recombinant proteins in transgenic plants: practical considerations, *Biotechnol. Bioeng.*, 1997, **56,** 473-484.
37. HATTI-KAUL, R., MATTIASSON, B., Downstream processing of proteins from transgenic plants, *in* Transgenic Plants: A Production System for Industrial and Pharmaceutical Proteins (M.R.L. Owen and J. Pen, eds.), John Wiley & Sons, Ltd., New York, 1996, pp. 115-147.
38. HOOD, E.E., JILKA J.M., Plant-based production of xenogenic proteins, *Curr. Opinion Biotechnol.*, 1999, **10**, 382-386.
39. PEN, J., Comparison of host systems for the production of recombinant proteins, *in* Transgenic Plants: A Production System for Industrial and Pharmaceutical Proteins (M. R.L. Owen and J. Pen, eds.), John Wiley & Sons, Ltd., New York, 1996, pp. 149-167.
40. PONSTEIN, A.S., VERWOERD, T.C., PEN, J., Production of enzymes for industrial use, *Annals of New York Acad. Sci.*, 1996, **792**, 1-98.

41. HERBERS, K., SONNEWALD, U., Production of new/modified proteins in transgenic plants, *Curr. Opinion Biotechnol.*, 1999, **10**, 163-168.
42. HOOD, E., WITCHER, D., MADDOCK, S., MEYER, T., BASZCZYNSKI, C., BAILEY, M., FLYNN, P., REGISTER, J., MARSHALL, L., BOND, D., Commercial production of avidin from transgenic maize: characterization of transformant, production, processing, extraction and purification, *Mol. Breed.*, 1997, **3**, 291-306.
43. WITCHER, D., HOOD, E., PETERSON, D., BAILEY, M., BOND, D., KUSNADI, A., EVANGELISTA, R., NIDOLOV, Z., WOOGE C., MEHIGH, R., Commercial production of β-glucuronidase (GUS): a model system for the production of proteins in plants, *Mol. Breed.*, 1998, **4**, 301-312.
44. AUSTIN, S., BINGHAM, E.T., KOEGEL, R.G., MATHEWS, D.E., SHAHAN, M.N., STRAUB, R.J., BURGESS, R.R., An overview of a feasibility study for the production of industrial enzymes in transgenic alfalfa, *Ann. New York Acad. Sci.,* 1994, **721,** 234-244.
45. GALLIE, D.R., The role of post-transcriptional control in transgenic gene design, *in* Transgenic Plants: A Production System for Industrial and Pharmaceutical Proteins (M.R.L. Owen and J. Pen, eds.), John Wiley & Sons, Ltd., New York, 1996, pp. 49-74.
46. GALLIE, D.R., Controlling gene expression in transgenics, *Curr. Opinion Plant Biol.*, 1998, **1**, 166-172.
47. FUTTERER, J., Expression signals and vectors, *in* Gene Transfer to Plants (I. Potrykus and G. Spangenberg, eds.), Springer, New York, 1995, pp. 311-324.
48. BENFEY, P.N., CHUA, N.-H., The cauliflower mosaic virus 35S promoter: combinatorial regulation of transcription in plants, *Science*, 1990, **250**, 959-966.
49. KAY, R., CHAN, A., DALY, M., MCPHERSON, J., Duplication of CaMV 35S promoter sequences creates a strong enhancer for plant genes, *Science*, 1987, **236**, 1299-1302.
50. GALLIE, D.R., SLEAT, D.E., WATTS, J.W., TURNER, P.C., WILSON, T.M.A., The 5'-leader sequence of tobacco mosaic virus RNA enhances the expression of foreign gene transcripts *in vitro* and *in vivo, Nucleic Acids Res.*, 1987, **15**, 3257-3273.
51. MATSUOKA, K., BEDNAREK, S.Y., Protein transport within the plant cell endomembrane system: an update, *Curr. Opinion Plant Biol.*, 1998, **1**, 463-469.
52. NAKAMURA, K., MATSUOKA, K., Protein targeting to the vacuole in plant cells, *Plant Physiol.*, 1993, **101**, 1-5.
53. OKITA, T.W., Compartmentation of proteins in the endomembrane system of plant cells, *Annu. Rev. Plant Physiol. Plant Mol. Biol.*, 1996, **47**, 327-50.
54. Staehelin, L.A., The plant ER: A dynamic organelle composed of a large number of discrete functional domains, *Plant J.*, 1997, **11**, 151-1165.
55. VITALE, A., RAIKHEL, N.V., What do proteins need to reach different vacuoles?, *Trends Plant Sci.*, 1999, **4**,149-155.
56. PELHAM, H.R.B., The retention signal for soluble proteins of the endoplasmic reticulum, *Trends Biochem. Sci.*, 1990, **15**, 483-486.
57. SEMENZA, J.C., HARDWICK, K.G., DEAN, N., PELHAM, H. R.B., ERD2, a yeast gene required for the receptor-mediated retrieval of luminal ER proteins from the secretory pathway, *Cell*, 1990, **61**, 1349-1357.

58. MUNRO, S., PELHAM, H.R.B., A C-terminal signal prevents secretion of luminal ER proteins, *Cell*, 1987, **48**, 899-907.
59. GOMORD, V., DENMAT, L.-A., FITCHETTE-LAINE, A.-C., SATIAT-JEUNEMAITRE, B., HAWES, C., FAYE, L., The C-terminal HDEL sequence is sufficient for retention of secretory proteins in the endoplasmic reticulum (ER) but promotes vacuolar targeting of proteins that escape the ER, *Plant J.*, 1997, **11**, 313-325.
60. HERMAN, E.M., TAGUE, B.W., HOFFMAN, L.M., KJEMTRUP, S.E., CHRISPEELS, M. J., Retention of phytohemagglutinin with carboxyterminal tetrapeptide KDEL in the nuclear envelope and the endoplasmic reticulum, *Planta*, 1990, **182**, 305-312.
61. WANDELT, C.I., KHAN, M.R.I., CRAIG, S., SCHROEDER, H.E., SPENCER, D.,HIGGINS, T.J.V., Vicilin with carboxy-terminal KDEL is retained in the endoplasmic reticulum and accumulates to high levels in the leaves of transgenic plants, *Plant J.*, 1992, **2**, 181-192.
62. HASELOFF, J., SIEMERING, K.R., PRASHER, D.C., HODGE, S., Removal of a cryptic intron and subcellular localization of green fluorescent protein are required to mark transgenic *Arabidopsis* plants brightly, *Proc. Natl. Acad. Sci. USA*, 1997, **94**, 2122-2127.
63. BOEVINK, P., SANTA CRUZ, S., HAWES, C., HARRIS, N., OPARKA, K.J., Virus-mediated delivery of the green fluorescent protein to the endoplasmic reticulum of plant cells, *Plant J.*, 1996, **10**, 935-941.
64. BASSHAM, D.C., RAIKHEL, N.V., Molecular aspects of vacuole biogenesis, *Adv. Bot. Res.*, 1997, **25**, 43-58.
65. MARTY, F., Plant vacuoles, *Plant Cell*, 1999, **11**, 587-599.
66. NEUHAUS, J.-M., ROGERS, J.C., Sorting of proteins to vacuoles in plant cells, *Plant Mol. Biol.*, 1998, **38**, 127-144.
67. WINK, M., The plant vacuole: a multifunctional compartment, *J. Exp. Bot.*, 1993, **44**, 231-246.
68. PARIS, N., STANLEY, C.M., JONES, R.L., ROGERS, J.C., Plant cells contain two functionally distinct vacuolar compartments, *Cell*, 1996, **85**, 563-572.
69. CHRISPEELS, M.J., RAIKHEL, N.V., Short peptide domains target proteins to plant vacuoles, *Cell*, 1992, **68**, 613-616.
70. MATSUOKA, K., MATSUMOTO, S., HATTORI, T., MACHIDA, Y., NAKAMURA, K., Vacuolar targeting and posttranslational processing of the precursor to the sweet potato tuberous root storage protein in heterologous plant cells, *J. Biol. Chem.*, 1990, **265**, 19750-19757.
71. MATSUOKA, K., NAKAMURA, K., Propeptide of a precursor to a plant vacuolar protein required for vacuolar targeting, *Proc. Natl. Acad. Sci. USA*, 1991, **88**, 834-838.
72. NEUHAUS, J.-M., STICHER, L., MEINS, F., Jr., BOLLER, T., A short C-terminal sequence is necessary and sufficient for the targeting of chitinases to the plant vacuole, *Proc. Natl. Acad. Sci. USA*, 1991, **88**, 10362-10366.
73. MATSUOKA, K., BASSHAM, D.C., RAIKHEL, N.V., NAKAMURA, K., Different sensitivity to wortmannin of two vacuolar sorting signals indicates the presence of distinct sorting machineries in tobacco cells, *J. Cell Biol.*, 1995, **130**, 1307-1317.

74. DI SANSEBASTIANO, G.-P., PARIS, N., MARC-MARTIN, S., NEUHAUS, J.-M., Specific accumulation of GFP in a non-acidic vacuolar compartment via a C-terminal propeptide-mediated sorting pathway, *Plant J.*, 1998, **15**, 449-457.
75. SMEEKENS, S., WEISBEEK, P., ROBINSON, C., Protein transport into and within chloroplasts, *Trends Biochem. Sci.*, 1990, **15**, 73-76.
76. NAWRATH, C., POIRIER, Y., SOMERVILLE, C., Targeting of the polyhydroxybutyrate biosynthetic pathway to the plastids of *Arabidopsis thaliana* results in high levels of polymer accumulation, *Proc. Natl. Acad. Sci. USA*, 1994, **91**, 12760-12764.
77. BOGORAD, L., Engineering chloroplasts: an alternative site for foreign genes, proteins, reactions, and products, *Trends Biotechnol.*, 2000, **18**, 257-263.
78. MCBRIDE, K.E., SCHAAF, D.J., STALKER, D.M., Controlled expression of plastid transgenes in plants based on a nuclear DNA-encoded and plastid-targeted T7 RNA polymerase, *Proc. Natl. Acad. Sci. USA*, 1994, **91**, 7301-7305.
79. ASPEGREN, K., MANNONEN, L., RITALA, A., PUUPPONEN-PIMIA, R., DURTEN, U., SALMENKALLIO-MARTTILA, M., KAUPPINEN, V., TEERI, T.H., Secretion of a heat-stable fungal β-glucanase from transgenic, suspension-cultured barley cells, *Mol. Breed.*, 1995, **1**, 91-99.
80. NUUTILA, A.M., RITALA, A., SKADSEN. R.W., MANNONEN, L., KAUPPINEN, V., Expression of fungal thermotolerant endo-1,4-β-glucanase in transgenic barley seeds during germination, *Plant Mol. Biol.*, 1999, **41**, 777-783.
81. IRWIN, D.C., SPEZIO, M., WALKER, L.P., WILSON, D.B., Activity studies of eight purified cellulases: specificity, synergism, and binding domain effects, *Biotechnol. Bioeng.*, 1993, **42**, 1002-1013.
82. WILSON, D.B., Biochemistry and genetics of actinomycete cellulases, *Crit. Rev. Biotechnol.*, 1992, **12**, 45-63.
83. ZIEGELHOFFER, T., WILL, J., AUSTIN-PHILLIPS, S., Expression of bacterial cellulase genes in transgenic alfalfa (*Medicago sativa* L.), potato (*Solanum tuberosum* L.) and tobacco (*Nicotiana tabacum* L.), *Mol. Breed.*, 1999, **5**, 309-318.
84. DAI, Z., HOOKER, B.S., QUESENBERRY, R. D., GAO, J., Expression of *Trichoderma reesei* exo-cellobiohydrolase I in transgenic tobacco leaves and calli, *Appl. Biochem. Biotechnol.*, 1999, **77-79**, 689-699.
85. MOHAGHEGHI, A., GROHMANN, K., HIMMEL, M., LEIGHTON, L., UPDEGRAFF, D. M., Isolation and characterization of *Acidothermus cellulolyticus* gen. nov., sp. nov., a new genus of thermophilic, acidophilic, cellulolytic bacteria, *Intl. J. Systematic Bacteriol.*, 1986, **36**, 435-443.
86. TUCKER, M.P., MOHAGHEGHI, A., GROHMANN, K., HIMMEL, M.E., Ultra-thermostable cellulases from *Acidothermus cellulolyticus*: comparison of temperature optima with previously reported cellulases, *Bio/Technology*, 1989, **7**, 817-820.
87. DAI, Z., HOOKER, B.S., ANDERSON, D.B., THOMAS S.R., Expression of *Acidothermus cellulolyticus* endoglucanase E1 in transgenic tobacco: biochemical characteristics and physiological effect, *Transgenic Res.*, 2000, **9**, 43-54.

88. HERBERS, K., WILKE, I., SONNEWALD, U., A thermostable xylanase from *Clostridium thermocellum* expressed at high levels in the apoplast of transgenic tobacco has no detrimental effect and is easily purified, *Bio/Technology*, 1995, **13**, 63-66.
89. VERWOERD, T.C., VAN PARIDON, P.A., VAN OOYEN, A.J.J., VANLENT, J.W.M., HOEKEMA, A., PEN, J., Stable accumulation of *Aspergillus niger* phytase in transgenic tobacco leaves, *Plant Physiol.*, 1995, **109**, 1199-1205.
90. ZIEGLER, M.T., THOMAS, S.R., DANNA, K.J., Accumulation of a thermostable endo-1,4-β-D-glucanase in the apoplast of *Arabidopsis thaliana* leaves, *Mol. Breed.*, 2000, **6**, 37-46.
91. CORNELISSEN, B.J.C., HOROWITZ, J., VAN KAN, J.A.L., GOLDBERG, R.B., BOL, J.F., Structure of tobacco genes encoding pathogenesis-related proteins from the PR-1 group, *Nucleic Acids Res.*, 1987, **15**, 6799-6811.
92. AN, G., High efficiency transformation of cultured tobacco cells, *Plant Physiol.*, 1985, **79**, 568-570.
93. WOOD, P.J., ERFLE, J.D., TEATHER, R.M., Use of complex formation between Congo Red and polysaccharides in detection and assay of polysaccharide hydrolases, *Meth. Enzymol.*, 1988, **160**, 59-75.
94. NIEVES, R.A., CHOU, Y.-C., HIMMEL, M.E., THOMAS, S.R., Quantitation of *Acidothermus cellulolyticus* E1 endoglucanase and *Thermomonospora fusca* E_3 exoglucanase using enzyme-linked immunosorbent assay (ELISA), *Appl. Biochem. Biotechnol.*, 1995, **51/52**, 211-223.
95. COUGHLAN, M.P., Staining techniques for the detection of the individual components of cellulolytic enzyme systems, *Meth. Enzymol.*, 1988, **160**, 135-144.
96. GIBEAUT, D.M., CARPITA, N.C., Biosynthesis of plant cell wall polysaccharides, *FASEB J.*, 1994, **8**, 904-915.
97. CHALFIE, M., TU, Y., EUSKIRCHEN, G., WARD, W., PRASNER, D., Green fluorescent protein as a marker for gene expression, *Science*, 1994, **263**, 802-805.
98. HASELOFF, J., AMOS, B., GFP in plants, *Trends Genet.*, 1995, **11**, 328-329.
99. INGRAM, L.O., GOMEZ, P.F., LAI, X., MONIRUZZAMAN, M., WOOD, B.E., YOMANO, L.P., YORK, S.W., Metabolic engineering of bacteria for ethanol production, *Biotechnol. Bioengin.*, 1998, **58**, 204-214.
100. MONIRUZZAMAN, M., DIEN, B.S., SKORY, C.D., CHEN, Z.D., HESPELL, R.B., HO, N.W.Y., DALE, B.E., BOTHAST, R.J., Fermentation of corn fibre sugars by an engineered xylose utilizing *Saccharomyces* yeast strain, *World J. Microbiol. Biotechnol.*, 1997, **13**, 341-346.
101. KRISHNAN, M.S., HO, N.W.Y., TSAO, G.T., Fermentation kinetics of ethanol production from glucose and xylose by recombinant *Saccharomyces* 1400 (pLNH33), *Appl. Biochem. Biotechnol.*, 1999, **77-9**, 373-388.
102. MOHAGHEGHI, A., EVANS, K. FINKELSTEIN, M., ZHANG, M., Cofermentation of glucose, xylose, and arabinose by mixed cultures of two genetically engineered *Zymomonas mobilis* strains, *Appl. Biochem. Biotechnol.*, 1998, **70-2**, 285-299.

103. KRISHNAN, M.S., XIA, Y., HO, N.W.Y. TSAO, G.T., Fuel ethanol production from lignocellulosic sugars – Studies using a genetically engineered *Saccharomyces* yeast, *Fuels and Chemicals from Biomass*, 1997, **666**, 74-92.
104. BOTHAST, R.J., NICHOLS, N.N., DIEN, B.S., Fermentation with new recombinant organisms, *Biotechol. Prog.*, 1999, **15**, 867-875.
105. ZVERLOV, V.V., FUCHS, K.P., SCHWARZ, W.H., VELIKODVORSKAYA, G.A., Purification and cellulosomal localization of *Clostridium thermocellum* mixed linkage β-glucanase *Lic*B β-1,3-1,4-(-D-glucanase), *Biotechnol. Lett.*, 1994, **16**, 29-34.
106. PHILLIPSON, B.A., Expression of a hybrid (1-3, 1-4)-β-glucanase in barley protoplasts, *Plant Sci.*, 1993, **91**, 195-206.
107. JENSEN, L.G., OLSEN, O., KOPS, O., WOLF, N., THOMSEN, K.K., VON WETTSTEIN, D., Transgenic barley expressing a protein-engineered, thermostable (1,3-1,4)-β-glucanase during germination, *Proc. Natl. Acad. Sci. USA*, 1996, **93**, 34887-3491.
108. PIRUZIAN, E.S., MONZAVI-KARBASSI, B., DARBINIAN, N.S., GOLDENKOVA, I.V., KOBETS, N.S., MOCHULSKY, A.V., The use of a thermostable β-glucanase gene from *Clostridium thermocellum* as a reporter gene in plants, *Mol. Gen. Genet.*, 1998, **257**, 561-567.
109. GREPINET, O., CHEBROU, M.-C., BEGUIN, P., Nucleotide sequence and deletion analysis of the xylanase gene (xynZ) of *Clostridium thermocellum*, *J. Bacteriol.*, 1988, **170**, 4582-4588.

Chapter Eleven

RESVERATROL GLUCOSIDE ENGINEERING: PLANT AND HUMAN HEALTH BENEFITS

Nancy L. Paiva* and John D. Hipskind[#]

Plant Biology Division
P.O.Box 2180
2510 Sam Noble Parkway
Samuel Roberts Noble Foundation
Ardmore, OK, 73401

Present address: Syngenta , 3054 Cornwallis Rd, P.O. Box 12257, Research Triangle Park, NC 27709

INTRODUCTION

Engineering resveratrol (3,5,4'-trihydroxystilbene) (Fig. 11.1) accumulation into crop plants is of interest for various practical reasons. Several species accumulate stilbenes as part of their defense against invasion by pathogens. Peanut, grapevine, and pine tree plants and cell cultures accumulate high levels of resveratrol and other stilbenes following fungal infection or treatment with elicitors.[1-3] These products possess antifungal activity in bioassays with the respective plant pathogens. Stilbenes may also be present prior to pathogen attack. Grape berries and roots of several medicinal plants contain high levels of resveratrol or derivatives, which may act as part of a preformed antimicrobial barrier.[4,5] Largely based on epidemiological studies, resveratrol was proposed to contribute to the improved cardiovascular health of populations consuming high levels of red wine, a food typically high in resveratrol and its glucoside. Later, *in vitro* assays and controlled animal studies confirmed that resveratrol decreased platelet aggregation, low-density lipoprotein oxidation, and had other activities consistent with these health claims.[4,6] Resveratrol has also been shown to have anticancer (chemopreventative) activity in several assays, possibly, in part, due to its strong antioxidant properties.[5] Thus, engineering constitutive accumulation of resveratrol in crop plants, particularly in the edible portions, may improve both the pathogen resistance of the plant and the nutritional value of the crop.

Resveratrol synthase (RS) (also known as stilbene synthase) catalyzes the condensation of one molecule of *p*-coumaroyl CoA and 3 molecules of acetyl CoA to form the 3,5,4'-trihydroxy stilbene resveratrol (Fig. 11.1). The same substrates are utilized by chalcone synthase (CHS), a structurally similar polyketide synthase, and by other biosynthetic pathways, such as lignin and lipid biosynthesis. While chalcone synthases have been found in the majority of vascular land plants, resveratrol synthase and other stilbene synthases have been identified in a smaller but diverse group of species; it has been proposed that stilbene synthases have evolved independently in these species from mutations in the endogenous CHS genes.[7] In species where stilbene synthases have not evolved, genetic engineering of resveratrol accumulation appears to be particularly straightforward, in that transfer of only a single gene is required. The required precursors for resveratrol synthesis are likely to be present in all plants during at least part of their development.

EARLY STUDIES OF STILBENE SYNTHASE EXPRESSION IN TRANSGENIC TOBACCO (*NICOTIANA TABACUM*)

The feasibility of introducing resveratrol synthesis into foreign plants for improving disease resistance was first confirmed by Hain and co-workers.[8,9] In a preliminary study, a stilbene synthase gene from peanut (*Arachis hypogaea*), including the promoter and complete coding region, was introduced into tobacco plants and cell cultures.[8] Expression of the stilbene synthase was inducible by fungal elicitor and UV treatments, while no expression was observed in uninduced tissues, consistent with the regulation of this gene in peanut.[8,10] Reverse phase HPLC was used to confirm the accumulation of resveratrol in elicitor-induced cell cultures from tranformed tobacco lines, while no resveratrol was observed in control lines. An ELISA assay for resveratrol was introduced by using polyclonal antisera raised against resveratrol conjugated to derivatized bovine serum albumin (BSA).[8] This assay was used to follow the kinetics of resveratrol accumulation following induction of cell cultures, as well as quantitate resveratrol levels. The maximum level observed was 50 ng/g fresh weight of transformed tobacco cell culture, less than 10% of the level that accumulated in elicited peanut cell cultures.[8] Surprisingly, control tobacco cultures registered the equivalent of 10 ng resveratrol/g fresh weight with the ELISA assay, although the authors state that no resveratrol was observed in the control cultures when HPLC was used for analysis; this indicates that the antiserum may not be absolutely specific for resveratrol.

In later studies, a fragment of grapevine (*Vitis vinifera*) genomic DNA containing two stilbene synthase genes (*Vst1* and *Vst2*) was used to transform tobacco plants.[9] Wild-type tobacco plants were susceptible to *Botrytis cinerea* infection, while much smaller lesions developed on transgenic tobacco plants that accumulated high levels of stilbene synthase mRNA following inoculation. This demonstrated the potential value of introducing resveratrol into foreign species, and represented the first example of directed phytoalexin engineering. Resveratrol levels were measured following *Botrytis* inoculation to induce the synthesis. The highest level reported was 1000 μg/g fresh weight, 5 days after infection. Two days after inoculation, the youngest leaves accumulated the highest resveratrol levels (400 μg/g), while the oldest (bottom) leaves accumulated little or none. Only the ELISA assay[8] was used for resveratrol analysis, and it was indicated that all values were corrected for the signal found in control tobacco extracts.[9]

The most recent study from this group involved the expression of a grapevine stilbene synthase gene (*Vst1*) under the control of a cauliflower mosaic virus (CaMV) 35S promoter with a duplicated enhancer, a strong constitutive promoter, in transgenic tobacco.[11] Transformants accumulating the highest levels of stilbene

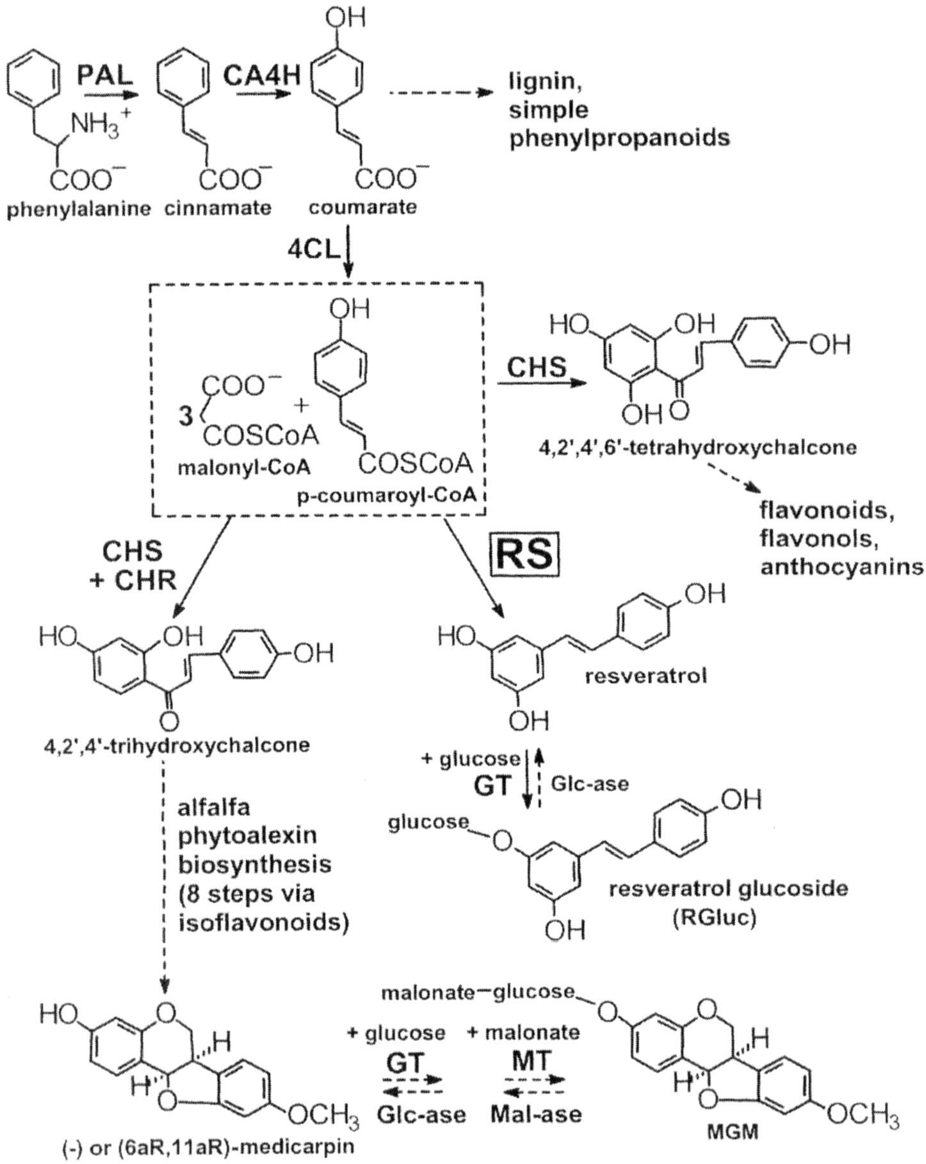
PAL
CA4H
lignin, simple phenylpropanoids
phenylalanine
cinnamate
coumarate
4CL
malonyl-CoA
p-coumaroyl-CoA
CHS
4,2',4',6'-tetrahydroxychalcone
flavonoids, flavonols, anthocyanins
CHS + CHR
RS
4,2',4'-trihydroxychalcone
resveratrol
+ glucose
GT
Glc-ase
resveratrol glucoside (RGluc)
alfalfa phytoalexin biosynthesis (8 steps via isoflavonoids)
(-) or (6aR,11aR)-medicarpin
+ glucose
+ malonate
GT
MT
Glc-ase
Mal-ase
malonate−glucose
MGM

Figure 11.1: Biosynthesis of resveratrol and *trans*-resveratrol-3-O-β-D-glucopyranoside in relation to other branches of phenylpropanoid metabolism. The enzymes are: PAL, phenylalanine ammonia lyase; CA4H, cinnamic acid 4-hydroxylase; 4CL, *p*-coumaric acid CoA-ligase; RS, resveratrol synthase; CHS, chalcone synthase; CHR, chalcone reductase; GT, glucosyl transferase; Glc-ase, glucosidase; MT, malonyl transferase. Wild-type tobacco contains only CHS (not CHR nor RS) while the wild-type legumes alfalfa and soybean contain both CHS and CHR (but not RS). The compound labeled resveratrol glucoside is also known as *trans*-resveratrol-3-O-β-D-glucopyranoside, piceid, polydatin and RGluc in the literature.

synthase mRNA had white flower petals and pollen incapable of germinating, resulting in complete male-sterility; wild-type plants had pink flowers and were completely male-fertile. Male-sterile lines contained as much as 250 μg resveratrol/g fresh weight of leaf, based solely on ELISA analysis. A few transformed lines that accumulated lower levels of stilbene synthase mRNA and lower levels of resveratrol (50-70 μg/g fresh weight leaf tissue), exhibited pale pink flowers, less pigmented than the parental line, but were completely male-fertile. Chalcone synthase activity assays indicated that CHS levels were unchanged in petal extracts, leading the authors to speculate that the decrease in anthocyanin levels in petals was due to stilbene synthase depleting precursors, not due to co-suppression. The male sterility was proposed to be caused by a similar depletion of flavonoid precursors in developing anthers, since tobacco has been demonstrated to require flavonols for pollination.[11,12] The fertility was partially restored (10 to 15%) by spraying the plants with kaempferol and isorhamnetin and the flavonoid precursor *p*-coumarate prior to flower opening, which the authors attributed to a replenishment of flavonols by these substances.[11] Surprisingly, spraying plants with resveratrol also restored pollination by 15%. The authors speculated that this was due to product inhibition of stilbene synthase, thereby blocking the competition for substrates. However, no resveratrol, flavonoid or enzyme analyses were carried out on the pollen or anthers, so direct evidence for these proposed mechanisms was not provided. The lack of pollen grain formation did not match phenotypes reported in other studies with petunia and maize mutants or transgenics that do not accumulate flavonols due to loss of chalcone synthase activity. In these studies, normal pollen grains were produced, but did not form normal germ tubes without the addition of flavonols.[13-16]

Whatever the mechanism by which pollen viability was destroyed, high expression of stilbene synthase was demonstrated as a novel approach for producing male sterile plants, and one that could have useful agronomic applications.[11] Despite the high levels of resveratrol reported, however, the disease resistance of these

CaMV 35S:stilbene synthase plants[11] was lower than tobacco plants containing the inducible grapevine construct.[9]

EXPRESSION OF REVERATROL SYNTHASE IN TRANSGENIC ALFALFA (*MEDICAGO SATIVA*)

Recently, we introduced a construct encoding a peanut resveratrol synthase under the constitutive control of an enhanced cauliflower mosaic virus (CaMV) 35S promoter into transgenic alfalfa plants via *Agrobacterium tumefaciens*-mediated transformation and somatic embryogenesis.[17] Unlike previously published studies in transgenic plants, HPLC analysis detected the accumulation of high levels of resveratrol as *trans*-resveratrol-3-O-β-D-glucopyranoside (Fig. 11.1), but no free resveratrol. The structure of the glucoside was determined by a combination of HPLC, UV, and ^{1}H- and ^{13}C-NMR analyses, the latter consistent with data reported for *trans*-resveratrol-3-O-β-D-glucopyranoside isolated from grapes and *Vitis* cell cultures.[4,18] We do not know if alfalfa contains a unique resveratrol glycosylation activity lacking in the species examined previously, or if any of the prior studies that used an ELISA assay for resveratrol were in error, in that they may have misidentified resveratrol glucoside as resveratrol due to possible antibody cross-reaction in the assay. Previous studies involving elicited or pathogen-infected transgenic plant tissues most likely were correct in reporting free resveratrol; rapid synthesis would likely produce the aglycone, and pathogen or host β-glucosidases would hydrolyze any conjugates formed, as is observed for isoflavonoid biosynthesis in elicited cell cultures.[19] However, the studies involving the CaMV 35S promoter driving the grapevine stilbene synthase[11] could be incomplete in that constitutive accumulation of resveratrol may have allowed conjugation reactions to occur.

In our studies, younger leaves contained the highest levels of resveratrol glucoside (more than 15 μg/g fresh weight), while the oldest leaves contained the lowest levels (approximately 5 μg/g fresh weight).[17] The opposite trend was observed in stem segments. No visible developmental phenotypes were observed. Flower color (purple) was unchanged, even though high levels of resveratrol glucoside (15-20 μg) was detected in floral racemes (unpublished results). Seed set following self-pollination was the same for lines accumulating high levels of resveratrol glucoside as for vector-transformed lines. It is not known if flavonols are required for alfalfa pollination. Although tobacco, petunia, and maize have been shown to require flavonols for pollen germination, *Arabidopsis* mutants completely lacking flavonoids were 100% self-fertile.[20] No resveratrol or resveratrol glucoside was detected in milligram amounts of transgenic pollen collected, but higher amounts of material may be required to reach the limit of detection. The roots of the transgenic plants were readily nodulated by *Rhizobium meliloti*, but only trace levels of resveratrol glucoside accumulate in these tissues, most likely due to the low level of expression driven by the CaMV 35S promoter in alfalfa roots.[17]

An alfalfa pathogen, *Phoma medicaginis*, was inhibited approximately 50% in preliminary agar plate bioassays[21] with resveratrol at 50 μg/ml of agar, compared to unsupplemented agar.[17] Resveratrol glucoside, purified from transgenic alfalfa, also inhibited mycelial growth approximately 50% when added at the same molar concentration.[17] *Phoma* showed a much greater reduction in symptom development on transgenic leaves accumulating only 20 μg/g fresh weight of the resveratrol glucoside relative to empty vector-transformed control plants.[17] The sizes of visible lesions were decreased 50% or more, depending on the level of fungal inoculum used. Following extended incubations to allow the formation of pycnidia (sporulation structures), trypan blue staining of fungal structures revealed lower mycelial growth and dramatically reduced sporulation in transgenic alfalfa leaves relative to controls lacking resveratrol glucoside. For example, out of 57 inoculated transgenic leaves, only 11 leaves formed any pycnidia, with an average of 8 per leaf, while 37 out of 50 inoculated control leaves formed pycnidia, with a average of 32 per leaf. The inhibition of sporulation is likely to be significant in a field setting, in that this will reduce infection of new leaves and plants, as well as reduce survival of the pathogen to the next year.

A number of factors may be involved in the striking *in planta* inhibition conferred by resveratrol glucoside. The foreign antifungal compound is, in effect, added to the natural suite of defense mechanisms (native phytoalexins, hydrolases, oxidative burst, etc.), and is present before the pathogen is applied, in contrast to the numerous defense responses that require induction or activation. The resveratrol glucoside is, therefore, present to inhibit all stages of fungal spore germination, initial leaf penetration, and subsequent development. By delaying any of these fungal processes, the host has more time to assemble its natural defenses.

We also tested resveratrol as an inhibitor of the mycelial growth of other important alfalfa pathogens (Table 11.1). Several pathogens were inhibited significantly, suggesting that resveratrol-accumulating alfalfa may also show increased resistance to these pathogens. Transgenic and control plants are now being challenged with selected leaf and stem pathogens. Stronger promoters for shoots may provide even greater protection, and stronger root promoters are necessary to evaluate the effect of resveratrol glucoside on root pathogens and symbionts.

Table 11.1: Inhibition of selected alfalfa pathogens by 0.5 mM resveratrol. Percent inhibition was calculated by comparing the net radial growth of fungal mycelia on agar media with and without resveratrol added. Growth measurements of duplicate plates were taken five or eight days after inoculation depending upon the fungal isolate. Media and assay design were similar to those used in previous studies.[21]

	% Inhibition	**Site of Infection**
Phoma medicaginis	64.6%	leaf/stem/crown
Fusarium oxysporum f.sp. *medicaginis*	53.1%	leaf/stem/crown/root
Leptosphaerulina medicaginis	43.4%	leaf
Colletotrichum trifolii ARLNW race 1	40.2%	leaf/stem
Phytophthora megasperma var. *medicaginis*	35.8%	root
Verticillium albo-atrum	35.3%	stem (vascular)
Colletotrichum trifolii SB-1 race 2	35.1%	leaf/stem

RESVERATROL GLUCOSIDE ACCUMULATION IN TRANSGENIC *GLYCINE MAX* CULTURES EXPRESSING A PEANUT STILBENE SYNTHASE

The identification of resveratrol glucoside in the transgenic alfalfa was unique among species transformed with RS at the time it was published.[17] In order to examine the phytochemical composition of another species transformed with the binary vector construct encoding resveratrol synthase from peanut (*Arachis hypogaea*) used in the previous alfalfa study,[17] the construct was introduced into transgenic soybean (*Glycine max*) cells. Stilbenes have not been reported to accumulate in untransformed soybean plants, seeds, cell cultures, or food products.

A fast *Agrobacterium rhizogenes*-mediated transformation system,[22] similar to many other "hairy root" co-transformation schemes,[23,24] was used to co-transform the expression construct into soybean cells. The CaMV-35S:RS binary construct or the pGA482 binary vector control plasmid from previous studies[17] was introduced into *Agrobacterium rhizogenes* strain K599[25] by using a standard freeze-thaw technique.[26] Soybean seeds (*Glycine max* cv. Harosoy 63) were surface-sterilized, and germinated on 0.5 % (w/v) sucrose-water-0.8 % agar (w/v) (SWA) plates. The cotyledons of nine to eleven-day-old soybean seedlings were wounded with a sharp scalpel previously dipped into transformed *Agrobacterium rhizogenes* cells growing on the surface of yeast extract-mannitol plates.[27] The inoculated seedlings were transferred to transparent plastic boxes containing SWA medium, with roots embedded in the medium to hold the seedlings upright. Roots emerged from the inoculation sites within 2 to 4 weeks. Plants wounded with a sterile scalpel (uninoculated control) did not form roots at the wound site. Individual hairy roots were transferred to Shenk and Hildebrand (SH) agar medium supplemented with phytohormones and carbenicillin (500 mg/l) to initiate callus from the roots and to eliminate any *A. rhizogenes*.[28] The hairy roots and newly initiated calli were transferred to fresh medium at weekly intervals.

From previous studies in our lab,[22] and consistent with other studies,[23,24] approximately 50 to 60% of the emerging hairy roots were "co-transformed" (contain both the Ri DNA from the genome of the *A. rhizogenes* and the T-DNA region of the introduced binary vector), based on the presence of neomycin phosphotransferase (NPT II). Calli induced by *A. rhizogenes* with no binary vector are negative for NPT II by ELISA analysis. These positive calli (and representative non-transformed calli) were then assayed by HPLC for resveratrol and its glucoside, and the presence of the CaMV-35S:RS construct, by using the PCR and genomic Southern techniques similar to those used for transgenic alfalfa.[17]

A comparison of the HPLC profiles of organic extracts of representative callus cultures revealed that several independent cultures containing the CaMV-35S:RS construct produced a peak (retention time 27.5 min, Fig. 11.2A) not present in control cultures (Fig. 11.2B). The retention time was identical to that of a *trans*-resveratrol-3-O-β-D-glucopyranoside standard purified from transgenic alfalfa harboring the same construct. The retention times of both peaks (novel soybean peak and standard) did not shift when 1% H_3PO_4 was added to the HPLC mobile phase, consistent with a lack of malonylation of the glucoside. Previous reports had shown that both soybean and alfalfa isoflavonoid glucoside malonates are more highly retained in the presence of acid than in its absence.[29]

The novel peak had a UV spectrum similar to those of resveratrol and *trans*-resveratrol-3-O-β-D-glucopyranoside, with a split maximum at 300-320 nm (Fig. 11.3), while none of the peaks in the control culture had UV scans resembling that of resveratrol. The novel compound was purified by two rounds of semi-preparative reverse-phase HPLC and subjected to treatment with β-glucosidase.[17] HPLC analysis indicated that the size of the novel peak (8 min) was reduced, while a later-eluting compound (15 min) was released. The released compound had a retention time and UV scan identical to resveratrol (Figs. 11.4A and 11.4B). Higher levels of β-glucosidase and longer incubation times resulted in complete hydrolysis of the compound (data not shown).

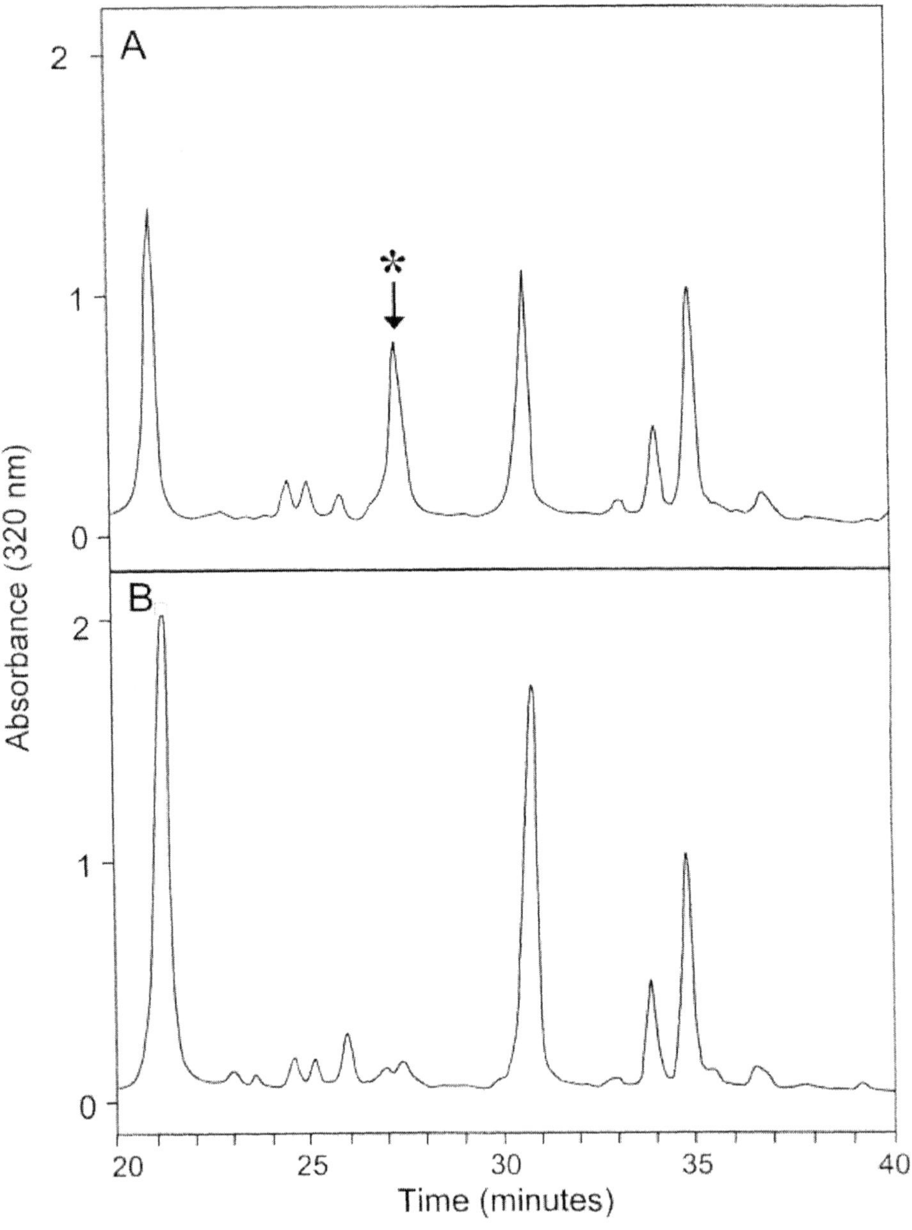
A
B
*
2
1
0
Absorbance (320 nm)
20
25
30
35
40
Time (minutes)

Figure 11.2: HPLC chromatograms of extracts of transgenic and control soybean callus. Extracts from (A) callus derived from roots co-transformed by *A. rhizogenes* with the CaMV35S:RS construct, encoding the peanut resveratrol synthase gene, and (B) control callus from roots induced by *A. rhizogenes* carrying no binary vector. The novel peak present in the transgenic extracts is marked with an *.

From one untransformed control line and one CaMV-35S:RS transgenic callus line, approximately 20 to 30 grams of soybean callus cells were homogenized in 150 ml acetone to extract the phenolic compounds. The mixture was centrifuged, the acetone extract taken to dryness, and the residue dissolved in 0.5 ml methanol. After centrifugation, the extracts were analyzed by an HPLC system similar to that described for alfalfa extracts.[17] To enhance the resolution of the resveratrol glucoside from the endogenous soybean cell culture phenolics in crude extracts, the slope of the gradient was reduced; the gradient for crude soy extracts was 90% water/10% CH_3CN initially, with a linear change to 25% CH_3CN in 45 minutes, flow rate 0.8 ml/min on a reverse phase column (4.6 x 250 mm Octadecyl silica, 5 μm, JT Baker).

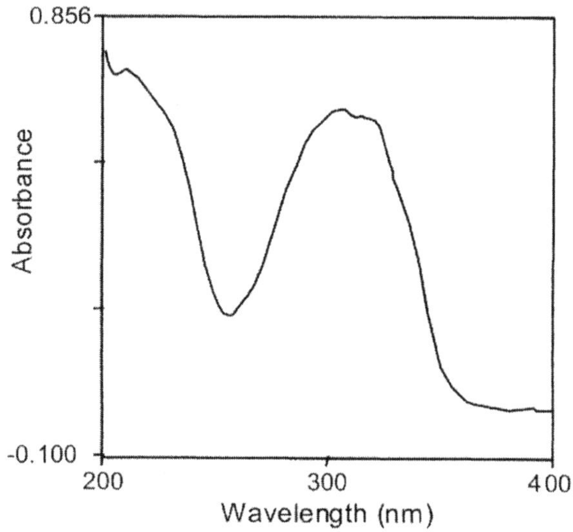

Figure 11.3: UV diode array scan of the novel peak present in soybean callus transformed with the CaMV-35S:RS construct, encoding the peanut resveratrol synthase gene. UV scan was recorded with a Beckman model 168 diode array detector.

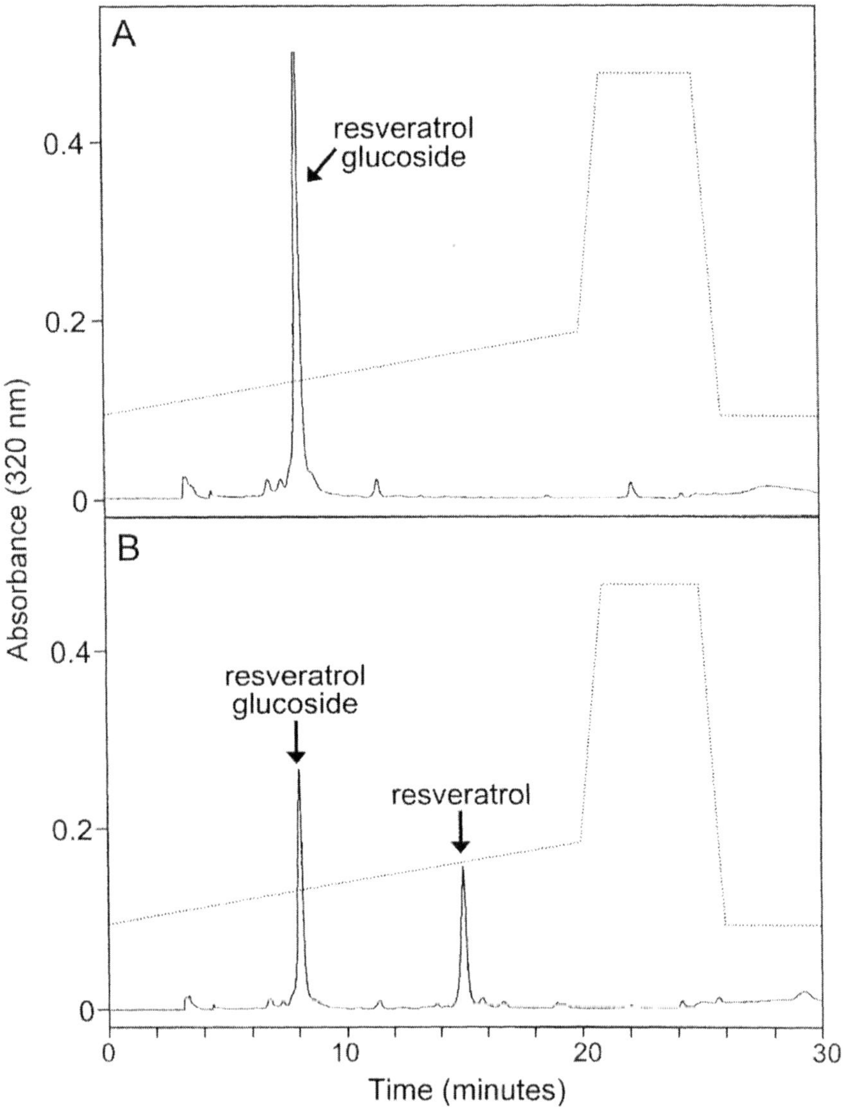

Figure 11.4: HPLC chromatograms of resveratrol glucoside partially purified from an extract of transgenic soybean callus, (A) without enzyme treatment, and (B) following partial hydrolysis by β-glucosidase and re-extraction into EtOAc. Digestion with β-glucosidase was performed as described previously.[17]

The partially purified compound was also subjected to LC/MS analysis, in comparison to a *trans*-resveratrol-3-O-β-D-glucopyranoside standard. The compounds had identical retention times in this system and produced identical MS fragmentation patterns. Each produced 2 major ions at m/z = 389 (85%) and m/z = 227 (100%), corresponding to m – 1 (parent ion) of the resveratrol glucoside and m-1-glucose, or the resveratrol aglycone fragment, respectively. Under the same conditions, a pure resveratrol aglycone standard (Sigma) produced only the major ion at m/z = 227 (m-1).

The novel compound from the transformed soy callus was indistinguishable from *trans*-resveratrol-3-O-β-D-glucopyranoside isolated from transgenic alfalfa, suggesting that the two species accumulate the same resveratrol conjugate when expressing the peanut resveratrol synthase. The compounds share identical UV scans, HPLC retention times in multiple systems, hydrolysis by β-glucosidase to release resveratrol, and LC/MS spectra. While from this analysis alone we cannot conclusively assign the position or stereochemistry of the glucose attachment in the soy-derived product, it is likely that the compound is the 3-β-O-glucoside; the 4'-glucoside of resveratrol has not been reported in nature, and 7-β-O-glucosides of daidzein, genistein, and other isoflavonoids are abundant in these same tissues. While the β-glucosidase preparation used may contain traces of other glycosidase activities,[30] the speed and extent of the enzymatic hydrolysis suggest that this is a glucose conjugate. Several studies have identified glucose conjugates of resveratrol (also known as piceid and polydatin),[4,18,31-34] but only one study has indicated conjugation with another sugar.[35] A resveratrol xyloside has only been identified in the bark of *Holodiscus discolor*.[35] Glucose conjugation is also a common mode of detoxification of xenobiotics such as herbicides in soybean, and such a conjugation system may be capable of glycosylating resveratrol. It is interesting that no malonylation of the resveratrol glucoside was observed. The majority of isoflavonoid glucosides from soybean roots and cell cultures are malonylated,[36,37] and *p*-nitrophenol added to soybean cell cultures was recovered as the glucose-malonate derivative.[38]

We, therefore, conclude that both transgenic alfalfa and transgenic soybean cultures that constitutively express a foreign RS gene accumulate a resveratrol glucoside, not free resveratrol. The maximum levels ($\leq$ 2 μg/g fresh weight) in soybean callus were less than 10 % of that observed in alfalfa leaves, perhaps, in part, due to the high water content of the callus, weak expression from the enhanced CaMV 35S promoter, and competition for substrates with the chalcone synthase(s) involved in isoflavonoid biosynthesis.

Preliminary agar plate bioassays (performed by using protocols established for alfalfa pathogens[21]) indicate that resveratrol at 100 μg/ml in 1/10-strength V8 agar is able to inhibit the growth of *Phytophthora soja* Race 1 by 30% and Race 35 by 25%, compared to growth on agar plates containing no resveratrol. These data suggest that this same gene may be used to engineer the synthesis of resveratrol glucoside into soybean plants, possibly increasing their resistance to this common soil pathogen, especially if a promoter driving strong root expression is used.

PHYTOCHEMICAL STUDIES OF RESVERATROL SYNTHASE IN TRANSGENIC TOBACCO

Other labs have reported only very high levels of free resveratrol accumulation and visible phenotypes associated with transgenic tobacco plant studies,[8,9,11] while our lab has observed only resveratrol glucoside in alfalfa shoots[17] and soybean cell cultures (above), with no phenotypic changes. To examine these differences we have begun studies with our binary vector introduced into tobacco plants. These plants are being analyzed with our HPLC system, to determine if the discrepancies among reports are a result of differences in gene constructs and species used, or due to differences in the assay methods (HPLC versus ELISA).

The binary vector containing the peanut resveratrol synthase driven by the enhanced CaMV 35S promoter was introduced into tobacco cv. Xanthi by using *Agrobacterium tumefaciens*-mediated transformation.[39] Regenerated plants were transferred to soil in the greenhouse. Young leaves were extracted with acetone, and the extracts subjected to HPLC analysis as previously described for alfalfa. Two peaks with UV spectra similar to that of resveratrol were observed in extracts of transformed plants that were not present in control extracts (N.L. Paiva and T. Flores, unpublished) (Fig. 11.5A and B). The earlier peak co-eluted with resveratrol glucoside from transgenic alfalfa (Fig. 11.5C and D), while the later peak co-eluted with free resveratrol. These peak intensities indicate a concentration of only 1-2 μg/g fresh weight, several hundred fold lower than that reported for the similar constitutive grapevine stilbene synthase expression construct measured with an ELISA assay for resveratrol.[11] However, despite the low levels of resveratrol detected, the flowers of higher accumulating lines show the same loss of anthocyanins (white flower petals) and lack of pollen grain dehiscence as the published study, suggesting that similar levels of resveratrol are being accumulated in the two sets of plants. From these preliminary results, it appears that tobacco may indeed accumulate the resveratrol glucoside as was observed for alfalfa and soybean, and that even low levels of free or conjugated resveratrol may have negative consequences, depending upon the species. Further studies are underway to determine conclusively the identity of these peaks, to measure the levels of resveratrol and the glucoside in leaves of various ages as well as different flower parts during flower development, and to analyze quantitatively the levels of endogenous flavonoids in developing anthers by using phytochemical methods.

It is unknown how conjugation to glucose would affect the quantitation of resveratrol conjugates, in that resveratrol glucoside may show a higher or lower response compared to free resveratrol, or a zero response, with the antibody used in the ELISA assay.[8] However, the above results suggest that the ELISA may overestimate resveratrol concentrations in the presence of the glucoside.

EXPRESSION OF STILBENE SYNTHASES IN OTHER PLANT SPECIES

Subsequent studies with tomato, rice, wheat, and barley have demonstrated an increased resistance to fungal pathogens in transgenic plants harboring variations of the grapevine gene construct (wherein RS expression is driven by the pathogen-inducible grapevine promoter).[40-42] In these studies, either resveratrol was quantitated by the ELISA assay,[41] or was not assayed.[40,42] No chromatographic techniques or other assays were used to determine conclusively or independently the presence of resveratrol in these tissues.

Recently, the coding regions of three stilbene synthase genes from the grapevine species *Vitis vinifera*, *Vitis labrusca, and Vitis riparia* under the control of a CaMV 35S promoter were introduced into kiwifruit vines (*Actinidia deliciosa*).[43] By using HPLC and parallel assays, it was determined that only a resveratrol beta-glucoside accumulated in the leaves of the vines, with no free resveratrol observed in unhydrolyzed extracts. Young leaves accumulated as much as 200 μg of resveratrol glucoside/g fresh weight, although the majority of the plants accumulated 50 μg/g fresh weight or less. Despite the high levels of accumulation, the plants did not show increased resistance to *Botrytis cinerea*, the same pathogen that was sensitive to resveratrol in the early tobacco studies.[9]

NUTRACEUTICAL STUDIES USING RESVERATROL-ACCUMULATING ALFALFA

In addition to its value in detering infection by plant pathogens, resveratrol has been proposed to be a valuable nutraceutical compound. Initially, epidemiological studies suggested that consumption of large amounts of red wine can protect humans from heart disease, despite the consumption of a high fat diet, a situation described as the "French paradox".[6] Strong antioxidants, including resveratrol, are likely candidates, but years of study have not identified the active component. One complicating factor in these studies was that initial analyses focussed solely on resveratrol aglycones (*trans* and *cis*), but later analyses revealed that some wines contained high levels of polydatin (resveratrol glucoside).[44,45]

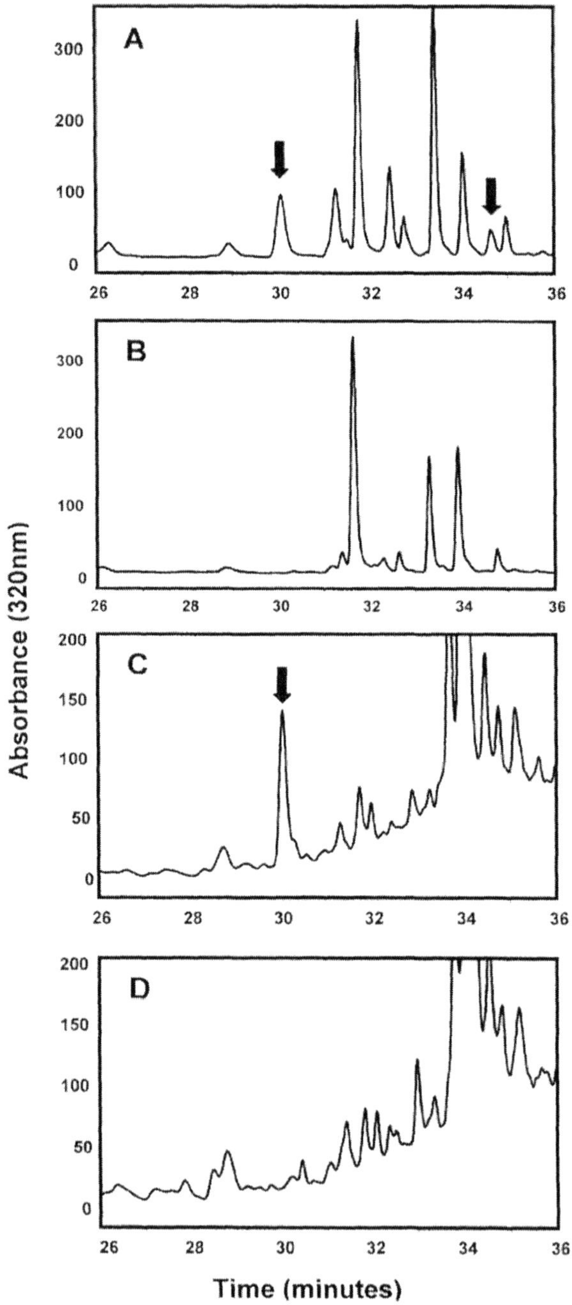

A
300
200
100
0
26
28
30
32
34
36
B
300
200
100
0
26
28
30
32
34
36
C
200
150
100
50
0
26
28
30
32
34
36
D
200
150
100
50
0
26
28
30
32
34
36
Absorbance (320nm)
Time (minutes)

Figure 11.5: HPLC chromatograms of extracts of transgenic and control tobacco and alfalfa leaves. Extracts from (A) tobacco leaf transformed via *A. tumefaciens* with the CaMV-35S:RS construct, encoding the peanut resveratrol synthase gene, (B) control (untransformed) tobacco leaf, (C) dried alfalfa leaf transformed via *A. tumefaciens* with the CaMV-35S:RS construct (low-accumulating line), and (D) dried control (untransformed) alfalfa leaf. The novel peaks with resveratrol-like UV scans present in the transgenic extracts are marked with arrows. In this extended HPLC method, resveratrol glucoside elutes at approximately 30 minutes, while free resveratrol elutes at 34.5 minutes.

More recently, resveratrol and its derivatives have also been shown to have activity as strong antioxidants and to act as suppressors of tumor formation,[5] inhibitors of tumor cell proliferation,[46] inducers of apoptosis of tumor cells,[47] phytoestrogens,[48] inhibitors of platelet aggregation,[49,50] and inhibitors of protein kinase C,[51] in addition to promoting good cardiovascular health.[6,44] Therefore, new dietary sources of resveratrol are being explored. However, little or no resveratrol accumulates in the edible portions of most crop plants. Currently, the main sources are grape-derived products such as red wine, but the levels vary greatly depending on growth, processing, and storage conditions,[6,44,45] and the cost and ethanol content may be prohibitive for the average consumer. Very low levels of resveratrol have been detected recently in peanut products,[52] but much of it is destroyed during common roasting procedures. Our studies with transgenic alfalfa and soybean tissues illustrate that constitutive expression of resveratrol synthase allows the accumulation of significant amounts of resveratrol glucoside. Theoretically, the same construct used in this study, or perhaps others with stronger leaf or seed-specific promoters, may be used to engineer the accumulation of resveratrol glucoside in leafy green vegetables or soybean seeds to allow increased human consumption of this potentially beneficial nutraceutical. Preliminary tests indicate that the glucoside is more stable in solution compared to free resveratrol, particularly when exposed to light (data not shown); presumably this is due to the glucose unit acting as a protecting group, reducing the rate of oxidation of resveratrol.

Alfalfa is a high quality animal feed[53] (Hansen, 1988), rich in high quality protein and vitamins, and is highly palatable in mouse diets, unlike the common model plants such as *Arabidopsis* and tobacco. We are currently carrying out studies to determine if diets prepared with transgenic resveratrol glucoside-accumulating alfalfa can reduce tumor rates in a mouse model system relative to diets containing wild-type alfalfa. In order to generate sufficient quantities of pesticide-free transgenic alfalfa for the mouse diet studies, plants were grown in a USDA-approved transgenic field test. Thousands of cuttings of primary transgenics and control plants were rooted in the greenhouse and transplanted to the field. High resveratrol glucoside-producing lines in the greenhouse were still high producing lines in the field (data not shown). Young leaf and stem material was harvested, washed, dried in a 55°C forage drying oven, and ground to 1 mm particles; little or no resveratrol glucoside was lost, yielding material containing 100 μg/g dry weight. This ground alfalfa was then blended with basal mouse diet ingredients.

Preliminary studies indicate that supplementation of basal diets with free resveratrol (100 μg/day) significantly reduces the numbers of aberrant crypt foci (ACF; pre-tumorous lesions) in colorectal tissues.[54] However, both alfalfa diets reduce ACF levels somewhat, with no significant difference between the diets containing dried wild-type and transgenic alfalfa (containing 100 μg resveratrol equivalents as the glucoside). This partial protection by either alfalfa diet may be due to natural phytochemicals and nutrients in alfalfa, making it more difficult to see a difference between the plant materials. The lack of difference between the diets could be because the bioavailability of the resveratrol glucoside may not be 100%. Future tests will investigate feeding higher levels or beginning diet supplementation prior to inducing tumor formation, and as well, will measure the bioavailability and uptake of resveratrol and resveratrol from the diets, as has been done for the nutraceutical soy isoflavones.[55,56]

SUMMARY

Genetic engineering of resveratrol accumulation in crop species may enhance their disease resistance as well as enhance their nutritional value, due to the antifungal and medicinal properties of resveratrol. In multiple species, expression of a resveratrol synthase gene has allowed the accumulation of resveratrol and/or resveratrol glucoside. Despite the use of similar stilbene synthase binary expression constructs, different product profiles and plant phenotypes have been reported. These may be due to differences in the endogenous glucosyl transferases, glucosidases, and/or differences in promoter strengths in different species. While it is clear that several species transformed to date have produced resveratrol, the possible derivatives that accumulate (free or glycosylated) may vary from species to species. Factors affecting the ratio of resveratrol aglycone to glucoside are being investigated, since glycosylation may affect the phytotoxicity, bioavailability, and stability of resveratrol.

Negative effects on pollination and flower color observed for tobacco have not been confirmed in other species, and resveratrol glucoside accumulation may pose no detrimental effects on other species. In the majority of the cases, an increase in resistance to at least one fungal pathogen was observed. At the same time, an increasing amount of dietary and pharmacological data suggest that increased consumption of resveratrol and its derivatives may lower the incidence of heart disease and cancer in humans. Future studies that use transgenic plants will allow researchers to evaluate directly the role of resveratrol and its derivatives in disease prevention. Thus, engineering edible crop species to accumulate resveratrol (free or glycosylated) may be beneficial to both plants and mankind.

ACKNOWLEDGEMENTS

We would like to acknowledge Dr. J. Tim Smith, Southeastern Oklahoma State University, Durant, Oklahoma, for acquiring the LC/MS data, Dr. Joachim Schröder, University of Freiburg, for providing the peanut RS cDNA and helpful discussion, Dr. Madan K. Bhattacharyya (SRNF) for providing the *P. soja* strains, Darla Boydston for assistance in preparing the chromatography figures. The transgenic alfalfa field test was conducted in Oklahoma with the assistance of Elizabeth Flatt, John Cooper and other SRNF personnel, while the animal feeding trials were conducted by Dr. Diane F. Birt, Angela Au and Dr. Jeanne W. Stewart at Iowa State University, Ames, Iowa. This research was supported by the Samuel Roberts Noble Foundation, Inc., Ardmore, OK.

REFERENCES

1. VORNAM, B., SCHON, H., KINDL, H., Control of gene expression during induction of cultured peanut cells: mRNA levels, protein synthesis and enzyme activity of stilbene synthase, *Plant Mol. Biol.*, 1988, **10**, 235-243.
2. LISWIDOWATI, MELCHIOR, F., HOHMANN, F., SCHWER, B., KINDL, H., Induction of stilbene synthase by *Botrytis cinerea* in cultured grapevine cells, *Planta*, 1991, **183**, 307-314.
3. FLIEGMANN, J., SCHRODER, G., SCHANZ, S., BRITSCH, L., SCHRODER, J., Molecular analysis of chalcone and dihydropinosylvin synthase from Scots pine (*Pinus sylvestris*), and differential regulation of these and related enzyme activities in stressed plants, *Plant Mol. Biol.*, 1992, **18**, 489-503.
4. WATERHOUSE, A.L., LAMUELA-RAVENTÓS, R.M., The occurrence of piceid, a stilbene glucoside in grape berries, *Phytochemistry*, 1994, **37**, 571-573.
5. JANG, M., CAI, L., UDEANI, G., SLOWING, K., THOMAS, C., BEECHER, C., FONG, H., FARNSWORTH, N., KINGHORN, D., MEHTA, R., MOON, R., PEZZUTO, J.M., Cancer chemopreventive activity of resveratrol, a natural product derived from grapes, *Science*, 1997, **275**, 218-220.

6. SOLEAS, G.J., DIAMANDIS, E.P., GOLDBERG, D.M., Wine as a biological fluid: history, production, and role in disease prevention, *J. Clin. Lab. Anal.*, 1997, **11**, 287-313.
7. TROPF, S., LANZ, T., RENSING, S., SCHRÖDER, J., SCHRÖDER, G., Evidence that stilbene synthases have developed from chalcone synthases several times in the course of evolution, *J. Mol. Evol.*, 1994, **38**, 610-618.
8. HAIN, R., BIESELER, B., KINDL, H., SCHRÖDER, G., STÖCKER, R., Expression of a stilbene synthase gene in *Nicotiana tobacum* results in synthesis of the phytoalexin resveratrol, *Plant Mol. Biol.*, 1990, **15**, 325-335.
9. HAIN, R., REIF, H.-J., KRAUSE, E., LANGEBARTELS, R., KINDL, H., VORNAM, B., WIESE, W., SCHMELZER, E., SCHREIER, P., STÖCKER, R., STENZEL, K., Disease resistance results form foreign phytoalexin expression in a novel plant, *Nature*, 1993, **361**, 153-156.
10. SCHRODER, G., BROWN, J.W.S., SCHRODER, J., Molecular analysis of resveratrol synthase: cDNA, genomic clones and relationship with chalconsynthase, *Eur. J. Biochem.*, 1988, **172**, 161-169.
11. FISCHER, R., BUDDE, I., HAIN, R., Stilbene synthase gene expression causes changes in flower colour and male sterility in tobacco, *Plant J.*, 1997, **11**, 489-498.
12. YLSTRA, B., TOURAEV, A., MORENO, R.M.B., STOGER, E., VAN TUNEN, A.J., VICENTE, O., MOL, J.N.M., HEBERLE-BORS, E., Flavonols stimulate development, germination, and tube growth of tobacco pollen, *Plant Physiol.*, 1992, **100**, 902-907.
13. MO, Y., NAGEL, C., TAYLOR, L.P., Biochemical complementation of chalcone synthase mutants defines a role for flavonols in functional pollen, *Proc. Natl. Acad. Sci. USA*, 1992, **89**, 7213-7217.
14. TAYLOR, L.P., Flavonols: effects on fertility and fecundity, *Crop Sci.*, 1995, **35**, 1521-1526.
15. DERKSEN, J., VAN WEZEL, R., KNUIMANN, B., YLSTRA, B., VAN TUNEN, A.J., Pollen tubes of flavonol-deficient *Petunia* show striking alterations in wall structure leading to tube disruption, *Planta*, 1999, **207**, 575-581.
16. NAPOLI, C.A., FAHY, D., WANG, H.Y., TAYLOR, L.P., White anther: a petunia mutant that abolishes pollen flavonol accumulation, induces male sterility, and is complemented by a chalcone synthase transgene, *Plant Physiol.*, 1999, **120**, 615-622.
17. HIPSKIND, J.D., PAIVA, N.L., Constitutive accumulation of a resveratrol glucoside in transgenic alfalfa increases resistance to *Phoma medicaginis*, *Molecular Plant-Microbe Interact.*, 2000, **13**, 551-62.
18. TEGUO, P.W., DECENDIT, A., VERCAUTEREN, J., DEFFIEUX, G., MÉRILLON, J-M., Trans-resveratrol-3-O-beta-glucoside (piceid) in cell suspension cultures of *Vitis vinifera*, *Phytochemistry*, 1996, **42**, 1591-1593.
19. DIXON, R.A., CHOUDHARY, A.D., DALKIN, K., EDWARDS, R., FAHRENDORF, T., GOWRI, G., HARRISON, M.J., LAMB, C.J., LOAKE, G.J., MAXWELL, C.A., ORR, J.D., PAIVA, N.L., Molecular biology of stress-induced phenylpropanoid biosynthesis in alfalfa, *in* Phenolic Metabolism in Plants (Stafford, H.A. and Ibrahim, R.K. eds.), Plenum Press, New York, 1992, pp. 91-138.

20. YLSTRA, B., MUSKENS, M., VAN TUNEN, A.J., Flavonols are not essential for fertilization in *Arabidopsis thaliana*, *Plant Mol. Biol.*, 1996, **32**, 1155-1158.
21. BLOUNT, J.W., DIXON, R.A., PAIVA, N.L., Stress responses in alfalfa (*Medicago sativa* L.) XVI. Antifungal activity of medicarpin and its biosynthetic precursors; implications for the genetic manipulation of stress metabolites, *Physiol. Mol. Plant Pathol.*, 1993, **41**, 333-349.
22. MAZAREI, M., MIAO, B., PAIVA, N., Cotransformed hairy roots of soybean: A rapid system for *in vivo* plant promoter analysis, *Phytopath.*, 1996, **86**, S93.
23. HATAMOTO, H., BOULTER, M.E., SHIRSAT, A.H., CROY, E.J., ELLIS, J.R., Recovery of morphologically normal transgenic tobacco from hairy roots co-transformed with *Agrobacterium rhizogenes* and a binary vector plasmid, *Plant Cell Rep.*, 1990, **9**, 88-92.
24. DAMIANI, F., ARCIONI, S., Transformation of *Medicago arborea* L. with an *Agrobacterium rhizogenes* binary vector carrying the hygromycin resistance gene, *Plant Cell Rep.*, 1991, **10**, 300-303.
25. SAVKA, M.A., RAVILLION, B., NOEL, G.R., FARRAND, S.K., Induction of hairy roots on cultivated soybean genotypes and their use to propagate the soybean cyst nematode, *Phytopath.*, 1990, **80**, 503-508.
26. AN, G., Development of plant promoter expression vectors and their use for analysis of differential activity of nopaline synthase promoter in transformed tobacco cells, *Plant Physiol.*, 1986, **81**, 86-91.
27. KRIEG, N.R., Enrichment and isolation, *in* Manual of Methods for General Bacteriology (P. Gerhardt, R.G.E. Murray, R.N. Costilow, E.W. Nester, W.A. Wood, N.R. Krieg and G.B. Phillips, eds.), American Society for Microbiology, Washington D C, 1981, pp. 112-142.
28. OOMMEN, A., DIXON, R., PAIVA, N.L., The elicitor-inducible alfalfa isoflavone reductase promoter confers different patters of developmental expression in homologous and heterologous transgenic plants, *Plant Cell*, 1994, **6**, 1789-1803.
29. KOSTER, J., STRACK, D., BARZ, W., High performance liquid chromatographic separation of isoflavones and structural elucidation of isoflavone 7-O-glucoside 6"-malonates from *Cicer arietinum* [Chickpeas, comparison with Leguminosae], *Planta Medica*, 1983, **48,** 131-135.
30. ESEN, A., β-Glucosidases: Biochemistry and Molecular Biology, American Chemical Society, Washington D C, 1993, 250 p.
31. HATA, K., KOZAWA, M., BABA, K., A new stilbene glucoside from Chinese crude drug "Hoshouwu" the roots of *Polygonum multiflorum* Thunb, *J. Pharm. Soc. Jap.*, 1975, **95**, 211-213.
32. NONAKA, G., MIWA, N., NISHIOKA, I., Stilbene glycoside gallates and proanthocyanidins from *Polygonum multiflorum*, *Phytochemistry*, 1982, **21**, 429-432.
33. HANAWA, F., TAHARA, S., MIZUTANI, J., Antifungal stress compounds from *Veratrum grandiflorum* leaves treated with cupric chloride, *Phytochemistry*, 1992, **31**, 3005-3007.

34. OSHIMA, Y., UENO, Y., Ampelopsins D, E, H and cis-ampelosin E, oligostilbenes from *Ampelopsis brevipedunculata* var. *hancei* roots, *Phytochemistry*, 1993, **33**, 179-182.
35. GONZALEZ-LAREDO, R. F., CHAIDEZ-GONZALEZ, J., AHMED, A. A., KARCHESY, J.J., A stilbene xyloside from *Holodiscus discolor* bark, *Phytochemistry*, 1997, **46**, 175 -176.
36. GRAHAM, T.L., Flavonoid and isoflavonoid distribution in developing soybean seedling tissues and in seed and root exudates, *Plant Physiol.*, 1991, **95**, 594-603.
37. BARZ, W., WELLE, R., Biosynthesis and metabolism of isoflavones and pterocarpan phytoalexins in chickpea, soybean, and phytopathogenic fungi, *in* Phenolic Metabolism in Plants (Stafford, H.A. and Ibrahim, R.K. eds.), Plenum Press, New York, 1992, pp. 139-164.
38. KNOPS, M., SCHUPHAN, I., SCHMIDT, B., Biotransformation of 4-nitrophenol by a fermenter grown cell suspension culture of soybean (*Glycine max*): isolation and identification of conjugates, *Plant Science*, 1995, **109**, 215-224.
39. HORSCH, R.B., JOYCE, F., HOFFMAN, N., NEIDERMEYER, J., ROGERS, S.G., FRALEY, R. T., Leaf disc transformation, *in* Plant Molecular Biology Manual 2nd Ed. (S.B. Gelvin and R.A. Schilperoort, eds.), Kluwer, Dordrecht, 1988, pp. A5: 1-9.
40. STARK-LORENZEN, P., NELKE, B., HÄNßLER, G., MÜHLBACH, H., THOMZIK, J.E., Transfer of a grapevine stilbene synthase gene to rice (*Oryza sativa* L.), *Plant Cell Rep.*, 1997, **16**, 668-673.
41. THOMZIK, J., STENZEL, K., STÖCKER, R., SCHREIER, P.H., HAIN, R., STAHL, D., Synthesis of a grapevine phytoalexin in transgenic tomatoes (*Lycopersican esculentum* Mill.) conditions resistance against Phytophthora infestans, *Physiol. Mol. Plant Path.*, 1997, **51**, 265-278.
42. LECKBAND, G., LÖRZ, H., Transformation and expression of a stilbene synthase gene of *Vitis vinifera* L. in barley and wheat for increased fungal resistance, *Theor. Appl. Genet.*, 1998, **96**, 1004-1012.
43. KOBAYASHI, S., DING, C.K., NAKAMURA, Y., NAKAJIMA, I., MATSUMOTO, R., Kiwifruits (*Actinidia deliciosa*) transformed with a *Vitis* stilbene synthase gene produce piceid (resveratrol-glucoside), *Plant Cell Rep.*, 2000, **19**, 904-910.
44. WATKINS, T.R., Wine: Nutritional and Therapeutic Benefits, American Chemical Society, Washington, DC, 1997, 296 p.
45. RIBEIRO-DE-LIMA, M.T., WAFFO-TEGUO, P., TEISSEDRE, P.L., PUJOLAS, A., VERCAUTEREN, J., CABANIS, J.C., MERILLON, J.M., Determination of stilbenes (*trans*-astringin, *cis*- and *trans*-piceid, and *cis*- and *trans*-resveratrol) in Portuguese wines, *J. Agric. Food Chem.*, 1999, **47**, 2666-2670.
46. MGBONYEBI O.P., RUSSO J., RUSSO I.H., Antiproliferative effect of synthetic resveratrol on human breast epithelial cells, *Int. J. Oncology*, 1998, **12**, 865-869.
47. CLEMENT, M.V., HIRPARA, J.L., CHAWDHURY, S.H., PERVAIZ, S., Chemopreventive agent resveratrol, a natural product derived from grapes triggers CD95 signaling-dependent apoptosis in human tumor cells, *Blood*, 1998, **92**, 996-1002.

48. GEHM, B.D., MCANDREWS, J.M., CHIEN, P.Y., JAMESON, J.L., Resveratrol, a polyphenolic compound found in grapes and wine, is an agonist for the estrogen receptor, *Proc. Natl. Acad. Sci. USA*, 1997, **94**, 14138-14143.
49. ORSINI F., PELIZZONI F., BELLINI B., MIGLIERINI G., Synthesis of biologically active polyphenolic glycosides (combretastatin and resveratrol series), *Carbohydr. Res.*, 1997, **301**, 95-109.
50. ORSINI, F., PELIZZONI, F., VEROTTA, L., ABURJAI, T., ROGERS, C.B., Isolation, synthesis, and antiplatelet aggregation activity of resveratrol 3-O- β-D-glucopyranoside and related compounds, *J. Nat. Prod.*, 1997, **60**, 1082-1087.
51. JAYATILAKE, G.S., JAYASURIYA, H., LEE, E.S., KOONCHANOK, N.M., GEAHLEN, R.L., ASHENDEL, C.L., MCLAUGHLIN, J.L., CHANG, C.J., Kinase inhibitors from *Polygonum cuspidatum*, *J. Nat. Prod.*, 1993, **56**, 1805-1810.
52. SOBOLEV, V.S., COLE, R.J., trans-Resveratrol content in commercial peanuts and peanut products, *J. Agric. Food Chem.*, 1999, **47**, 1435-1439.
53. HANSON, A.A., BARNES, D. K., HILL, R.R., Alfalfa and Alfalfa Improvement (Monograph no. 29), American Society of Agronomy, Madison, 1988, 1084 p.
54. AU, A., STEWART, J.W., PAIVA, N.L., BIRT, D.F., Reduced formation of azoxymethane induced aberrant crypt foci in mice fed purified resveratrol aglycone but not resveratrol glucoside in transgenic alfalfa., *Proc. Annu. Meet. Am. Assoc. Cancer Res.*, 2000, **41**, A5371.
55. XU, X., HARRIS, K.S., WANG, H.J., MURPHY, P.A., HENDRICH, S., Bioavailability of soybean isoflavones depends upon gut microflora in women, *J. Nutr.*, 1995, **125**, 2307-2315.
56. XU, X., WANG, H.J., MURPHY, P.A., COOK, L., HENDRICH, S., Daidzein is a more bioavailable soymilk isoflavone than is genistein in adult women, *J. Nutr.*, 1994, **124**, 825-832.

Chapter Twelve

CROP ALLELOPATHY: ENHANCEMENT THROUGH BIOTECHNOLOGY

Brian E. Scheffler, Stephen O. Duke, Franck E. Dayan, and Eiji Ota[1]

USDA, ARS,
Natural Products Utilization Research Unit,
University of Mississippi,
University, MS 38677, USA

[1]*Kubota Corporation, Ryugasaki, Japan*

INTRODUCTION

Weed management in agriculture has become increasingly dependent on synthetic herbicides due to many factors, including an increased emphasis on reduced tillage cultural practices, the introduction of herbicide-resistant crops, and the availability of a complete arsenal of chemical weed killers. The same is not true for dependence on insecticides, nematocides, and fungicides, as crops can be made resistant to these pests by conventional breeding or use of pest resistance transgenes.

At the same time, current concerns over the potential environmental and toxicological impacts of pesticides have led many developed nations to re-assess their pesticide regulatory policies. In the United States, the Food Quality Protection Act (FQPA) of 1996 has revolutionized the way pesticides are registered. As a result, many commercially available pesticides will be removed from the market, and greater than 50% of herbicides registered for use in minor crops in the U.S. will be lost according to the IR-4 program.[1] For the same reason, the increased cost associated with pesticide re-registration and review will have a negative impact on the development of new and environmentally friendly pest control products.

Novel approaches to weed control, including biocontrol and the use of phytotoxic natural products, have not yet been developed to replace or complement the impending void. Attempts have been made to breed crops that are more competitive with weeds through allelopathy, but the results have not been commercially acceptable. Clearly, the work with cucumbers, barley, rice, and sorghum demonstrates that germplasm for allelochemical production exists within some crop species, however, the level of allelopathic activity is not sufficient to provide adequate weed management in the field.[2-5] In the case of rice, herbicide use can be reduced with allelopathic varieties, but these varieties are not yet commercially available.[6] Transgene technology has been used to make crops resistant to insects, pathogens, and herbicides, but has not been used to make crops repress growth of weeds by either increasing their competitive nature and/or increasing the production of allelochemicals. In this chapter, we discuss genetic engineering strategies that might be used to impart or enhance allelopathic traits in crops. Success with such an endeavor in major crops could drastically reduce pesticide use in agriculture. This review is tempered by a discussion of anticipated difficulties and environmental risks.

STRATEGIES

Enhancement of the Production of Existing Allelochemicals

Imparting or enhancing allelopathic traits to crops represents a major challenge to molecular biology, as the steps from gene identification to isolation and the eventual transfer into other species are complicated. In most instances, the allelopathic potential associated with particular weeds is based on observation in the field, but the presence of allelopathic compound(s) or their identifications are not known. In the few cases where phytotoxins have been identified, there is limited knowledge about their biosynthetic pathways. The lack of basic information at the biochemical level has also hindered studies that use traditional genetic means or the identification of specific mutations affecting these biosynthetic routes. Therefore, there are limited avenues available for a molecular biologist to clone the genes required to introduce these pathways in target crops. Even the multinational plant genomic programs may not shed light on such interesting traits, as each occurrence of allelopathy is probably unique to a given species or closely related species, and in most instances the species will not have been the focus of intensive molecular biology investigations. In many respects, the elucidation of the genes responsible for the synthesis of allelochemicals is a challenging task for the molecular biologist.

Elucidation of the biosynthetic pathways associated with the production of an allelochemical is vital to characterizing genes. Access to the biochemical information opens one avenue to isolate the target genes by traditional methods for cloning a gene through enzyme purification. The protein can be used as an antigen to raise antibodies or to sequence the protein. Antibodies can be used to screen a cDNA expression library for a clone expressing the protein of interest. Furthermore, the protein sequence can be used to design degenerative oligonucleotides that can be used in PCR reactions to amplify the desired segment of DNA. The major advantage of such approaches is that a gene is isolated by using a protein with a function known to be associated with the pathway of interest. However, the biochemical studies leading to the identification of enzymes of the secondary metabolism are quite laborious due to the small amount of enzyme normally present in the tissues. These approaches are being used by some laboratories, including ours, for isolating genes involved in the synthesis of allelochemicals, but the substantial research effort required may have a relatively small pay-off if no other approaches are used concomitantly.

More importantly, biochemical studies provide the main key to the identification of a gene's function. Although almost every gene of an organism can be cloned by using standard techniques, the clones are of limited value until their functions have been discovered. The putative function of a cloned gene can be assigned by comparison of its DNA sequence to DNA/protein databases. This is particularly useful if the function corresponds to a step expected to occur in the biochemical pathway. Therefore, a large effort must be dedicated to the isolation of genes.

One of the most successful methods used to isolate genes is through mutagenesis by using transposable elements. Transposable elements are mobile segments of DNA that can move within the genome. When they insert into an active gene, they can disrupt normal gene function, thus creating an allele with a mutant phenotype. In most cases, these alleles would exhibit a recessive phenotype. In the case of the modified systems from maize, the transposable elements have been altered to allow for maximum detection of mutagenic events.[7] Unfortunately, one must have a method to screen for a mutation in the biosynthetic pathway of the allelochemical, and this must be performed on thousands of individual plants. It might be possible to use this technique to isolate allelopathic genes in rice and sorghum, but there are several reasons why it will probably not be applied to isolating genes associated with allelopathy in other species. In order to use transposon tagging it is necessary to have an active and characterized transposable element system. The most efficient way to do that is to introduce a modified system from *Zea mays* into the desired plant *via* transformation.[7] The Ac/Ds transposable element system from maize has already been transformed into rice, and this will probably be achieved in sorghum.[8,9] Unfortunately, many other species that exhibit allelopathy are not the subjects of major molecular biology research programs and will not receive the necessary input of resources to achieve this goal in the near future.

Differential screening techniques can be used to generate a pool of candidate clones for testing to see if they are related to any step in the production of the allelochemical. There are numerous procedures that use differential screening as their selection criteria, but the basis for all of them is mRNA isolated from tissue that is and is not expressing the allelochemical. Comparisons are then made between the two pools to determine what mRNAs are unique to the tissue expressing the allelochemical. These can then be cloned and characterized. Two of the common techniques are differential display (DD) and subtractive hybridization (SH).

The "general concept" for DD is that mRNA is converted to double stranded DNA, which is called cDNA.[10] The two pools of cDNAs are used in PCR experiments to amplify the DNA. During this process a radioactive or other detectable marker is incorporated into the DNA. The PCR samples are then separated by acrylamide-gel-electrophoresis conditions that are sensitive enough to detect size differences of a single base pair between PCR bands. Each band would correspond to mRNA produced from a gene. The intensity of a band is a relative, but not 100% accurate, indication of the amount of transcript produced by a gene. The intensity and size of the bands are compared between the two mRNA pools. In most instances, there is no detectable difference between the two pools. The working hypothesis is that the rare differences represent variances in gene expression. The bands of interest are then cloned and sequenced, and this information is used in further experiments to help characterize their corresponding genes using standard molecular biology techniques. There are numerous variations of conducting DD, but they basically follow this procedure.

In comparison, for SH there are numerous procedures that are so different from each other that it is difficult to give a detailed description that fits all of these techniques. The concept behind SH is that one mRNA pool is subtracted from the other pool. In most instances, a given gene will be expressed at a similar level in either mRNA pool and will be removed due to the subtraction procedure. The remainder of the mRNA pool is then cloned and characterized. Therefore, a mRNA pool derived from tissue expressing an allelochemical is subtracted using a pool that is not expressing the compound. In theory, the remainder represents only those genes associated with the production of the allelochemical. Naturally, this is only true if this is the only difference between the two tissue sources of mRNA. It should be noted that it might be wise to do the subtraction procedure twice, once in each direction. Genes that are associated with structural steps in a biochemical pathway will generally be expressed in the tissue that contains the allelochemical. On the other hand, regulatory loci for the pathway (activators and suppressors) may be expressed in either tissue source.[11]

Neither DD or SH are perfect techniques, but DD has many shortcomings, and the artifacts generated by this method have misled many researchers.[12,13] SH requires more skill, but there are commercial kits available that greatly enhance this procedure and make it slightly easier to perform. No matter what procedure is used, many problems can be avoided by testing any isolated clones to see if they are truly differentially expressed.

The major problem with differential screening is that the two appropriate tissues must be identified. As most allelochemicals will be expressed in the root tissue, this is an obvious choice, but the non-expressing tissue represents a bigger challenge. Potential tissues must be tested to assure the allelochemical is not produced. Another critical requirement is that the non-expressing tissue should be similar to the tissues expressing the allelochemical. Obviously, roots and leaves would have similar housekeeping genes being expressed for general cellular functions, but there will also be a large number of differences that will be difficult to sort through to determine which specific genes are associated with the allelochemcial. Furthermore, the root represents a special problem for differential screening, as it is a unique plant tissue. Nonetheless, through careful physical and biochemical examination of allelochemical production, it may be possible to identify differences within root tissues that may be used as sources of expressing and non-expressing mRNA. Differences may be found at various developmental stages, within diverse root types, or within substructures of the root.

The alternative would be to use genetic or environmental conditions where the allelochemical is expressed in one plant and not the other, and compare the mRNA of the appropriate tissue from both plants. In this instance, it would be best to use inbred lines to reduce genetic differences between the two plants.

Whether using transposon-tagging, differential screening, or related techniques, there is always the problem of proving that a particular clone truly represents a key step in the biochemical pathway of an allelochemical. As previously mentioned, comparison to DNA and protein databases may provide useful leads but this is not definitive proof. Proving a clone is associated with a specific step of a biochemical pathway can be approached by many methods. The ideal method is through direct proof. This can be accomplished by expressing the clone, isolating the produced protein, and demonstrating its activity on precursor(s) of the allelochemical. This is only feasible if the clone represents a structural and not a regulatory gene, and if it can be expressed in active form.

If independent mutations exist for a given locus (such as those generated by transposon tagging or mutagenesis with EMS), the DNA corresponding to the clone in question can be isolated from each mutant and be compared. If the clone represents the same gene that was mutated, then base pair alterations will be detected in the DNA sequence isolated from the mutant, and the difference should represent something that would alter its expression or cause a frameshift mutation. A variation of this approach would be to make an anti-sense or over-expression construct of the clone in question and transform it back into the plant. When a plant is transformed with a functional anti-sense or over-expression construct, the expression of the corresponding gene is altered. With anti-sense constructs, the targeted gene is down regulated, while over-expression constructs can result in up or down (known as co-suppression) regulation.[14] In either case, the production of the allelochemical should be directly affected.

DIMBOA

The dissection of the DIMBOA (2,4-dihydroxy-7-methoxy-1,4-benzoxazin-3-one) pathway (Fig. 12.1) in *Zea mays* best represents how genetic, molecular biological, and biochemical techniques can be applied in combination to isolate the various genes involved in a specific pathway.[15,16] There is significant evidence that DIMBOA is a phytochemical and possibly an allelochemical.[17] In grasses, the compounds DIBOA and DIMBOA also form part of the defense system against insect and microbial pathogens. DIMBOA helps confer resistance to European corn borer (*Ostrinia nubilas*), northern corn leaf blight (*Helminthosporium turcicum*), stalk rot (*Diplodia maydis*), and interestingly, the herbicide atrazine.[18] Mendelian genetics was used to isolate a mutation at the Bx1 gene. The resulting homozygous recessive plants (*bx1/bx1*) did not produce DIMBOA. Plants, homozygous recessive for *bx1*, were then used in a directed transposon-tagging program with the *Mutator* transposable element. Isolated mutations were then screened by using genetic-segregation analysis. The screening process looks for the co-segregation of the mutated locus and the *Mutator* transposable element. With this approach, it was possible to isolate the *Bx1* locus.[16] The isolated gene had DNA homology to tryptophan synthase alpha (TSA), which catalyzes the conversion of indole-3-glycerol phosphate to indole (Fig. 12.1). Through genetic complementation tests and biochemical analysis, it was possible to show that the *Bx1* locus is responsible for the production of indole in the DIMBOA pathway.[16]

Subtractive screening of high versus low DIMBOA producing plants resulted in the isolation of a cDNA clone encoding a P-450-dependent monooxygenase. Through DNA homology, three other P-450-dependent monooxygenase loci were cloned. The four loci (*Bx2-5*) were expressed in a yeast expression system. The expressed proteins were then tested on the various intermediates of DIMBOA to determine which gene was responsible for each biochemical step.[16]

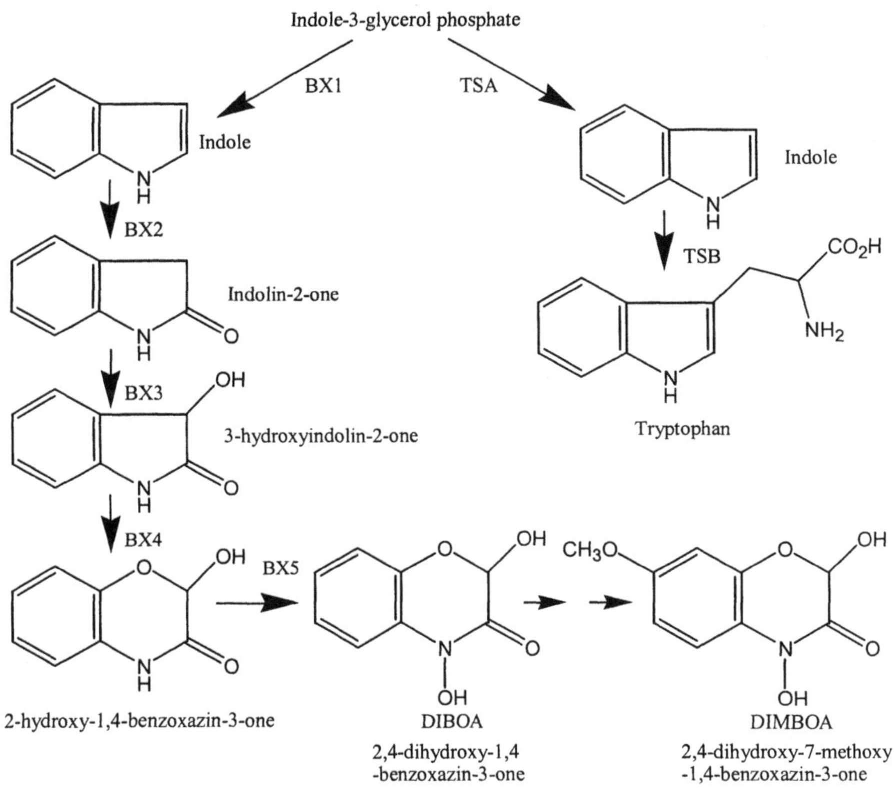

Figure 12.1: DIMBOA and tryptophan biochemical pathways of *Zea mays*.

Producing Transgenic Plants

While the isolation of genes associated with an allelopathic trait is difficult, it does not represent the final and perhaps most tricky hurdle. The ultimate goal of such a project is to increase the production of an allelochemical or to introduce the trait into an agronomic crop. In most instances, it will not be possible to transfer directly an allelopathic biosynthetic pathway to another organism simply by direct transformation, as the genes would often be cloned as cDNAs. This means it will be necessary to make DNA constructs where the cDNA is under the control of a tissue-specific promoter. For safety to the plant and for effective expression, it would be desirable to express the genes solely in the roots or root hairs. Such promoters are not readily available in the public sector.

Through the determination of a biochemical pathway and additional biochemical analysis, it is possible to identify which enzymatic step is rate limiting. If it were possible to remove this limitation, one would expect more product to be produced. This approach is appealing for crops known to be allelopathic, such as rice and sorghum. The simplest way to do this is to change the promoter of the rate-limiting gene so that it is expressed at a slightly increased level, thus avoiding the phenomenon of co-suppression associated with high levels of expression.[19] In reality, the overall gain in allelochemical output would be limited in most cases, as a different step in the pathway will become rate limiting. Therefore, gains will be measured in small percentage points and not leaps and bounds. Fortunately, in the case of allelopathy, it is probably more desirable to increase slightly the production of an allelopathic compound than to have massive amounts being produced per gram of tissue. An alternative approach to increasing output would be to change the expression of a regulatory gene instead of the structural genes. If there is a regulatory gene that controls several steps in a biosynthetic pathway, an enhancement in its expression level can cause the output from the controlled genes also to increase and, therefore, there will be a boost in the final product.[20,21]

Sorgoleone: A Test Case

The use of sorghum (*Sorghum bicolor*) and its ability to reduce weed growth *via* allelopathic interactions is one of the most compelling examples of the potential for enhancing crop allelopathy through biotechnology. Sorghum is often selected as a summer annual cover crop because of its rapid growth and ability to suppress weeds.[22] Sorghum *(Sorghum bicolor*) cover crops are known to exude oily droplets from living root hairs that are composed of the allelopathic compound sorgoleone (Fig. 12.2) and many other minor constituents that are potent inhibitors of photosynthesis, respiration, and plastoquinone synthesis in higher plant systems (Meazza *et al*., unpublished data).[23,24] Root exudates of sorghum contain between

85-90% pure sorgoleone, and the concentration of this natural product in soils in which sorghum is grown can reach 10^{-4} to 10^{-5} M.[25]

Sorgoleone is a potent phytotoxin that represses the growth of large crabgrass seedlings (*Digitaria sanguinalis*) with a GR_{50} of 10 μM for shoot and root growth.[26] Inhibition of shoot and root growth of velvetleaf (*Abutilon theophrasti*) and barnyardgrass (*Echinochloa crus-galli*) also have been observed at higher concentrations ranging from 10 to 200 μM, but ivyleaf morningglory shoots and roots (*Ipomoea hederaceae*) are tolerant to this natural product.

Inhibition of photosynthetic electron transport in isolated spinach (*Spinacea oleracea*) thylakoids as well as competitive binding to spinach, pigweed (*Amaranthus retroflexus*), and potato (*Solanum tuberosum*) thylakoids have been investigated for sorgoleone in comparison to other PSII inhibitors.[26] Sorgoleone inhibits photosynthetic electron transport more than diuron (DCMU), with an I_{50} of 0.09 μM compared to 0.12 for diuron, by competing for the same Q_B binding site of the D_1 protein.

The quinone portion of sorgoleone is probably derived from pathways associated with *p*-benzoquinone natural products, such as the homogentisate pathway or the polyketide pathway.[27,28] Fate *et al.* demonstrated that a small amount of ^{13}C-acetate could be incorporated in the quinone head of sorgoleone.[28] However, the incorporation was less than 1% of the total labeled acetate applied. Protocols that use *in vitro* tissue cultures (of embryo or excised roots) that will enable better feeding experiments are being developed according to protocols already existing for the production of plant pharmaceuticals, and more specifically for the production of natural quinones in roots.[29-32] The other approach taken in our laboratory is the use of differential display as described above.

Ultimately, one would like to identify the biosynthetic pathway at the biochemical level, isolate the genes responsible, and introduce part of the machinery into a related crop, such as corn, to obtain a new cultivar that requires lower input of synthetic herbicides.

CAVEATS

Autotoxicity

In the case of allelopathy, a plant is producing a potentially potent toxin that acts as a natural herbicide, and therefore, autotoxicity is a major concern. Autotoxicity occurs when a compound is toxic to the organism that produces it. An organism must have mechanisms to prevent poisoning itself and, in some instances, other plants within its species when it produces an allelochemical. Autotoxicity can be avoided through sequestration, secretion, resistance at the target site, or metabolic

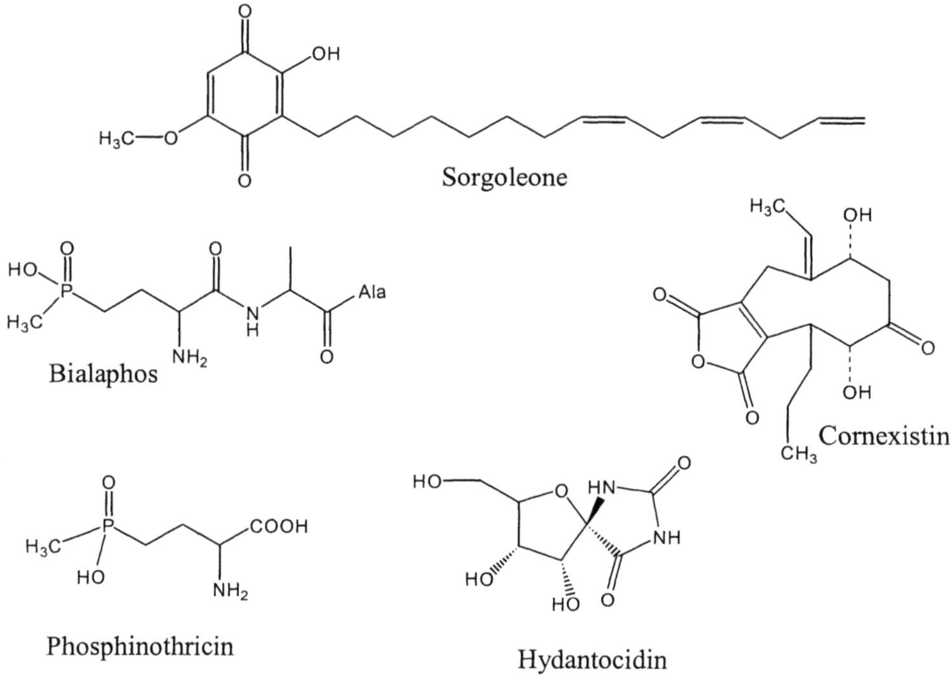

Figure 12.2: Structures of the natural products mentioned in the text.

inactivation. Many phytotoxic compounds, although most are not allelopathic, are sequestered from biological processes by placement within the vacuole and/or in organs like glandular trichomes. In living plants, most allelopathic compounds involved in plant-plant interactions are probably secreted into the environment through the roots, and although they are no longer within the plant, they can still pose a threat if they can be reabsorbed. Therefore, secretion without an additional level of defense may be of limited use. A simple method to avoid autotoxicity is for the plant to have resistance at the molecular target site. This may be found in several different forms ranging from a modified binding site all the way to total absence of the molecular target. We are not aware of any examples of this. However, the molecular target site of most allelochemicals is unknown. The most elegant forms of autotoxicity avoidance are through metabolic inactivation. Of these, the simplest is modification of the phytotoxin through degradation or chemical modification, such as oxidation with P-450 monooxygenase enzymes. A more sophisticated system is when the plant produces the phytotoxin in an inactive form called a prophytotoxin, which becomes an active phytotoxin only after modification. This is common with phytotoxic natural compounds produced by microbes. For example, the prophytotoxins cornexistin, hydantocidin, and bialaphos (Fig. 12.2) are all converted to phytotoxins in the plant.[33-35]

Autotoxicity avoidance is a major concern when developing a new allelopathic plant through genetic engineering. The absence of the molecular target site is the most desirable situation, while sequestration is probably not a viable option, as the goal in most projects would be to have the allelopathic compound secreted into the environment. As mentioned, secretion by itself is insufficient if the plant can reabsorb the compound. Unless the plant already has natural resistance to the phytotoxin, it might be difficult to engineer resistance by using metabolic inactivation. In order to genetically engineer this trait into a plant, a metabolic inactivation mechanism must be discovered, characterized, the corresponding gene(s) cloned, and then transferred in a desirable form into the plant. Similar limitations would be encountered if the goal were to confer resistance at the molecular target site.

Engineering a plant to produce a natural phytotoxin to which it has total resistance is not beyond technical limitations. For example, many crops have been altered transgenically with the bar or pat gene to be resistant to the natural phytotoxin phosphinothricin (Fig. 12.2).[34] This has been done to render the crop resistant to the synthetic version of phosphinothricin, glufosinate, or in order to use glufosinate as a selection agent for successful transformations. The genes for the bialaphos biosynthesis are available.[36] Bialaphos is a protoxin, which is converted to phosphinothricin in the target weed. It may be possible to introduce these genes into roots of resistant crops, so that the plant exudes this potent proherbicide into the rhizophere. This herbicide does not have a high level of soil activity, but the crop would be producing it on a continuous basis.

Autotoxicity avoidance may also be a problem when the goal is to enhance the production of an existing allelopathic compound. Although the plant would have an inherent mechanism for such avoidance, the level of protection might be insufficient when the quantity of the allelochemical is increased. This situation might be analogous to that with commercial selective herbicides with which a low level of phytotoxicity is tolerated. In such cases, increases in dosage above the tolerated levels can cause unacceptable phytotoxicity.

The Metabolic Cost of Secondary Products

Another potential problem in producing a highly allelopathic crop is that the metabolic cost associated with the production of the allelochemical may reduce yield. In a detailed review, Gershenzon concludes that the metabolic cost of most secondary compounds is higher than that of most compounds found in primary metabolism.[37] It follows that this expenditure of resources toward bioactive secondary metabolites should provide a significant survival benefit to the plant in a natural ecosystem. There are several reasons for this, including the extensive chemical reduction of many of the compounds, the lack of shared enzymes in the biosynthetic pathways, and the costs of compartmentalizing and storing the compounds. Whether this tradeoff will be acceptable in an agroecosystem in which yield, rather than propagation, is the primary concern will probably depend on the species and the allelochemical. Certainly, the higher the unit activity of the allelochemical the less concern there would be with this resource-partitioning dilemma. Metabolic imbalance is always a concern when producing transgenic plants. The redirection of plant resources can deregulate the physiological equilibrium of the cells and cause shortages of needed compounds and/or lead to the production of toxic byproducts.[38]

Evolution of Resistance to Allelochemicals

Just as resistance to some synthetic herbicides has evolved, resistance to some allelochemicals produced by crops also will evolve. We are unaware of any proven examples of evolved resistance to an allelochemical. However, a recent paper suggests that diffuse knapweed is much more allelopathic toward plant species of North America than to those of its geographical origin because plants from its native habitat have evolved in response to its allelochemicals.[39] The speed with which it will happen is dependent on a number of factors. In general, the stronger the selection pressure, the more likely resistance is to evolve. The fact that the selection pressure invoked by the allelochemical would be continuous and widespread, rather than transient and localized, increases the odds of resistance evolving. However, the relatively low level of effect of most known allelochemicals in field situations would decrease the chances of evolution of resistance. Another strong factor that influences evolution of resistance is the molecular target site of the allelochemical. We know from the numerous cases of emerging resistance to synthetic herbicides that evolution of resistance is much more rapid with certain molecular target sites (*e.g.*, acetolactate synthase, ACCase) than to those with other sites of action (*e.g.*, EPSP synthase, protoporphyrinogen oxidase).[40] Evolution of resistance to a phytotoxin with more than one molecular target site is likely to be slow. We know little about molecular sites of allelochemicals and nothing about resistance that might have evolved in natural ecosystems. Some estimate of the probability of the evolution of resistance can be obtained by treating a large population of mutagenized seeds with the lowest concentration of the toxin required for 100% mortality to see the frequency of resistance. This has been done with synthetic herbicides, but not with natural phytotoxins.[41,42] We expect there would be little or no difference in results between natural and synthetic phytotoxins.

Transgenic Risk Assessment

In the production of transgenic plants, there are always two major concerns that must be addressed before they can be released into the environment as agricultural crops. The first is the direct risk to humans, mainly in the form of potential allergens, and the second is an alteration of the environment. In the latter case, the main concern is with horizontal gene transfer of the transgene into related wild species, thus giving the recipient plant a competitive edge over other native plants.[43]

The enhancement of allelopathic traits in known edible crops, such as sorghum and rice, or their transfer to other crops would probably pose little risk to human health, as they are already widely consumed cereals, and there are no known adverse effects to human consumption. In addition, crops properly engineered to produce an allelochemical pose little health risk because the compound is excreted to the rhizosphere and would not be present in the edible portion of the plant as long as root crops are not used.

Horizontal gene transfer of engineered traits in transgenic crops to wild species poses little risk to the environment, because few of the traits would give the wild species any foreseeable competitive advantage over native plants. In comparison, transfer of an allelopathic trait could give the recipient plant some competitive advantage. Nonetheless, there are numerous reasons why there is little environmental threat from developing this technology. 1) There are several wild plant species that are known to be allelopathic, and there is no indication that these plants have led to the extinction of other plant species. 2) For most major crops in the USA, there are no wild weed species present for cross-pollination to occur, thus, horizontal transfer is not likely. 3) Most natural forms of allelopathy are probably multigenic traits. Therefore, to genetically modify a crop to produce an allelopathic compound would require the transfer of many genes, which, in most instances, would indicate that the transgenes are located at different locations within the genome. The probability of all these genes being simultaneously incorporated in a wild species genome is infinitesimal. 4) Even if horizontal gene transfer did occur, the recipient plant would also have to deal with the problem of autotoxicity avoidance.

If a crop could be made allelopathic with transgenes, the probability and impact of horizontal gene transfer would need to be evaluated. Whether horizontal gene flow to a weedy relative occurs at all is a first critical question. Many crops will interbreed with one or more related species in some part of the world.[44] For some crop/weed relative pairs, estimation of whether introgression will occur is difficult. Extrapolation of laboratory introgression experiments to the field is often unrealistic.[45,46] In the field, introgression can be safely determined by growing the crop and weedy relative together monitoring the transfer of a gene conferring resistance to an herbicide or phytotoxin that will not be used with the crop. Introgression will be readily detected by testing the seed of the weedy relative with the herbicide or phytotoxin. Determination of the potential risk of introgressed allelopathy-conferring or -enhancing transgenes is more difficult than determining whether introgression will occur. The value of the transgene(s) in an ecosystem will be dependent on both the physical and biotic environment.

If a significant risk of horizontal gene flow is identified, physical or biological barriers to gene flow should be utilized to prevent such gene flow. Numerous methods to prevent or eliminate pollen flow should be considered.[47] Genetic barriers or transgene mitigating strategies can be used. These include: using the transgene only in sterile hybrids, using the transgene in the plastid genome in species that do not transmit plastome genes via pollen, flanking the gene with a transgenic mitigation gene or with transgene constructs that are unlikely to have significant environmental impact.[45,48,49]

CONCLUSION

The development of allelopathy in crops represents a major challenge to the agriculture and biotechnology industries. It is a value-added trait that is difficult to measure in monetary terms and will probably never totally replace other weed management practices, as management and costs are dependent on the crop, predominant types of weeds, and environmental conditions. In fact, farmers in developed countries might continue to use herbicides to some degree even if a crop contained an allelopathic trait. The use of genetic engineering to develop allelopathic crops will require substantial investment of time and money. Ultimately, the benefits of this investment would be substantially greater than the cost.

Whether it is the use of manual labor or herbicides, weed management practices are an absolute requirement in crop production. While insect or plant pathogen infestations may vary in intensity from year to year, the absence of weed control results in drastic reduction in yield every year. In 1997, 70% of all pesticides used in the United States were herbicides, and even with the advent of the development of the environmentally friendly herbicide glyphosate, additional reduction in herbicide applications are financially and environmentally desirable.[50]

Allelopathic crops would reduce herbicide usage in the United States, resulting in financial gains for the farmer and would also address environmental concerns. In developing countries, the gains might be more substantial. Some developing countries can afford to use various forms of herbicides, but these are often environmentally unfriendly and have already been banned for use in developed countries. More importantly, a reduction in the backbreaking work of weeding would result in economic and social gains. In some countries, weeding is the responsibility of the women and children; individuals who can reduce their investment of time and energy in one area can redirect those resources to other important endeavors such as education!

REFERENCES

1. Interregional Research Project #4 WWW page: <http://aesop.rutgers.edu/~ir4/>
2. PUTNAM, A.R., DUKE, W.B., Biological suppression of weeds: evidence for allelopathy in accessions of cucumber, *Science*, 1974, **185**, 370-372.
3. LOVETT, J.V., HOULT, A.H.C., Allelopathy and self-defense in barley, *in*: Allelopathy: Organisms, Processes, and Applications, Amer. Chem. Soc., Washington, D.C., 1995, pp. 170-183.
4. OLOFSDOTTER, M., WANG, D., NAVAREZ, D., Allelopathic rice for weed control, *in* Recent Advances in Allelopathy, Vol. I. A Science for the Future (F.A. Macias, J.C.G. Galindo, J.M.G. Molinillo, and H.G. Cutler eds.), University of Cadiz, Spain, 1999, pp. 383-390.
5. NIMBAL, C.I., YERKES, C.N., WESTON, L.A., WELLER, S.C., Herbicidal activity and site of action of the natural product sorgoleone, *Pestic. Biochem. Physiol.*, 1996, **54**, 73-83.
6. CHAVEZ, R.S.C., GEALY, D.R., BLACK, H.L., Reduced propanil rates and naturally suppressive cultivars for barnyardgrass control in drill-seeded rice, *in* Arkansas Agricultural Experiment Station, B. R. Wells Rice Research Studies-1998, (R.J. Norman and T.H. Johnson, eds.), Arkansas, 1999, pp. 43-50.
7. BANCROFT, I., BHATT, A.M., SJODIN, C., SCOFIELD, S., JONES, J.D.G., DEAN, C., Development of an efficient two-element transposon tagging system in *Arabidopsis thaliana*, *Mol. Gen. Genet.*, 1992, **233**, 449-461.
8. IZAWA, T., OHNISHIK, T., NAKANO, T., ISHIDA, N., ENOKI, H., HASHIMOTO, H., ITOH, K., TERADA, R., WU, C., MIYAZAKI, C., Transposon tagging in rice, *Plant Mol. Biol.*, 1997, **35**, 219-229.
9. SHIMAMOTO, K., MIYAZAKI, C., HASHIMOTO, H., IZAWA, T., ITOH, K., TERADA, R., INAGAKI, Y., IIDA, S., Trans-activation and stable integration of the maize transposable element Ds cotransfected with the Ac transposase gene in transgenic rice plants, *Mol. Gen. Genet.*, 1993, **239**, 354-360.
10. ROUX, C., BILANG, J., THEUNISSEN, B.H, PERROT-RECHENMANN, C., Identification of new early auxin markers in tobacco by mRNA differential display, *Plant Mol Biol.*, 1998, **37**, 385-389.
11. BURR, F.A., BURR, B., SCHEFFLER, B.E., BLEWITT, M., WIENAND, U., MATZ E.C., The maize repressor-like gene intensifier1 shares homology with the r1/b1 multigene family of transcription factors and exhibits missplicing, *Plant Cell*, 1996, **8**, 1249-1259.
12. SOMPAYRAC, L., JANE, S., BURN, T.C., TENEN, D.G., DANNA, K.J., Overcoming limitations of the mRNA differential display technique, *Nuc. Acids Res.*, 1995, **23**, 4738-4739.

13. VON DER KAMMER, H., ALBRECHT, C., MAYHAUS, M., HOFFMANN, B., STANKE, G., NITSCH, R.M., Identification of genes regulated by muscarinic acetylcholine receptors: Application of an improved and statistically comprehensive mRNA differential display technique, *Nuc. Acids Res.*, 1999, **27**, 2211-2218.
14. JORGENSEN, R.A., CLUSTER, P.D., ENGLISH, J., QUE, Q., NAPOLI, C.A., Chalcone synthase cosuppression phenotypes in petunia flowers: comparison of sense vs. antisense constructs and single-copy vs. complex T-DNA sequences, *Plant Mol. Bio.*, 1996, **31**, 957-973.
15. FREY, M., KLIEM, R., SAEDLER, H., GIERL, A., Expression of a cytochrome P450 gene family in maize, *Mol. Gen. Genet.*, 1995, **246**, 100-109.
16. FREY, M., CHOMET, P., GLAWISCHNIG, E., STETTNER, C., GRUN, S., WINKLMAIR, A., EISENREICH, W., BACHER, A., MEELEY, R.B., BRIGGS, S.P., Analysis of a chemical plant defense mechanism in grasses, *Science*, 1997, **277**, 696-699.
17. WU, H., PRATLEY, J., LEMERLE, D., HAIG, T., Crop cultivars with allelopathic capability, *Weed Res.*, 1999, **39**, 171-180.
18. NIEMEYER, H.M., Hydroxamic acids (4-hydroxy-1,4-benzoxazin-3-ones), defense chemicals in the Gramineae, *Phytochemistry*, 1988, **27**, 3349-3358.
19. QUE, Q., WANG, H.Y., ENGLISH, J.J., JORGENSEN, R.A., The frequency and degree of cosuppression by sense chalcone synthase transgenes are dependent on transgene promoter strength and are reduced by premature nonsense codons in the transgene coding sequence, *Plant Cell*, 1997, **9**, 1357-1368.
20. BRADLEY, J.M., DAVIES, K.M., DEROLES, S.C., BLOOR, S.J., LEWIS, D.H., The maize Lc regulatory gene up-regulates the flavonoid biosynthetic pathway of Petunia, *Plant J.: for Cell and Mol. Bio.*, 1998, **13**, 381-392.
21. SCHEFFLER, B., FRANKEN, P., SCHUTT, E., SCHRELL, A., SAEDLER, H., WIENAND, U., Molecular analysis of C1 alleles in *Zea mays* defines regions involved in the expression of this regulatory gene, *Mol. Gen. Genet.*, 1994, **242**, 40-48.
22. FORNEY, D.R., FOY, C.L., Phytotoxicity of products from rhizospheres of a sorghum-sudangrass hybrid (*S. bicolor* x *S. sudanese*), *Weed Sci.*, 1985, **33**, 597-604.
23. GONZALEZ, V., NIMBAL, C.I., WESTON, L.A., CHENIAE, G.M., Inhibition of a photosystem II electron transfer reaction by sorgoleone, a natural product, *J. Agric. Food Chem.*, 1997, **45**, 1415-1421.
24. RIMANDO, A.M., DAYAN, F.E., CZARNOTA, M.A., WESTON, L.A., DUKE, S.O., A new photosystem II electron transfer inhibitor from *Sorghum bicolor*, *J. Nat. Prod.*, 1998, **61**, 927-930.
25. NETZLEY, D.H., REOPEL, J.L., EJETA, G., BUTLER, L., Germination stimulants of witchweed (*Striga asiatica*) from hydrophobic root exudate of Sorghum (*Sorghum bicolor*), *Weed Sci.*, 1988, **36**, 441-446.

26. NIMBAL, C.I., PEDERSON, J.F., YERKES, C.N., WESTON, L.A., WELLER, S.C., Phytotoxicity and distribution of sorgoleone in grain sorghum germplasm., *J. Agric. Food Chem.*, 1996, **44**, 1343-1347.
27. LUCKNER, M., Secondary Metabolism in Microorganisms, Plants and Animals, Springer-Verlag, Berlin, 1984, 576 p.
28. FATE, G., CHANG, M., LYNN, D.G., Control of germination in *Striga asiatica*: Chemistry of spatial definition, *Plant Physiol.*, 1990, **93**, 201-207.
29. MEDINA-BOLIVAR, F., FLORES, H., Selection for hyoscyamine and cinnamoyl putrescine overproduction in cell and root cultures of *Hyoscyamus muticus*, *Plant Physiol.*, 1995, **108**, 1553-1560.
30. FLORES, H.E., MEDINA-BOLIVAR, F., Root culture and plant natural products: Unearthing the hidden half of plant metabolism, *Plant Tissue Culture Biotech.*, 1995, **1**, 59-74.
31. FUKUI, H., HASAN, A.F.M., UEOKA, T., KYO, M., Formation and secretion of a new brown benzoquinone by hairy root cultures of *Lithospermum erythrorhizon*, *Phytochemistry*, 1998, **47**, 1037-1039.
32. FUKUI, H., HASAN, A.F.M., KYO, M., Formation and secretion of a unique quinone by hairy root cultures of *Lithospermum erythrorhizon*, *Phytochemistry*, 1999, **51**, 511-515.
33. AMAGASA, T., PAUL, R.N., HEITHOLT, J.J., DUKE, S.O., Physiological effects of cornexstin on *Lemna pausicostata*, *Pesticide Biochem. Physiol.*, 1994, **49**, 37-52.
34. SIEHL, D.L., SUBRAMANIAN, M.V., WALTERS, E.W., LEE, S.F., ANDERSON, R.J., TOSHI, A.G., Adenylosuccinate synthetase: Site of action of hydantocidin, a microbial phytotoxin, *Plant Physiol.*, 1996, **110**, 753-758.
35. LYDON, J., DUKE, S.O., Inhibitors of glutamine biosynthesis, *in* Plant Amino Acids, Biochemistry and Biotechnology, (B.K. Singh ed.), Marcel Dekker, New York, 1996, pp. 445-464.
36. LYDON, J., The molecular genetics of bacterial phytotoxins, *Plant Growth Regul. Soc. Amer. Quarterly*, 1996, **24**, 111-139.
37. GERSHENZON, J., Metabolic costs of terpenoid accumulation in higher plants, *J. Chem. Ecol.*, 1994, **20**, 1281-1328.
38. MILLAR, A.A., SMITH, M.A., KUNST, L., All fatty acids are not equal: discrimination in plant membrane lipids, *Trends Plant Sci.*, 2000, **5**, 95-101.
39. CALLAWAY, R.M., ASCHEHOUG, E.T., Invasive plants versus their new and old neighbors: A mechanism for exotic invasion, *Science*, 2000, **290**, 521-523.
40. DEVINE, M.D., SHUKLA, A., Altered target sites as a mechanism of herbicide resistance, *Crop Protec.*, 2000, **19**, 881-889.
41. JUSTUM, A.R., HEANEY, S.P., PERRIN, B.M., WEGE, P.J., Pesticide resistance: assessment of risk and the development and implementation of effective management strategies, *Pestic. Sci.*, 1998, **54**, 435-446.
42. TONNEMAKER, K.A., AULD, D.L., THILL, D.C., MALLORY-SMITH, C.A., ERICKSON, D.A., Development of sulfonylurea-resistant rapeseed using chemical mutagenesis, *Crop Sci.*, 1992, **32**, 1387-1391.

43. DALE P.J., SCHEFFLER, J.A., Gene dispersal from transgenic crops, *in* Transgenic Organisms and Biosafety, (E.R. Schmidt and T. Hankeln eds.), Springer-Verlag, Berlin, Germany, 1996, pp. 85-93.
44. HANCOCK, J.F., GRUMET, R., HOKANSON, S.C., The opportunity for escape of engineered genes from transgenic crops, *HortSci.*, 1996, **31**, 1080-1085.
45. GRESSEL, J., Tandem constructs; preventing the rise of superweeds, *Trends Biotech.*, 1999, **17**, 361-366.
46. GRESSEL, J., Molecular biology of weed control, *Transgenic Res.*, 2000, **9**, 355-382.
47. Proceedings of a workshop on: Ecological effects of pest resistance genes in managed ecosystems, (P.L. Traynor and J.H. Westwood eds.), Blacksburg, VA., 1999, WWW page: <ftp://www.isb.vt.edu/pub/nbiap/Proceedings/>
48. DANIELL, H., DATTA, R., VARMA, S., GRAY, S., LEE, S-B., Containment of herbicide resistance through genetic engineering of the chloroplast genome, *Nature Biotech.*, 1998, **16**, 345-348.
49. SENIOR, I.J., DALE., P.J., Molecular aspects of multiple transgenes and gene flow to crops and wild relatives, *Brighton Crop Protection Conf. Symp. Proc.*, 1999, 225-231.
50. ANONYMOUS, US ACPA members' sales up 5.5% in 1997, *Agrow World Crop Protection News*, 1998, **304**, 16.

Chapter Thirteen

MOLECULAR REGULATION OF MONOTERPENOID INDOLE ALKALOID BIOSYNTHESIS

Frédérique Hilliou, Leslie van der Fits, and Johan Memelink*

Institute of Molecular Plant Sciences
Clusius Laboratory
Wassenaarseweg 64
2333 AL Leiden, The Netherlands

**Author to whom correspondence should be addressed*
e-mail: memelink@rulbim.leidenuniv.nl

INTRODUCTION

Primary metabolites include amino acids, nucleotides, sugars, fatty acids and the polymers derived from them (such as proteins, RNA, DNA, polysaccharides, lipids). According to an early definition, secondary metabolites are naturally occurring substances that are not vital to the organism that produces them. However, it now becomes clear that secondary metabolites serve key functions during the plant's life cycle in the natural environment.

In contrast to primary metabolites, the range of secondary metabolites found in different plant species varies enormously. Secondary metabolites are synthesized at specific developmental stages within specialized cell types and in specific tissues and organs.[1,2] This suggests that sophisticated control mechanisms exist to regulate their synthesis during plant development. In addition, many secondary metabolites that are constitutively present in certain cell types and/or during development, are produced in other tissues in response to various stress- or pathogen-related signals.

Alkaloids are a prominent class of plant secondary metabolites, consisting of low molecular weight nitrogen-containing compounds with a heterocyclic structure.[1,3] About 20 % of plant species synthesize a wide range of alkaloids that are species-specific. Biosynthetic pathways leading to economically important alkaloids have been the focus of research for many years in several laboratories because of the important medicinal uses of alkaloids[3,4] and the fact that many accumulate only in trace amounts in the plant source.[5]

This contribution reviews aspects of the regulation of terpenoid indole alkaloid (TIA) biosynthesis in *Catharanthus roseus* (Madagascar periwinkle), with emphasis on stress-responsive expression of the tryptophan decarboxylase (*Tdc*) and strictosidine synthase (*Str*) genes, encoding enzymes acting early in TIA biosynthesis. These two genes were instrumental in the recent isolation of central regulators of the TIA biosynthetic pathway, belonging to the AP2/ERF-domain family of transcription factors.

TERPENOID INDOLE ALKALOIDS

TIAs have a mixed biosynthetic origin, and are derived from L-tryptophan and terpenoid precursors. TIAs are found in a limited number of plant species belonging to the plant families Apocynaceae, Loganiaceae, Rubiaceae, and Nyssaceae. The TIA group of compounds is rich in biologically active substances, and includes the anti-tumor agent camptothecin from the Chinese "happy tree" *Camptotheca acuminata* and the anti-malarial compound quinine from *Cinchona ledgeriana*.[3]

Madagascar periwinkle, a perennial evergreen herb from the Apocynaceae, is able to produce more than a hundred TIAs.[6-8] Some are used in modern medicine including ajmalicine and serpentine used as antihypertensive agents, and vinblastine and vincristine, two dimeric TIAs used in the treatment of leukemia and Hodgkin's disease.

Vinblastine and vincristine accumulate only in trace amounts in *C. roseus* (0.003% dry weight), making their extraction difficult and expensive. Approximately half a ton of dry leaves is necessary to obtain 1 g of vinblastine,[9] and the world-wide requirement for vinblastine is on the order of 5 kg per year.[10] Because of the pharmaceutical importance of vinblastine and vincristine, together with their low levels of accumulation in *C. roseus*, much research has focused on elucidation of the pathway and development of alternative sources of production. Attempts to produce dimeric TIAs in *C. roseus* cell cultures have failed, and the only source of vinblastine and vincristine still remains the *C. roseus* plant.

The TIA biosynthetic pathway is complex, with more than twenty enzymatic steps leading to vinblastine and vincristine (Fig. 13.1).[3,11,12] A number of enzymes have been localized in different subcellular compartments, or their localization has been inferred from the presence of targeting signal peptides in their amino acid sequences.[5,11] In addition, there is evidence that certain (unknown) metabolic intermediates are transported between different specialized cell types.[1,13] Although intra- and intercellular transport of metabolic intermediates must be an important aspect of TIA biosynthesis, nothing is known about transport proteins or their regulation.

The amino acid L-tryptophan, providing the indole moiety of TIAs, is derived from the shikimate pathway, whereas the terpenoid precursor for secologanin is derived from the Rohmer or MEP (2-C-methyl-D-erythritol 4-phosphate) pathway (Fig. 13.1).[14] Both precursor pathways are localized in plastids and chloroplasts. L-tryptophan is decarboxylated by the cytoplasmic enzyme tryptophan decarboxylase (TDC; EC 4.1.1.28) to form tryptamine. Conversion of the terpenoid geraniol into secologanin is thought to occur in the vacuole. The vacuolar enzyme strictosidine synthase[15] (STR; EC 4.3.3.2) catalyzes the stereospecific condensation of secologanin and tryptamine to form 3α(*S*)-strictosidine, the universal precursor of TIAs in plants. Additional enzymes bound to the endoplasmic reticulum,[16,17] as well as chloroplastic[18] and cytoplasmic[19,20] enzymes, are involved in the following steps of TIA biosynthesis (Fig. 13.1). The dimerization of monomeric TIAs is catalyzed by a vacuolar peroxidase (POD).[21]

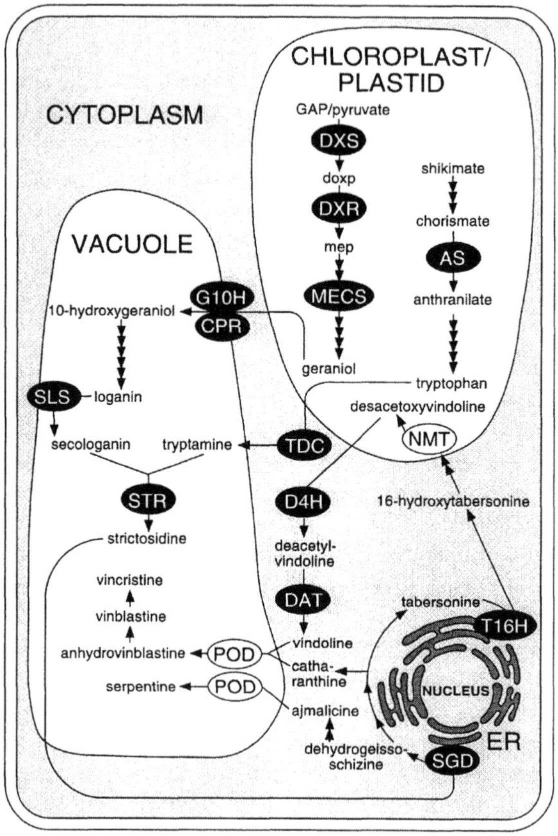

Figure 13.1: Compartmentalization of the TIA biosynthetic pathway Enzymes with black background circles have been cloned. Although the entire pathway is depicted in a single cell, some reactions in the pathway are likely to occur in different cells or during different stages of development.

DXS, D-1-Deoxyxylulose 5-phosphate synthase; DXR, D-1-Deoxyxylulose 5-phosphate reductoisomerase; MECS, Methylerythritol 2,4-cyclodiphosphate synthase; AS, Anthranilate synthase; CPR, Cytochrome P450-reductase; G10H: Geraniol 10-hydroxylase; SLS, Secologanin synthase; TDC: Tryptophan decarboxylase; STR: Strictosidine synthase; SGD: Strictosidine β-D-glucosidase; T16H: Tabersonine 16-hydroxylase; NMT: 16-Methoxy-2,3-dihydro-3-hydroxytabersonine-*N*-methyltransferase; DAT: Deacetylvindoline 4-*O*-acetyltransferase; D4H: Desacetoxyvindoline 4-hydroxylase; POD: Peroxidase; ER: endoplasmic reticulum.

TIAs as Plant Defense Compounds

It is assumed that TIAs function in plants as defense compounds against herbivores and microbial pathogens.[22] In addition, given their UV-absorbing properties and their inducibility by UV light,[23] TIAs also may provide protection against UV radiation. Physiological concentrations of *Cinchona* or *Catharanthus* alkaloids have antifeeding activity or can delay growth of insect larvae, fungi and microbes.[24-26] The *Catharanthus* alkaloids ajmalicine and vincamine have been shown experimentally to act as feeding deterrents for polyphagous *Syntomis* (Lep.).[22] TIAs, or their precursors, may also affect other aspects of insect physiology. The reproduction of whitefly was reduced on transgenic tobacco plants accumulating tryptamine as a result of the over-expression of TDC.[27]

The localization in the plant also provides support for the notion that alkaloids may deter feeding of insects and herbivores. Alkaloids accumulate in those developmental stages and parts of *Cinchona* plants where herbivorous and insect attacks would most affect plant fitness. In *Cinchona* seedlings, alkaloids and TDC and STR enzyme activities increase transiently after germination.[28] In older plants, alkaloid contents and STR activity are highest in young developing leaves and roots.[29]

Alkaloid levels usually correlate well with levels of biosynthetic enzymes and corresponding mRNA levels, indicating that TIAs are synthesized and stored on site. Control of biosynthetic gene expression appears to be the main mechanism to regulate alkaloid levels. The probable function of TIAs in defense is reflected in the regulation of the biosynthetic genes, as illustrated for the *Tdc* and *Str* genes in the next section.

COORDINATED REGULATION OF *TDC* AND *STR*

Tdc and *Str* cDNA and genomic clones have been isolated from *C. roseus*,[30-34] and both genes are present as single copies in the haploid genome.[31,33,3] At the plant level, both genes are expressed in all tissues, with the highest mRNA levels detected in roots.[33] RNA *in situ* hybridization experiments localized *Tdc* and *Str* mRNA in the upper and lower epidermis of *C. roseus* leaves, with higher levels of both transcripts detected in the distal part of the young leaf and other plant tissues such as stems and flower primordia.[1,13] In older leaves, the cell-specific localization is maintained, but transcript levels are lower compared with young leaves.[13] In periwinkle leaves, *Tdc* and *Str* gene expression is enhanced by a pulse of UV-B light.[23] Whether UV light stimulates expression in the epidermal leaf cells, or induces expression in other cell types is not known.

Studies using cell cultures have uncovered important aspects of stress-responsive *Tdc* and *Str* gene expression. *Tdc* and *Str* mRNA accumulate in suspension-cultured cells after auxin starvation.[33,35] Concurrent with the effect on gene expression, auxin starvation results in arrest of cell division and increased alkaloid production.[36]

Fungal extracts, including an extract from baker's yeast, also induced *Tdc* and *Str* expression in cell cultures.[33,37] Yeast extract contains several components that can elicit plant defense responses, including chitin, *N*-acetylglucosamine oligomers, β-glucan and ergosterol. However, a different component is responsible for *Tdc* and *Str* induction in periwinkle cells.[38,39] This elicitor has not been purified to homogeneity but it has a molecular mass of around 2000 and is most likely a peptide, based on its protease sensitivity and chromatographic behavior on anion exchange columns.[39]

Elicitor-responsive expression of *Tdc* and *Str* depends on jasmonate as a secondary signal.[38] MeJa and jasmonic acid (JA), collectively called jasmonates, are lipid-based plant hormones synthesized *via* the octadecanoid (ODA) pathway in response to stress.[40,41] Yeast elicitor was shown to induce accumulation of endogenous JA, and exogenously added (Me)JA was shown to induce *Tdc* and *Str* gene expression.[38] In addition, blocking JA biosynthesis abolished elicitor-induced expression of biosynthetic genes.[38] Concurrent with its effect on gene expression MeJA also was shown to enhance alkaloid production in seedlings[42] and in cell suspension cultures.[36,43] Elicitor-induced jasmonate biosynthesis requires protein kinase activity[38] and an elevation of cytosolic calcium concentration[39] (Fig. 13.2).

Since TDC and STR are encoded by single-copy genes in *C.roseus*, all these signals act on the same promoter regions. The observations that *Tdc* and *Str* mRNAs coordinately accumulate in response to fungal elicitors, jasmonates, UV light, and auxin starvation, clearly indicate that the *Tdc* and *Str* genes are controlled by common regulators. Since all TIA biosynthetic genes tested were induced by MeJA,[44] this suggests that a common jasmonate-responsive regulator controls many and possibly all TIA structural genes. Such regulators could prove to be useful to engineer TIA biosynthesis, and the following sections discuss two successful strategies for their isolation.

YEAST ONE-HYBRID SCREENING FOR ISOLATION OF TRANSCRIPTIONAL REGULATORS

The yeast one-hybrid system is a genetic screening method for the isolation of proteins that bind to characterized promoter elements. Multimers of the promoter element of interest and a yeast TATA box region are fused to form an artificial promoter. A yeast strain, carrying a selection gene driven by this artificial promoter is then transformed with a plasmid library of plant cDNAs cloned in a translational fusion with the yeast GAL4 activation domain. Transformed yeast cells are selected for presence of a library plasmid and for expression of the selection gene. Thereby cDNA-encoded proteins are selected for binding to the promoter of interest.

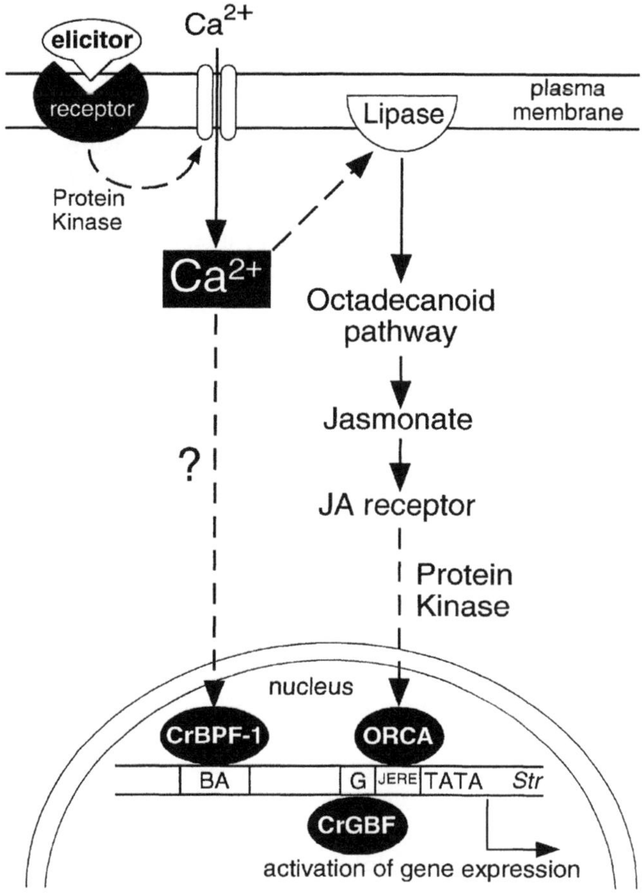

Figure 13.2: Model for elicitor signal transduction leading to *Str* expression. The model shows the positions of CrBPF-1 in a JA-independent, and ORCAs in a JA-dependent elicitor signal transduction pathway. Protein phosphorylation and calcium influx are required for elicitor-induced jasmonate biosynthesis, as well as for the induction of *CrBPF-1*, *Orca2*, and *Orca3* gene expression. *Orca* gene expression and ORCA protein activation additionally depend on jasmonate biosynthesis. The existence of a lipase and of receptors for elicitor and JA is postulated. The positions of the TATA box, the BA region, the jasmonate- and elicitor-responsive element (JERE), and the G-box within the *Str* promoter are indicated. The G-box interacts with GBFs 1 and 2 from *C. roseus*.

The *Str* promoter has been studied in great detail to identify elicitor- and jasmonate-responsive sequences (Fig. 13.2).[34,45,46] The so-called BA region (-339 to -208; Fig. 13.2) contains the main quantitative enhancer elements.[34,45] The *Str* promoter also contains the sequence CACGTG (-108 to –103; Fig. 13.2) or G-box.[34,45] This promoter element occurs in many unrelated genes and has been implicated in response to a variety of environmental signals, such as exposure to visible and UV light, dehydration stress, cold stress, abscisic acid, and sucrose (see Ref. 47 and references therein). In addition, the G-box confers tissue-specific gene expression (Ref. 47 and references therein). The *Str* G-box interacts with two G-box binding factors (GBFs) from *C. roseus* (Fig. 13.2).[48] The GBFs have a repressing effect on *Str* gene expression,[48] but do not seem to be involved in stress-responsive expression since mutation or deletion of the G-box did not change elicitor or MeJa responsiveness.[34,45] An important regulatory sequence is present in the *Str* promoter close to the TATA box and is called the jasmonate- and elicitor-responsive element (JERE). Mutation or removal of the JERE results in an inactive and unresponsive *Str* promoter derivative.[45] A tetramer of the JERE fused to a minimal promoter confers JA-responsive gene expression onto a reporter gene,[45] showing that the JERE is an autonomous JA-responsive sequence.

Use of the BA region in a one-hybrid screen resulted in the isolation of a *C. roseus* homolog of the MYB-like factor BPF-1 from parsley.[46] Like its parsley homolog,[49] transcription of periwinkle *CrBPF-1* is induced by fungal elicitor. *CrBPF-1* elicitation depends on protein kinase activity and an increase in cytosolic calcium concentration, but is independent of jasmonate biosynthesis (Fig. 13.2).[46]

Yeast one-hybrid screening with the JERE as a bait identified two transcription factors of the AP2/ERF-domain family that were named ORCA1 and ORCA2 (Octadecanoid-Responsive *Catharanthus* AP2).[45] *Orca2* expression was induced by MeJa and elicitor.[45] Furthermore, ORCA2 showed sequence-specific binding to the JERE and *trans*-activated *Str* gene expression.[45] In contrast to *Orca2*, *Orca1* expression is not induced by MeJA and elicitor, indicating that ORCA1 is not involved in the regulation of JA- and elicitor-induced *Str* expression.[45]

Figure 13.2 shows a model for elicitor signal transduction leading to *Str* expression. Upon perception of the elicitor, protein phosphorylation is required to induce a calcium influx. This transient increase in cytosolic $[Ca^{2+}]$ is necessary for activation of the ODA pathway. Jasmonate induces *Orca* gene expression[44,45,50] and activates pre-existing ORCA proteins.[50] Then, activated ORCAs switch on *Str* expression via interaction with the JERE. In addition, calcium influx is required for JA-independent induction of *CrBPF-1* mRNA accumulation.[46] CrBPF-1 is putatively involved in regulation of the *Str* gene via interaction with the jasmonate- and elicitor-responsive BA region. In summary, the *Str* promoter contains the JERE, which forms an ORCA-dependent switch for elicitor- and jasmonate-responsive activity, and other *cis*-acting elements with their cognate binding factors (such as BA/CrBPF-1), which may be important for enhancing the elicitor and/or jasmonate response.

T-DNA ACTIVATION TAGGING FOR ISOLATION OF TRANSCRIPTIONAL REGULATORS

Gene tagging techniques developed for gene isolation were used in our laboratory in addition to yeast one-hybrid experiments to identify regulators of the terpenoid indole alkaloid pathway. Gene tagging consists of the insertion of a defined piece of DNA in the vicinity of a target gene. The insertion of foreign DNA alters the expression of flanking genes and results in either a gain or loss of function responsible for the production of a screenable phenotype. The piece of DNA used as a tag should be able to insert randomly into the host genome, and insertion of the tag should result in the production of a screenable phenotype. The tag should also contain elements for the rescue of the flanking DNA sequences after insertion. Plant transposable elements were first used in gene tagging experiments[51,52] allowing for instance the isolation of genes involved in flower[53,54] and leaf development.[55]

The tumor-inducing soil bacterium *Agrobacterium tumefaciens* is another tool used in gene tagging experiments. This bacterium contains a large plasmid, the Tumor-inducing (Ti) plasmid, which is required for tumor induction. After infection of plant cells, proteins encoded by the virulence region of the Ti plasmid catalyze the production of a single-strand transferred DNA (T-DNA). The T-DNA is transferred to the plant cell where it inserts preferentially into transcribed regions of the plant genome by illegitimate recombination.[56] Two imperfect repeat sequences of 25 base pairs (bp) form the respective borders of the T-DNA on the Ti-plasmid. Between these borders the T-DNA can be modified without interfering with the process of transfer to the plant and insertion into the plant genome.[56]

Several types of vectors have been engineered for T-DNA tagging experiments. Passive tagging vectors (or promoter trap vectors) contain a marker gene that lacks a promoter sequence or is linked to a minimal promoter unable to initiate transcription. After insertion into a plant genome, such T-DNAs might be flanked by DNA plant sequences able to direct gene expression. This flanking DNA sequence will, thus, drive the expression of the marker gene. Passive vectors have been used to isolate promoters responsible for tissue-specific expression.[57] Active vectors with T-DNA containing transcriptional enhancers, for example derived from the Cauliflower Mosaic virus (CaMV) *35S* promoter, have been developed in order to activate the expression of flanking plant DNA sequences. This technique produces dominant mutations, and is called activation tagging.

Activation tagging was successfully used for the isolation of a histidine kinase involved in cytokinin signal transduction.[58] Recently, activation tagging with T-DNA vectors carrying multimerized transcriptional enhancers from the *CaMV 35S* promoter was applied to *Arabidopsis* plants and led to the isolation of several morphological mutants.[59] From this experiment, several characteristics were observed such as random insertion of the T-DNA into the *Arabidopsis* genome, activation of gene expression both upstream or downstream of the *CaMV 35S* enhancer, and a close position of the *CaMV 35S* enhancer to the overexpressed gene (380 bp to 3.6 kilo bp, kb). T-DNA activation tagging in *Arabidopsis* appears to be as efficient as the use of *Dissociation* (*Ds*) element carrying a *CaMV 35S* promoter for the screening of morphological mutants.[59]

T-DNA Activation Tagging of Periwinkle Cells

C. roseus cell suspensions were transformed with an *Agrobacterium* strain carrying an activation tagging construct (Fig. 13.3). The T-DNA activation construct contains several key elements between its right and left border sequences. The hygromycin phosphotransferase (*Hpt*) gene conferring resistance to hygromycin was inserted near the left border to allow the selection of stably transformed cells. The T-DNA construct also contains the *Escherichia coli* pUC21 vector with unique restriction sites that can be used to rescue the plant genomic DNA that flanks the integration site of the T-DNA. Finally, directly flanking the right border the T-DNA construct carries multimerized *CaMV 35S* enhancer sequences without a TATA box or ATG start codon.

Agrobacterium-mediated transformation of *C. roseus* cell suspensions[60] with the activation tagging construct resulted in roughly 4 to 5 $x10^5$ transformants that were screened for resistance to 0.4 mM 4-mT.[61] Resistance to 4-mT can result from different mechanisms such as an increase in tryptophan production, impaired 4-mT uptake, or increased TDC enzyme activity. TDC can also use L-tryptophan analogs as substrates, such as 4-methyl-tryptophan (4-mT).[62] 4-mT is a toxic compound to plant cells, but becomes de-toxified by conversion into 4-methyl-tryptamine.[63]

Selection on 4-mT yielded 180 resistant cell lines. To identify mutant lines in which the T-DNA tag activated a regulator of *Tdc* gene expression, 4-mT resistant lines were screened for *Tdc* expression levels by Northern blot analysis. Among 20 lines with increased *Tdc* expression, 6 lines also showed increased expression of a second gene involved in TIA production, *Str*, suggesting that the tag activated a regulator of multiple genes involved in TIA biosynthesis.[61]

From one of the lines showing high *Tdc* and *Str* expression (line 46), 1.6 kb of plant DNA flanking the T-DNA insertion site was isolated by plasmid rescue.[44] An open reading frame was located approximately 600 bp downstream of the T-DNA, which predicted a protein of 203 amino acids with a conserved AP2/ERF domain.

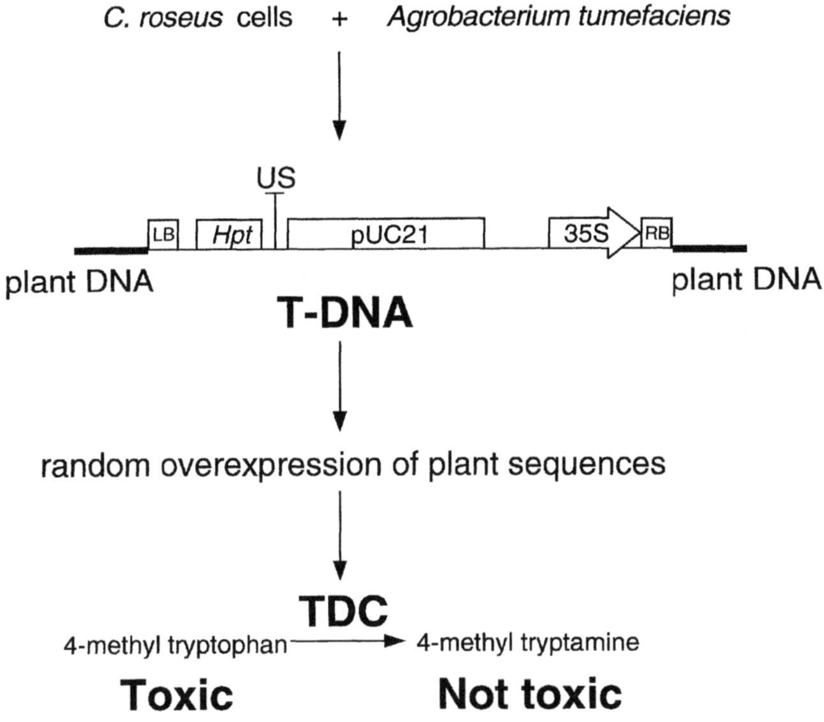

Figure 13.3: T-DNA activation tagging approach. The T-DNA activation tagging construct contains: *Hpt*, Hygromycin resistance marker driven by a deletion derivative of the Figwort Mosaic Virus (FMV) 34S promoter; pUC21, *E. coli* pUC21 vector; LB and RB, Left and right T-DNA border sequences; 35S, CaMV *35S* promoter; US, Unique restriction sites that can be used for plasmid rescue. TDC: Tryptophan decarboxylase.

The protein encoded by the rescued plant DNA was named ORCA3 because of its similarity to ORCA2. Expression of *Orca3* is activated by the T-DNA tag (Fig. 13.4).[44] The overexpressed transcript and the endogenous mRNA have identical sizes, indicating that full-length mRNA and protein are produced, which is consistent with the tagged gene structure.[44]

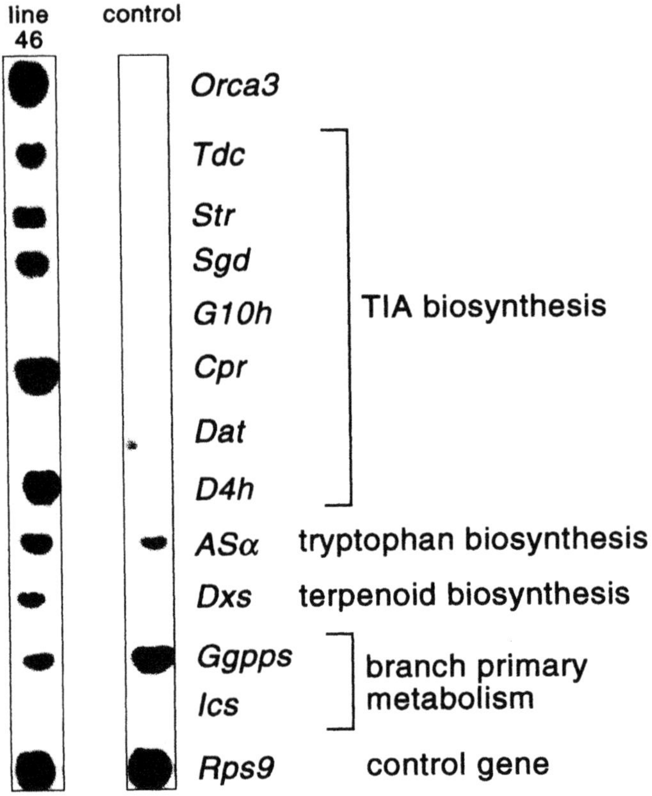

Figure 13.4: Expression of metabolic genes in line 46 overexpressing ORCA3. cDNAs of the genes indicated were used as probes on RNA extracted from tagged cell line 46 and from a representative control line. *Tdc*: Tryptophan decarboxylase; *Str:* Strictosidine synthase; *Sgd:* Strictosidine β-D-glucosidase; *G10h*: Geraniol 10-hydroxylase, *Cpr*: Cytochrome P450-reductase; *Dat*: Deacetylvindoline 4-*O*-acetyltransferase; *D4h*: Desacetoxyvindoline 4-hydroxylase; *Asα*: Anthranilate synthase; *Dxs:* D-1-Deoxyxylulose 5-phosphate synthase; *Ggpps*: Geranylgeranyl pyrophosphate synthase; *Ics*: Isochorismate synthase; *Rps9:* 40S ribosomal protein S9.

ORCA3 is a Master Regulator of Metabolism

Expression of TIA biosynthetic genes *Tdc*, *Str*, *Sgd* (strictosidine β-D-glucosidase), *Cpr* (cytochrome P450-reductase), and *D4h* (desacetoxyvindoline 4-hydroxylase) was increased in the *Orca3*-tagged line (Fig. 13.4).[44] TIA biosynthetic genes *G10h* (geraniol 10-hydroxylase) and *Dat* (deacetylvindoline 4-*O*-acetyltransferase) were not induced, suggesting that these genes are not controlled by ORCA3. Genes encoding the alpha subunit of anthranilate synthase (ASα) and D-1-deoxyxylulose 5-phosphate synthase (DXS), enzymes involved in primary metabolism leading to TIA precursor synthesis, were also induced by *Orca3* overexpression, whereas two other primary metabolic genes not involved in production of TIA precursors (geranylgeranyl pyrophosphate synthase, *Ggpps*, and isochorismate synthase, *Ics)* were not regulated by ORCA3 (Fig. 13.4). Except for *Sgd*, which shows variable expression in control cell lines, expression patterns observed in cell lines independently transformed by particle bombardment with the rescued *Orca3* gene or cDNA under control of *CaMV 35S* promoter elements were identical to that of line 46.[44] ORCA3 transcriptionally activates at least three genes (*Tdc,Str,Cpr*) in TIA metabolism by binding directly to their promoters.[44] ORCA3 binds specifically to the JERE in the *Str* promoter, and activates gene expression *via* an acidic N-terminal amino acid domain.[50]

These results indicate that ORCA3 is a regulator of primary as well as secondary metabolite biosynthetic genes involved in TIA biosynthesis. Although previous studies report differential regulation of genes involved in early (*i.e., Tdc* and *Str*) and late (*i.e. D4h*) steps in vindoline biosynthesis in periwinkle plants,[1,13] genes of both classes are controlled by *Orca3* in suspension-cultured cells. Expression of *Orca3*,[44,50] as well as of all other genes used in this study,[44] was induced by MeJA. However, not all these genes were induced by *Orca3* overexpression. Combinations of different jasmonate-responsive transcription factors, such as ORCA2 and ORCA3, may regulate TIA biosynthesis by selective activation of target genes.

AP2/ERF-DOMAIN PROTEINS IN PLANT STRESS RESPONSES

The AP2 domain is a plant DNA-binding domain first identified in the APETALA2 protein.[64] The *APETALA2* gene from *Arabidopsis* controls important processes during flower development.[64] The AP2-domain family can be divided into two sub-families based on whether the proteins contain one or two AP2 domains. The AP2-like proteins containing two AP2 domains (such as AINTEGUMENTA and Glossy15) act as key regulators in developmental processes, whereas the AP2/ERF-domain proteins with a single domain (such as the ERF/EREBP proteins in tobacco and *Arabidopsis*) appear to be involved in responses to environmental stress.[65] The AP2/ERF domain is highly conserved within the AP2/ERF family.[65] The domain spans a 58-59 amino acid region and consists of a three-stranded anti-parallel β-sheet and an α-helix packed approximately parallel to the β-sheet.[66]

AP2/ERF-domain transcription factors occupy central positions in regulation of plant stress responses. ERF1, for example, is an ethylene-responsive AP2/ERF-domain protein (Fig. 13.5). Its overexpression resulted in induction of several ethylene-responsive defense genes in *Arabidopsis*.[67] In tomato, Pto is a receptor kinase that is involved in the recognition of the *Pseudomonas syringae* avirulence gene product AvrPto. Yeast two-hybrid screening identified three Pto-interacting (Pti4, 5 and 6) tomato proteins, each of which contained an AP2/ERF-domain.[68] Pti4, 5 and 6 are thought to be involved in elicitor-induced activation of defense genes in tomato (Fig. 13.5). Another family of AP2/ERF-domain proteins, represented by DREB2A and DREB2B, is involved in drought-responsive gene expression, whereas the CBF/DREB1 family includes central regulators of cold-regulated gene expression in *Arabidopsis* (Fig. 13.5).[69] Upon overexpression of the AP2/ERF-domain proteins CBF1 or DREB1A, expression of a set of cold-induced genes was induced, thereby increasing freezing tolerance.[69] In periwinkle, expression of TIA biosynthetic genes is coordinately induced by jasmonates.[17,38,44] A considerable number of these genes are controlled by ORCA3.[44] Therefore, the spectrum of defense genes regulated by AP2/ERF-domain transcription factors includes genes involved in jasmonate-responsive secondary metabolism (Fig. 13.5).

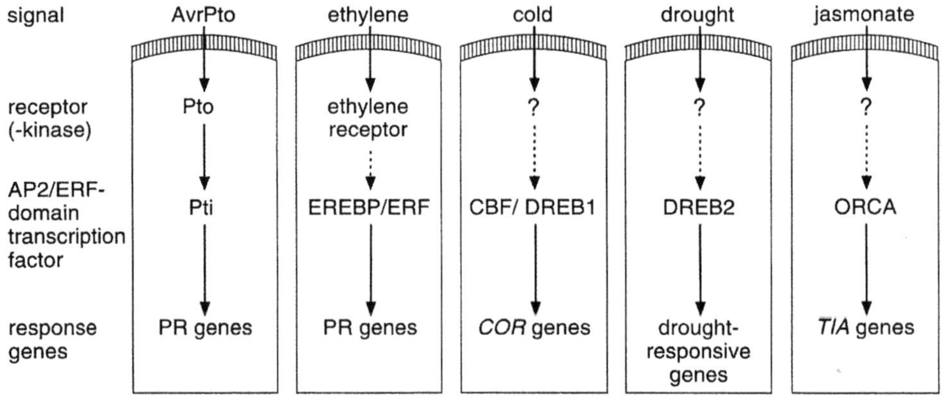

Figure 13.5: Involvement of AP2/ERF-domain transcription factors in plant stress responses. Stress signals are perceived by specific receptors or receptor-kinases. Induction of *de novo* synthesis and/or modulation of pre-existing AP2/ERF-domain transcription factors activates gene expression. Unidentified signal transduction components are indicated with a question mark. Dashed arrows indicate the possible involvement of multiple signaling steps.

CONCLUSIONS

Molecular techniques such as the yeast one-hybrid system and T-DNA activation tagging have been successfully used in our laboratory for the isolation of transcriptional regulators of the terpenoid indole alkaloid pathway. The yeast one-hybrid system with *Str* promoter elements as bait was used to isolate *Orca1*, *Orca2*,[45] and *CrBPF-1* clones.[46] These clones encode transcription factors of the AP2/ERF–domain and MYB families, respectively. The corresponding mRNAs are induced by MeJA (*Orca2*) [45] and fungal elicitors (*Orca2* and *CrBPF-1*).[46]

ORCA3 was isolated using T-DNA activation tagging *via* selection of *C. roseus* lines with elevated *Tdc* and *Str* mRNA levels. *Orca*3 transcription is induced by MeJA. Lines overexpressing ORCA3 have increased mRNA levels of several genes of the TIA pathway and accumulate higher levels of alkaloids than control lines.[44] All the genes induced by ORCA3 are also inducible by MeJA. ORCAs are, therefore, central regulators of TIA biosynthetic gene expression, and are responsible for MeJA-responsive gene expression in *C. roseus.*

Metabolic engineering of the TIA pathway by overexpression of *Tdc*[70] and/or *Str* (F. Hilliou, unpublished data) only resulted in transgenic lines with higher levels of tryptamine compared to controls. Two-fold increases of strictosidine levels were observed in transgenic lines overexpressing *Tdc* and *Str* compared to controls when fed with loganin. However, despite higher enzymatic levels of TDC and/or STR in transgenic lines, no increase in strictosidine-derived TIAs was observed, suggesting further block(s) downstream of strictosidine in metabolic flux toward TIA endproducts.

These observations suggest that manipulation of TIA biosynthesis in transgenic *C. roseus* cell cultures will not only require the over-expression of TDC and STR, but also the manipulation of the iridoid pathway that supplies secologanin, and manipulation of enzyme activities downstream of strictosidine biosynthesis. Therefore, transcription factors regulating expression of multiple genes of the TIA biosynthetic pathway may provide a better approach for metabolic engineering. Transgenic *C. roseus* cell cultures overexpressing *Orca3* and fed with loganin accumulate more tryptamine and TIAs (including strictosidine) than control lines.[44] *Orca3* overexpression causes induction of expression of some genes involved in the iridoid pathway (*Dxs*, *Cpr*),[44] however, the induction of these genes is not sufficient to trigger secologanin biosynthesis. *Orca2* overexpression, unlike *Orca3,* causes induction of *G10h* (unpublished data), which has been shown to be a cornerstone gene of the iridoid pathway. Overexpression of the transcriptional regulators *Orca2* together with *Orca3* may result in transgenic *C. roseus* lines with increased TIA biosynthesis in absence of loganin feeding. However, the TIA biosynthetic pathway is highly compartmentalized (Fig. 13.1), and metabolite trafficking might form a bottleneck. Metabolites are not always synthesized in the same compartment as the enzyme using them as substrates. On the other hand, it is possible that the (unknown) genes encoding metabolite transporters are also controlled by ORCAs.

Transcription factors have been used previously for successful engineering of secondary metabolism in plants. Increases in the flux through the phenylpropanoid/flavonoid pathway were obtained by overexpressing MYB and/or bHLH (basic helix-loop-helix) proteins in either homologous or heterologous plant species.[71-74] Expression of *C. roseus* ORCA proteins in other plant species might result in induction of jasmonate-responsive metabolic pathways and increased metabolite production. With the isolation of an increasing number of metabolic master regulators, new opportunities for engineering of plant secondary metabolism are offered by inter-species exchange of transcription factors.

ACKNOWLEDGEMENTS

The authors thank Ward de Winter for assistance with tissue culturing and P. Hock for preparing the figures. F.H. is supported by a Marie Curie research training grant from the European Community (MCFI-1999-01097), whereas LvdF was financially supported by the Ministry of Economic Affairs, the Ministry of Education, Culture and Science, the Ministry of Agriculture, Nature Management and Fishery in the framework of an industrial relevant research program of the Netherlands Association of Biotechnology Centres in the Netherlands (ABON).

REFERENCES

1. DE LUCA, V., ST-PIERRE, B., The cell and developmental biology of alkaloid biosynthesis, *Trends Plant Sci.*, 2000, **5**, 168-173.
2. HRAZDINA, G., JENSEN, R.A., Spatial organization of enzymes in plant metabolic pathways, *Annu. Rev. Plant Physiol. Plant Mol. Biol.*, 1992, **43**, 241-267.
3. KUTCHAN, T.M., Alkaloid biosynthesis-the basis for metabolic engineering of medicinal plants, *Plant Cell*, 1995, **7**, 1059-1070.
4. SCHMELLER, T., WINK, M., Utilization of alkaloids in modern medicine, *in* Alkaloids, Biochemistry, Ecology and Medicinal Applications (M.F. Roberts and M. Wink, eds.), Plenum Press, New York, 1998, pp. 435-461.
5. VERPOORTE, R., VAN DER HEIJDEN, R., MEMELINK, J., Engineering the plant cell factory for secondary metabolite production, *Transgenic Res.*, 2000, **9**, 323-343.
6. FARNSWORTH, N., BLOMSTER, N., DAMRATOSKI, D., MEER, W., CAMMARATO, L., Studies on *Catharanthus* alkaloids. VI. Evaluation by means of thin-layer chromatography and ceric ammonium sulfate spray reagent, *Lloydia*, 1964, **27**, 302-314.

7. SCOTT, I.A., LEE S-L., CULVER, M., WAN, W., HIRATA, T., GUERITTE, F., BAXTER, R., NORDLOV, H., DORSCHEL, C., MIZUKAMI, H., MACKENZIE, N., Indole alkaloid biosynthesis, *Heterocycles*, 1981, **15**, 1257-1274.
8. VAN DER HEIJDEN, R., VERPOORTE, R., TEN HOOPEN, H., Cell and tissue cultures of *Catharanthus roseus* (L.) G. Don: A literature survey, *Plant Cell Tiss. Org. Cult.*, 1989, **18**, 231-280.
9. NOBLE, R.L., The discovery of the vinca alkaloids-chemotherapeutic agents against cancer, *Biochem. Cell Bio.*, 1990, **68**, 1344-1351.
10. ROBERTS, M.F., Production of alkaloids in plant cell culture, *in* Alkaloids, Biochemistry, Ecology and Medicinal Applications (M.F. Roberts and M. Wink, eds.), Plenum Press, New York, 1998, pp. 159-238.
11. MEIJER, A.H., VERPOORTE, R., HOGE, J.H.C., Regulation of enzymes and genes involved in terpenoid indole alkaloid biosynthesis in *Catharanthus roseus*, *J. Plant Res.*, 1993, **3**, 145-164.
12. KUTCHAN, T.M., Molecular genetics of plant alkaloid biosynthesis, *in* The Alkaloids (G.A. Cordell, ed.), Academic Press, San Diego, 1998, pp. 257-316.
13. SAINT-PIERRE, B., VAZQUEZ-FLOTA, F.A., DE LUCA, V., Multicellular compartmentation of *Catharanthus roseus* alkaloid biosynthesis predicts intercellular translocation of a pathway intermediate, *Plant Cell*, 1999, **11**, 887-900.
14. CONTIN, A., VAN DER HEIJDEN, R., LEFEBER, A.W.M., VERPOORTE, R., The iridoid glucoside secologanin is derived from the novel triose phosphate/pyruvate pathway in *Catharanthus roseus* cell culture, *FEBS Lett.*, 1998, **434**, 413-416.
15. MCKNIGHT, T.D., BERGEY, D.R., BURNETT, R.J., NESSLER, C.L., Expression of enzymatically active and correctly targeted strictosidine synthase in transgenic tobacco plants, *Planta*, 1991, **185**, 148-152.
16. SCHRÖDER, G., UNTERBUSCH, E., KALTENBACH, M., SCHMIDT, J., STRACK, D., DE LUCA, V., SCHRÖDER, J., Light-induced cytochrome P450-dependent enzyme in indole alkaloid biosynthesis: tabersonine 16-hydroxylase, *FEBS Lett.*, 1999, **458**, 97-102.
17. GEERLINGS, A., MARTINEZ-LOZANO IBANEZ, M., MEMELINK, J., VAN DER HEIJDEN, R., VERPOORTE, R., Molecular cloning and analysis of strictosidine β-D-glucosidase, an enzyme in terpenoid indole alkaloid biosynthesis in *Catharanthus roseus*, *J. Biol. Chem.*, 2000, **275**, 3051-3056.
18. DE LUCA, V., CUTLER, A.J., Subcellular localization of enzymes involved in indole alkaloid biosynthesis in *Catharanthus roseus*, *Plant Physiol.*, 1987, **85**, 1099-1102.
19. VAZQUEZ-FLOTA, F., DE CAROLIS, E., ALARCO, A.-M., DE LUCA, V., Molecular cloning and characterization of desacetoxyvindoline-4-hydroxylase, a 2-oxoglutarate dependent-dioxygenase involved in the biosynthesis of vindoline in *Catharanthus roseus* (L.), *G. Don. Plant Mol. Biol.*, 1997, **34**, 935-948.
20. SAINT-PIERRE, B., LAFLAMME, P., ALARCO, A.-M., DE LUCA, V., The terminal O-acetyltransferase involved in vindoline biosynthesis defines a new class of proteins responsible for Coenzyme A-dependent acyl transfer, *Plant J.*, 1998, **14**, 703-713.
21. SOTTOMAYOR, M., DE PINTO, M.C., SALEMA, R., DICOSMO, F., PEDRENO, M.A., ROS BARCELLO, A., The vacuolar localization of a basic peroxidase isoenzyme responsible for the synthesis of alpha-3',4'-anhydrovinblastine in *Catharanthus roseus* (L.) G. leaves, *Plant, Cell Environment*, 1996, **19**, 761-767.

22. WINK, M., Chemical ecology of alkaloids, *in* Alkaloids, Biochemistry, Ecology and Medicinal Applications (M.F. Roberts and M. Wink, eds.), Plenum Press, New York, 1998, pp. 265-298.
23. OUWERKERK, P.B.F., HALLARD, D., VERPOORTE, R., MEMELINK, J., Identification of UV-B light-responsive regions in the promoter of the tryptophan decarboxylase gene from *Catharanthus roseus*, *Plant Mol. Biol.*, 1999, **41**, 491-503.
24. AERTS, R.J., SNOEIJER, W., AERTS-TEERLINK, O., VAN DER MEIJDEN, E., VERPOORTE, R., Control and biological implications of alkaloid biosynthesis in *Cinchona* seedlings, *Phytochemistry*, 1991, **30**, 3571-3577.
25. AERTS, R.J., STOKER, A., BEISHUIZEN, M., JAARSMA, I., VAN DE HEUVEL, M., VAN DER MEIJDEN, E., VERPOORTE, R., Detrimental effects of *Cinchona* leaf alkaloids on larvae of the polyphagous insect *Spodoptera exigua*, *J. Chem. Ecol.*, 1992, **18**, 1955-1964.
26. LUIJENDIJK, T.J.C., Strictosidine Glucosidase in Indole Alkaloid Biosynthesis, PhD Thesis, Leiden University, Leiden, 1995.
27. THOMAS, J.C., ADAMS, D.G., NESSLER, C.L., BROWN, J.K., BOHNERT, H.J., Tryptophan decarboxylase, tryptamine, and reproduction of the white fly, *Plant Physiol.*, 1995, **109**, 717-720.
28. AERTS, R.J., VAN DER LEER, T., VAN DER HEIJDEN, R., VERPOORTE, R., Developmental regulation of alkaloid production in *Cinchona* seedlings, *J. Plant Physiol.*, 1990, **136**, 86-91.
29. AERTS, R.J., DE WAAL, A., PENNINGS, E.J.M., VERPOORTE, R., The distribution of strictosidine-synthase activity and alkaloids in *Cinchona* plants, *Planta*, 1991, **183**, 536-541.
30. DE LUCA, V., MARINEAU, C., BRISSON, N., Molecular cloning and analysis of cDNA encoding a plant tryptophan decarboxylase: comparison with animal dopa decarboxylases, *Proc. Natl. Acad. Sci. USA*, 1989, **86**, 2582-2586.
31. GODDIJN, O.J.M., LOHMAN, F.P., DE KAM, R.J., SCHILPEROORT, R.A., HOGE, J.H.C., Nucleotide sequence of the tryptophan decarboxylase gene of *Catharanthus roseus* and expression of *tdc-gusA* fusions in *Nicotiana tabacum*, *Mol. Gen. Genet.*, 1994, **242**, 217-225.
32. MCKNIGHT, T.D., ROESSNER, C.A., DEVAGUPTA, R., SCOTT, A.I., NESSLER, C.L., Nucleotide sequence of cDNA encoding the vacuolar protein strictosidine synthase from *Catharanthus roseus, Nucleic Acids Res.*, 1990, **18**, 4939.
33. PASQUALI, G., GODDIJN, O.J.M., DE WAAL, A., VERPOORTE, R., SCHILPEROORT, R.A., HOGE, J.H.C., MEMELINK, J., Coordinated regulation of two indole alkaloid biosynthetic genes from *Catharanthus roseus* by auxin and elicitors, *Plant Mol. Biol.*, 1992, **18**, 1121-1131.
34. PASQUALI, G., ERVEN, A.S.W., OUWERKERK, P.B.F., MENKE, F.L.H., MEMELINK, J., The promoter of the strictosidine synthase gene from periwinkle confers elicitor-inducible expression in transgenic tobacco and binds nuclear factors GT1 and GBF, *Plant Mol. Biol.*, 1999, **39**, 1299-1310.
35. GODDIJN, O.J.M., DE KAM, R.J., ZANETTI, A., SCHILPEROORT, R.A., HOGE, J.H.C., Auxin rapidly down-regulates transcription of the tryptophan decarboxylase gene from *Catharanthus roseus*, *Plant Mol. Biol.*, 1992, **18**, 1113-1120.

36. GANTET, P., IMBAULT, N., THIERSAULT, M., DOIREAU, P., Inhibition of alkaloid accumulation by 2,4-D in *Catharanthus roseus* cell suspension is overcome by methyl jasmonate, *Acta Bot. Gallica*, 1997, **144**, 501-508.
37. ROEWER, I.A., CLOUTIER, N., NESSLER, C.L., DE LUCA, V., Transient induction of tryptophan decarboxylase (TDC) and strictosidine synthase (SS) genes in cell suspension cultures of *Catharanthus roseus*, *Plant Cell Rep.*, 1992, **11**, 86-89.
38. MENKE, F.L.H., PARCHMAN, S., MUELLER, M.J., KIJNE, J.W., MEMELINK, J., Involvement of the octadecanoid pathway and protein phosphorylation in fungal elicitor-induced expression of terpenoid indole alkaloid biosynthetic genes in *Catharanthus roseus*, *Plant Physiol.*, 1999, **119**, 1289-1296.
39. MENKE, F.L.H., Elicitor Signal Transduction in *Catharanthus roseus* Leading to Terpenoid Indole Alkaloid Biosynthetic Gene Expression, PhD Thesis, Leiden University, Leiden, 1999.
40. MÜLLER, M.J., Enzymes involved in jasmonic acid biosynthesis, *Physiol. Plant.*, 1997, **100**, 653-663.
41. GUNDLACH, H., MÜLLER, M.J., KUTCHAN, T.M., ZENK, M.H., Jasmonic acid is a signal transducer in elicitor-induced plant cell cultures, *Proc. Natl. Acad. Sci. USA*, 1992, **89**, 2389-2393.
42. AERTS, R.J., GISI, D., DE CAROLIS, E., DE LUCA, V., BAUMANN, T., Methyl jasmonate vapor increases the developmentally controlled synthesis of alkaloids in *Catharanthus* and *Cinchona* seedlings, *Plant J.*, 1994, **5**, 635-643.
43. GANTET, P., IMBAULT, N., THIERSAULT, M., DOIREAU, P., Necessity of a functional octadecanoid pathway for indole alkaloid synthesis by *Catharanthus roseus* cell suspensions cultured in an auxin-starved medium, *Plant Cell Physiol.*, 1998, **39**, 220-225.
44. VAN DER FITS, L., MEMELINK, J., ORCA3, a jasmonate-responsive transcriptional regulator of plant primary and secondary metabolism, *Science*, 2000, **289**, 295-297.
45. MENKE, F.L.H., CHAMPION, A., KIJNE, J.W., MEMELINK, J., A novel jasmonate- and elicitor-responsive element in the periwinkle secondary metabolite biosynthetic gene *Str* interacts with a jasmonate- and elicitor-inducible AP2-domain transcription factor, ORCA2, *EMBO J.*, 1999, **18**, 4455-4463.
46. VAN DER FITS, L., ZHANG, H., MENKE, F.L.H., DENEKA, M., MEMELINK, J., A *Catharanthus roseus* BPF-1 homologue interacts with an elicitor-responsive region of the secondary metabolite biosynthetic gene *Str* and is induced by elicitor via a jasmonate-independent signal transduction pathway, *Plant Mol. Biol.*, 2000, **44**, 675-685.
47. OUWERKERK, P.B.F., MEMELINK, J., A G-box element from the *Catharanthus roseus* strictosidine synthase (*Str*) gene promoter confers seed-specific expression in transgenic tobacco plants, *Mol. Gen. Genet.*, 1999, **261**, 635-643.
48. SIBÉRIL, Y., BENHAMRON, S., MEMELINK, J., GIGLIOLI-GUIVARC'H, N., THIERSAULT, M., BOISSON, B., DOIREAU, P., GANTET, P., *Catharanthus roseus* G-box binding factors 1 and 2 act as repressors of strictosidine synthase gene expression in cell cultures, *Plant Mol. Biol.*: in press, 2001.
49. DA COSTA E SILVA, O., KLEIN, L., SCHMELZER, E., TREZZINI, G.F., HAHLBROCK, K., BPF-1, a pathogen-induced DNA binding protein involved in the plant defense response, *Plant J.*, 1993, **4**, 125-135.

50. VAN DER FITS, L., MEMELINK, J., The jasmonate-inducible AP2/ERF-domain transcription factor ORCA3 activates gene expression via interaction with jasmonate-responsive promoter element, *Plant J.*, 2001, **25**, 43-53.
51. SHEPERD, N., Transposable elements and gene tagging, *in* Plant Molecular Biology: A Practical Approach (C.H. Shaw ed.), IRL Press, Oxford, 1988, pp. 187-220.
52. DORING, H.-P., Tagging genes with maize transposable elements, *Maydica*, 1989, **34**, 73-88.
53. SOMMER, H., BELTRON, J.-P., HUIJSER, R., PAPE, M., LONNING, W.-E., SAEDLER, H., SCHWARZ-SOMMER, Z., *Deficiens*, a homeotic gene involved in the control of flower morphogenesis in *Antirrhinum majus*: the protein shows homology to transcriptional factors, *EMBO J.*, 1990, **9**, 605-615.
54. CARPENTER, R., COHEN, E., Floral homeotic mutations produced by transposon-mutagenesis in *Anthirrhinum majus*, *Genes Dev.*, 1990, **4**, 1483-1493.
55. VOLLBRECHT, E., VEIT, B., SINHA, N., HAKE, S., The developmental gene *knotted-1* is a member of a maize homeobox gene family, *Nature*, 1991, **350**, 241-243.
56. GELVIN, S.B., *Agrobacterium* and plant genes involved in T-DNA transfer and integration, *Annu. Rev. Plant Physiol. Plant Mol. Biol.*, 2000, **51**, 223-256.
57. TEERI, T.H., HERRERA-ESTRELLA, L., DEPICKER, A., VAN MONTAGU, M., PALVA, E.T., Identification of plant promoters *in situ* by T-DNA mediated transcription fusions to the NPTII gene, *EMBO J.*, 1986, **5**, 1755-1760.
58. KAKIMOTO, T., CKI1, a histidine kinase homolog implicated in cytokinin signal transduction, *Science*, 1996, **274**, 982-985.
59. WEIGEL, D., AHN, J.H., BLAZQUEZ, M.A., BOREVITZ, J.O., CHRISTENSEN, S.K., FANKHAUSER, C., FERRANDIZ, C., KARDAILSKY, I., MALANCHARUVIL, E.J., NEFF, M.M., NGUYEN, J.T., SATO, S., WANG, Z.-Y., XIA, Y., DIXON, R.A., HARRISON, M.J., LAMB, C.J., YANOFSKY, M.F., CHORY, J., Activation tagging in *Arabidopsis*, *Plant Physiol.*, 2000, **122**, 1003-1013.
60. VAN DER FITS, L., DEAKIN, E.A., HOGE, J.H.C., MEMELINK, J., The ternary transformation system: Constitutive *virG* on a compatible plasmid dramatically increases *Agrobacterium*-mediated plant transformation, *Plant Mol. Biol.*, 2000, **43**, 495-502.
61. VAN DER FITS, L., HILLIOU, F., MEMELINK, J., T-DNA activation tagging as a tool to isolate regulators of a metabolic pathway in a genetically non-tractable plant species, *Transgenic Res.*: in press, 2001.
62. SASSE, F., BUCHHOLZ, M., BERLIN, J., Site of action of growth inhibitory tryptophan analogues in *Catharanthus roseus* cell suspension cultures, *Z. Naturforsch.*, 1983, **38C**, 910-915.
63. GODDIJN, O.J.M., VAN DER DUIJN SCHOUTEN, P.M., SCHILPEROORT, R.A., HOGE, J.H.C., A chimeric tryptophan decarboxylase gene as a novel selectable marker in plant cells, *Plant Mol. Biol.*, 1993, **22**, 907-912.
64. JOFUKU, K., DEN BOER, B.G.W., VAN MONTAGU, M., OKAMURO, J.K., Control of *Arabidopsis* flower and seed development by the homeotic gene *APETALA2*, *Plant Cell*, 1994, **6**, 1211-1225.
65. RIECHMANN, J.L., MEYEROWITZ, E.M., The AP2/EREBP family of plant transcription factors, *Biol. Chem.*, 1998, **379**, 633-646.

66. ALLEN, M.D., YAMASAKI, K., OHME-TAKAGI, M., TATENO, M., SUZUKI, M., A novel mode of DNA recognition by a β-sheet revealed by the solution structure of the GCC-box binding domain in complex with DNA, *EMBO J.*, 1998, **17**, 5484-5496.
67. SOLANO, R., STEPANOVA, A., CHAO, Q., ECKER, J.R., Nuclear events in ethylene signalling: A transcriptional cascade mediated by ETHYLENE-INSENSITIVE3 and ETHYLENE-RESPONSIVE-FACTOR1, *Genes Dev.*, 1998, **12**, 3703-3714.
68. ZHOU, J., TANG, X., MARTIN, G.B., The Pto kinase conferring resistance to tomato bacterial speck disease interacts with proteins that bind a *cis*-element of pathogenesis-related genes, *EMBO J.*, 1997, **16**, 3207-3218.
69. SHINOZAKI, K., YAMAGUCHI-SHINOZAKI, K., Molecular responses to dehydration and low temperature: differences and cross-talk between two stress signaling pathways, *Curr. Opin. Plant Biol.*, 2000, **3**, 217-223.
70: GODDIJN, O.J.M., PENNINGS, E.J.M., VAN DER HELM, P., SCHILPEROORT, R.A., VERPOORTE, R., HOGE, J.H.C., Overexpression of a tryptophan decarboxylase cDNA in *Catharanthus roseus* crown gall calluses results in increased tryptamine levels but not in increased terpenoid indole alkaloid production, *Transgenic Res.*, 1995, **4**, 315-323.
71. LLOYD, A.M., WALBOT, V., DAVIS, R.W., *Arabidopsis* and *Nicotiana* anthocyanin production activated by maize regulators *R* and *C1*, *Science*, 1992, **258**, 1773-1775.
72. MOONEY, M., DESNOS, T., HARRISON, K., JONES, J., CARPENTER, R., COEN, E., Altered regulation of tomato and tobacco pigmentation caused by the *delila* gene of *Antirrhinum*, *Plant J.*, 1995, **7**, 333-337.
73. BRADLEY, J.M., DAVIES, K.M., DEROLES, S.C., BLOOR, S.J., LEWIS, D.H., The maize *Lc* regulatory gene up-regulates the flavonoid biosynthetic pathway of *Petunia*, *Plant J.*, 1998, **13**, 381-392.
74. GROTEWOLD, E., CHAMBERLIN, M., SNOOK, M., SIAME, B., BUTLER, L., SWENSON, J., MADDOCK, S., ST.CLAIR, G., BOWEN, B., Engineering secondary metabolism in maize cells by ectopic expression of transcription factors, *Plant Cell*, 1998, **10**, 721-740.

Chapter Fourteen

MODELS FOR VACUOLAR SEQUESTRATION OF ANTHOCYANINS

Lukas A. Mueller[1] and Virginia Walbot[2]

[1]*Carnegie Institution*
260 Panama Street
Stanford, CA 94305 USA

[2]*Stanford University*
Dept of Biological Sciences
385, Serra Mall
Stanford, CA 94305 USA

INTRODUCTION

The flavonoids, including the colored anthocyanins, are a large family of compounds of diverse and complex structures,[1-4] that occur mainly in the vacuole, but are also found in other cellular structures, such as the cell wall[5] and the plastids.[6,7] Flavonoids and anthocyanins carry different modifications according to their site of storage; in environments such as the cell wall or in cuticular wax, they are often prenylated[1,2,8] or methylated,[5,9] whereas vacuolar anthocyanins carry modifications such as rich glycosylation and acylation.[8] The anthocyanin pathway is arguably the best understood pathway in plant secondary metabolism.[3,4,10] It is surprising, therefore, that so little is known about how anthocyanins are transported from their site of synthesis to their final subcellular storage locations. No single transport mechanism can be expected to account for all transport phenomena, as the structures and final destinations are diverse.

In this chapter, we focus on the transport of flavonoids and anthocyanins from the cytoplasm to the vacuole, which contains by far the largest storage pool of anthocyanins. Therefore, vacuolar sequestration affords an excellent model for the study of transport phenomena of these compounds in the cell. It is now generally accepted that anthocyanin biosynthesis occurs in the cytoplasm;[11,12] the biosynthetic enzymes are located within or associated with the endoplasmatic reticulum[3,13-15] The anthocyanins, therefore, have to cross the cytoplasm and have to be transported across the tonoplast membrane for sequestration. The questions we ask here are: (1) What is the nature of the pump in the tonoplast involved in anthocyanin transport? (2) Are there additional cytoplasmic factors required for import?

For the purpose of our analysis, we distinguish two phases in vacuolar import of a metabolite. In the first, anthocyanin is "prepared" for transport, such as through the attachment of special "tags." In the second, import *per se* occurs, presumably through a specific pump located in the tonoplast membrane. Additional factors may be required in the vacuole to stabilize or trap the pigment;[16,17] little is known about such factors and we do not discuss them further. The answers will involve a combination of chemistry, genetics, and molecular biology. As a basis for our discussion, we present three models, which we have found useful when thinking about vacuolar anthocyanin import. In the later sections, we describe in more detail some recent ideas on the function of cytoplasmic factors required for the sequestration of anthocyanins in model systems such as *Zea mays* (maize) and *Petunia hybrida* (petunia).

Pump Characteristics

In the process of vacuolar sequestration, anthocyanins must cross the tonoplast membrane. This step is assumed to be performed by specific pumps or channels in the tonoplast.[18] To date, the exact nature of the pump or channel is unknown, but two different energetic requirements have been proposed: (1) a ΔpH dependence and (2) an ATP-dependent pump, such as an ATP binding cassette (ABC) transporter.[19-21] In support for the former, Klein *et al.*[22] measured import of the flavonoid glucoside isovitexin and compared it to an herbicide glucoside (hydroxyprimisulfuron-glucoside) in isolated barley vacuoles. Transport of the flavonoid glucoside was ΔpH-dependent, whereas the herbicide glucoside was ATP-dependent. They concluded that an ABC transporter was required for the import of the herbicide glucoside but that a different type of transporter was required for the flavonoid glucoside.[18] In contrast, Hopp and Seitz[23] found that ATP did not stimulate import in an *in vitro* assay. In maize protoplasts, on the other hand, anthocyanin sequestration is vanadate-sensitive, pointing to an ABC transporter for import.[24] It is possible that individual plant species use different pumps for the sequestration of anthocyanin and other flavonoids. More efforts are needed to identify the anthocyanin pump through mutational analysis; detailed knowledge of the pump would permit the development of tests for some of the hypotheses put forward in this chapter.

Additional Cytoplasmic Factors

In addition to the pump, cytoplasmic factors may be required to assist import. We have summarized the present knowledge about vacuolar import by presenting three simple models (Fig. 14.1):

Model 1: Free Anthocyanin Transport

The simplest model assumes that no additional cytoplasmic or vacuolar factors are required for proper anthocyanin import. Import may happen spontaneously as a result of some intrinsic anthocyanin property, such as its affinity to a pump. Free anthocyanin transport has indeed been observed in experiments such as those performed by Klein *et al.*[22] These authors showed that unmodified flavonoid glucosides are readily imported into isolated barley vacuoles; import did not require the addition of soluble protein factors. These *in vitro* experiments are difficult to confirm *in vivo*, and should, therefore, be interpreted with caution. The model remains, however, a formal possibility for many plants.

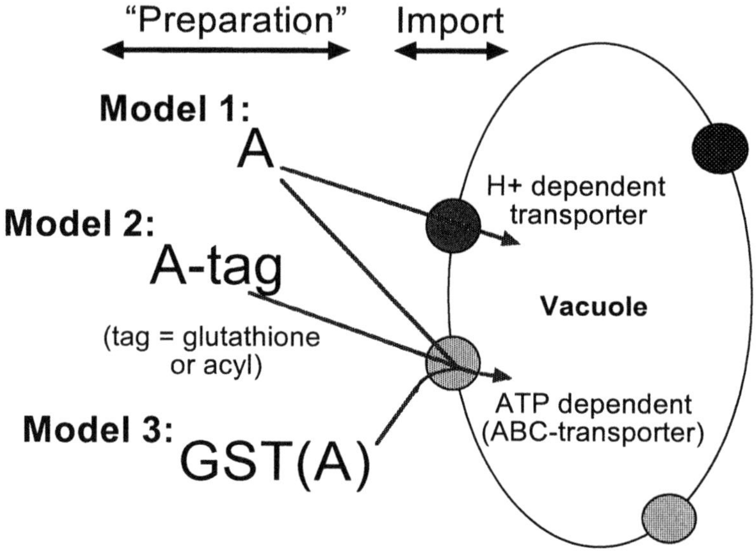

Figure 14.1: Models for anthocyanin sequestration into vacuoles. The first model postulates that unmodified flavonoid glucosides are readily imported into vacuoles.[22] No modifications are necessary, and no soluble protein factors are required. According to Model 2, anthocyanins and flavonoids require a special "tag", such as glutathionation, for vacuolar import.[24] Marrs *et al.* suggested that BZ2 enzyme catalyzes glutathionation of anthocyanin and that this conjugate is recognized by an ABC-transporter in the tonoplast membrane and transported into the vacuole.[26] Besides glutathionation, acylation of anthocyanins has been implicated as a prerequisite for vacuolar import.[23,28] Model 3 also builds on the observation that a GST is required, but maintains that the mechanism of the import is different. Instead of catalyzing a conjugate, the GST is proposed to be a binding protein. Binding could have multiple functional implications: It could prevent oxidation of the anthocyanin, or the GST could deliver the "cargo" directly to the pump, effectively acting as a carrier enzyme.[31]

Model 2: Conjugation Required for Transport

In genetically well characterized species such as maize and petunia, several lines of evidence point to the involvement of additional cytoplasmic factors. In

Model 2, anthocyanins and flavonoids require a special "tag", such as glutathionation or acylation, for vacuolar import.

Glutathionation. It has long been known that the *Bronze2* (*Bz2*) locus is required for vacuolar sequestration of maize anthocyanins.[25] Mutant *bz2* aleurone tissue is brown instead of the usual deep purple color, because anthocyanin accumulates in the cytoplasm where it polymerizes and oxidizes. Marrs *et al.*[24] suggested that BZ2 enzyme catalyzes glutathionation of anthocyanin and that this conjugate is recognized by an ABC-transporter[19] in the tonoplast membrane and transported into the vacuole.[26] Another example is the phytoalexin medicarpin, which is an *in vitro* substrate for glutathione *S*-transferase (GST)-mediated glutathionation, and is assumed to be imported by an ABC-transporter.[27]

Acylation. Besides glutathionation, acylation of anthocyanins has been implicated in vacuolar import.[23,28] Anthocyanins of many plants are acylated,[29] and acylation is a cytoplasmic modification.[30] Hopp *et al.*[23] prepared a mixture of radiolabeled anthocyanins and tested import with vacuoles isolated from a carrot cell line. De-acylated flavonoids were not efficiently taken up, whereas acylated forms were readily imported. The mixture of acylated, labeled compounds used in the carrot experiment, however, was isolated from the carrot cell culture, and only vacuoles prepared from the carrot line readily imported these anthocyanins. Vacuoles prepared from other plant species did not import the acylated anthocyanins.[23] The mechanism, thus, does not seem to be general, and may be limited to the species used in these experiments. A concern with all *in vitro* uptake experiments is that the *in vitro* assay may not reflect *in vivo* conditions. For example, the pH optimum of import into carrot vacuoles was in the range of pH 7.5-8.0, and little import was observed at the normal cytoplasmic pH of 7.0. Another concern is that at the small scale of these assays the chemical nature of the imported molecule(s) is difficult to verify, as label-exchange reactions can occur that result in misidentification of the sequestered compounds.

Model 3: Carrier Protein-Mediated Sequestration

This model also postulates that additional protein factors are required, in particular, glutathione *S*-transferase (GST), but proposes a different mechanism: Instead of catalyzing conjugate formation, the GST is proposed to be a binding protein. Binding could have multiple functional implications. It could prevent oxidation of the anthocyanin, or the GST could deliver the "cargo" directly to the pump, effectively acting as a carrier protein.[31] This model will be discussed in more detail in later sections.

Interpretation of Current Models

The three models we described above are difficult to reconcile. One possible resolution is that different plants utilize different vacuolar import mechanisms for anthocyanins and other flavonoids. Current data, however, are limited and may not adequately describe *in vivo* conditions. Clearly, more information is required to understand anthocyanin sequestration.

Both models 2 and 3 have been proposed for maize and petunia, species for which there is both genetic and biochemical data. For the remainder of the discussion, we will, therefore, focus on these two models. We also exclude acylation from the discussion as it has been implicated only in a few model systems, such as carrot cell lines.

In maize, *Bz2* encodes a protein required after formation of the mature cyanidin 3-glucoside. Similarly, *An9*, which encodes a GST enzyme, is required in petunia *hybrida*.[32] The isolation of these mutants suggests that efficient anthocyanin transport *in vivo* requires either a modification (Model 2) or a carrier protein (Model 3) in these monocot and dicot examples. *Bz2* was cloned by using transposon-tagging; somatic excision of *Mu1* from the *bz2-mu1* allele late in development restores anthocyanin pigmentation in aleurone tissues, demonstrating that the encoded function is required at the time of sequestration.[33] This is also true for the mutable *an9* allele.[32]

ROLE OF GSTS IN ANTHOCYANIN VACUOLAR SEQUESTRATION: DISTINGUISHING MODELS 2 & 3

To differentiate Models 2 and 3, detailed biochemical analyses are necessary.[24,31,32] Analyzing the BZ2 protein appeared to be the most promising route for understanding the *in vivo* requirements for sequestration, because there is a clear mutant phenotype.[24] Furthermore, the gene has been cloned and characterized.[33-37] However, when expressed in bacteria, BZ2 solubility and GST activity against the model substrate 1-chloro 2,4-dinitrobenzene (CDNB) are low.[24,31] More successful biochemical analysis were obtained with AN9. This *P. hybrida* enzyme is active towards model GST substrates, and it is much more soluble.[31,32] AN9 complements *bz2* tissue in particle bombardment assays.[32]

Glutathione S-Transferases (GST) in Herbicide Detoxification and Endogenous Metabolism

Based on sequence homology and the ability to form glutathione conjugates with model substrates, *Bz2* and *An9* are glutathione *S*-transferases (EC 2.5.1.18).[38] The well characterized GSTs[39-43] comprise a large family of ubiquitous, usually soluble enzymes found in both procaryotes and eukaryotes. GSTs from non-plant sources are classified as sigma, alpha, mu, pi, and theta types. GSTs not only form homo-dimers but also hetero-dimers within members of the same class; the monomer has a molecular mass of 20-30 kDa. Collectively, they constitute about 1% of the soluble protein in photosynthetic plant cells.[44] Until recently, all plant GSTs were considered to be members of the heterogeneous class theta GSTs. Plant GSTs were traditionally further subdivided into types I – IV. A more recent nomenclature recognized that there are, in fact, two GST types specific to photosynthetic organisms and that the theta designation obscured the deep phylogenetic integrity of the plant-specific GSTs.[45] Theta type I enzymes are now classified as Phi (F) enzymes, a plant-specific group present in photosynthetic microbes and green plants. The former theta type II are now Zeta (Z), a class also present in mammals. Type III are now Tau (U), a second plant-specific group, and theta type IV are members of the now narrower Theta (T). This nomenclature also proposes a new systematic convention for naming individual loci. For example, *Bz2* is *ZmGSTU4* (*Zea mays* GST Tau 4, for the fourth Tau locus in this species) according to the new rules.[45]

GSTs are required for detoxification of diverse exogenous substrates including many herbicides.[46,47] Because of their economic importance, herbicides have been the major focus of substrate analysis with plant GSTs. These well-studied detoxification reactions ultimately lead to vacuolar sequestration of substrates conjugated to GSH. The detoxification of many xenobiotic organic compounds involves three distinct stages.[46,48] In phase 1, cytochrome p450 mono-oxygenases hydroxylate toxic compounds to produce reactive sites and render the compounds soluble in the cytoplasm. In phase 2, transferases add sugar residues or amino acids to these newly formed reactive sites; in many cases, a GST covalently attaches a tripeptide glutathione (γ-glutamic acid-cysteine-glycine) residue forming a C-S bond. The resulting glycosylated and glutathionated conjugates are substrates for phase 3, in which they are transferred into the vacuole *via* a Multi-drug Resistance-associated Protein (MRPs).[19,49] The MRPs are part of the ABC superfamily of transporters.[20]

Parallels between Xenobiotics Detoxification and Sequestration of Natural Compounds

Like herbicides and other xenobiotics, many toxic natural products are stored in the vacuole. Biosynthesis often involves both cytochrome p450s and glycosyl transferases, suggesting metabolism parallel to xenobiotic detoxification. This parallel has been strengthened by the demonstration that a GST enzyme is required after glycosylation to facilitate anthocyanin sequestration into maize and petunia vacuoles.[24]

It is tempting to draw parallels to the detoxification of xenobiotics and conclude that GSTs also glutathionate their natural substrates as a pre-requisite for vacuolar sequestration. Glutathionated anthocyanin, however, has never been identified in plant extracts.[1,2] The vacuolar forms of anthocyanin in both maize and petunia lack sulfur, indicating either complete removal of glutathione or that conjugates do not form. Sulfated anthocyanins have been identified in only a few plant species, and these are synthesized by specific sulfotransferases that are unrelated to GSTs.[50] The only biochemical data indicating that a glutathionated intermediate might form came from co-localization of anthocyanin and ^{35}S from glutathione on a two-dimensional thin layer chromatographic analysis; this result was achieved in plant protoplasts induced to synthesize a high level of anthocyanin.[24] The chemical nature of the putative conjugate was not established, and it remained a possibility that a glutathionated compound other than anthocyanin co-migrated with the pigment in the analysis. Neither cyanidin nor cyanidin 3-glucoside contain any obvious electrophilic centers that would be the predicted site of glutathionation. It is a characteristic of xenobiotic substrates that they form spontaneous GS-conjugates.[48] No spontaneous anthocyanin-glutathione conjugates are observed.

Biochemical Analysis of AN9

Recombinant AN9 was unable to catalyze conjugates between glutathione and anthocyanins or flavonoids, although it readily conjugated glutathione to the model substrate CDNB.[31] Hence, a GST involved in anthocyanin sequestration *in vivo* has GST activity *in vitro*, but does not glutathionate anthocyanins.

AN9 does bind flavonoids, as shown by an inhibition assay. Most flavonoids inhibit its GST activity towards the substrate CDNB with C_{50}–values (inhibitor concentrations yielding half-maximal activities) on the order of 1 – 100 μM.[31] This indicates that many of the flavonoids have a higher affinity for AN9 than the model substrate CDNB. Furthermore, equilibrium dialysis has been used to measure the affinity of flavonoid substrates to AN9.[31] Our current model for *An9* and *Bz2* action, therefore, is that they act as binding proteins. Binding could serve several purposes:

protection of the bound anthocyanin from oxidation; protection of cellular constituents from flavonoids; and delivery of the flavonoid to a vacuolar pump. These hypothesis are consistent with Model 3.

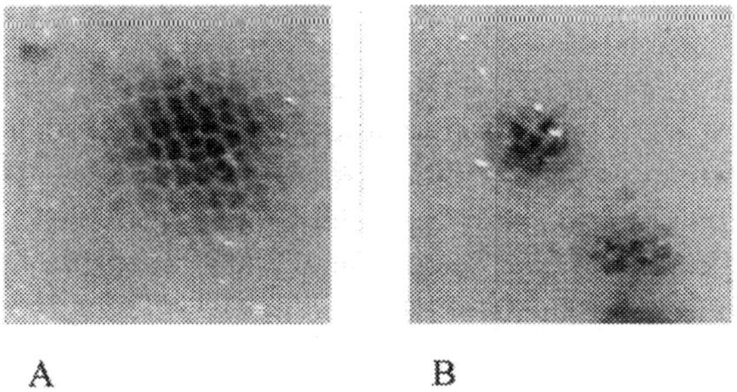

Figure 14.2: Model 2 and 3 can be distinguished by eliminating the transferase function of the GST, while retaining the binding function, and testing for *in vivo* complementation. Preliminary results from complementation experiments in *bz2* tissue by wildtype *Bz2* (A) and mutant *Bz2S9A* (B) are shown. Plasmids were bombarded into *bz2* aleurone tissue by using a BioRad (Hercules, CA) Biolistics device. The mutant, *Bz2S9A,* has the catalytically important serine 9 replaced by alanine, resulting in strongly reduced transferase activity.[52] This mutant plasmid can still complement *bz2* in this *in vivo* assay for anthocyanin sequestration. The implication is that the transferase function of the enzyme is not important for anthocyanin sequestration, which is consistent with biochemical data.[31] We conclude that no anthocyanin-glutathione conjugate is implicated in vacuolar import.

Is Transferase Function Necessary for Pigment Sequestration?

To further test Model 3, we designed an experiment to separate transferase and sequestration functions. We asked if mutant enzymes that have the ability to bind ligands but cannot catalyze the glutathione transferase reaction could still complement *bz2* aleurone tissue in a particle bombardment assay.[32] If Model 2 is correct, mutated enzyme should not complement, but if Model 3 is correct, complementation should still occur.

Several amino acid residues are crucial for transferase activity in animal GSTs. In class mu GSTs, a tyrosine residue near the amino terminus is critical,[51] because it activates glutathione to the thiolate form required for a nucleophilic attack on the substrate. If this tyrosine is mutated to phenylalanine, the transferase activity is reduced to less then 1% of wild-type activity.[51] In all plant GSTs characterized to date, there is a serine residue in the position occupied by a tyrosine in most animal GSTs. The serine hydoxyl plays an analogous role in activation of the SH of glutathione to a thiolate ion. In a theta type enzyme from the Australian sheep blowfly (*Lucilia cuprina*), the serine residue was mutated to an alanine. This change reduced conjugating activity by a factor of over 280-fold.[52]

Site-directed mutagenesis of *Bz2* was conducted to convert serine-9 to alanine; the mutation was verified by DNA sequencing. The mutant gene was then tested for complementation of vacuolar sequestration in *bz2* aleurone tissue by assaying for dark purple cells after particle bombardment. If complementation occurs after expression of the mutant enzyme, we can conclude that the transferase reaction can be eliminated or greatly reduced without impairing vacuolar sequestration. Plasmid p35S::Bz2i contains a fully functional *Bz2* gene, including the single intron, with the 35S viral promoter. In p35S::Bz2iS9A, serine 9 was replaced by alanine. Bombardment with the wild type vector resulted in deep purple spots several cells wide (Fig.14.2A). The spot size reflects diffusion into neighboring cells, presumably movement of the BZ2 protein.

Transformation with the mutant construct p35S::Bz2iS9A also resulted in the formation of purple cells (Fig. 14.2B). This definitive result indicates that the transferase function is unlikely to be important in the process of anthocyanin sequestration into the vacuole. This experimental evidence is consistent with Model 3 and with the biochemical analysis, where we concluded that AN9 and BZ2 act as carrier proteins.

In future experiments, we will introduce similar mutations in AN9. We expect that the mutant enzyme will be unable to conjugate glutathione to CDNB but will retain *in vivo* complementation activity.

CONCLUSIONS

We have presented three models for anthocyanin sequestration into vacuoles. Model 1 postulates the import of unmodified anthocyanins into vacuoles. Model 2 states that anthocyanins require modifications, such as glutathione tags, prior to import. Model 3 is based on additional cytoplasmic protein factors, which act as carrier enzymes. Other models are formally possible, but we have found little evidence for other possibilities in the literature. We did not consider evidence, however, on anthocyanin transport to subcellular locations other than the vacuole.

For vacuolar sequestration in maize and petunia, ample biochemical and genetic data indicate that free anthocyanins are not substrates for the tonoplast pump. Additional soluble, cytoplasmic proteins, such as *Bz2* and *An9*, are required. At least in these two species direct import as presented in Model 1 is considered unlikely. Model 2 has the shortcoming that the tags suggested are either not a general feature of anthocyanins in maize, or are not detectable. In view of the large amount of biochemical data supporting Model 3, we conclude that it best accounts for all available observations in maize and petunia.

What is the purpose of the substrate binding by AN9 and BZ2? It could prevent oxidation and crosslinking of the ligand and/or be directly involved in transporting the anthocyanins to a tonoplast pump. No data, however, are presently available to distinguish the two modes of action.

Transposon-inactivated and other mutant alleles produce a dramatic effect: when these particular GSTs are absent, the pigment does not reach its final destination, the vacuole, and forms a brownish color in the cytoplasm. Even single cell somatic revertants resulting from transposon excision are bright purple, indicating that BZ2 and AN9 can restore vacuolar sequestration late in tissue development. Similarly, particle bombardment into mature, brown *bz2* aleurone tissue results in purple cells; this suggests that some free anthocyanin remains in the cytoplasm and can be sequestered.

Several GSTs have been described that bind low molecular weight metabolites. GSTs were first described as "ligandins", which were binding proteins for large anionic metabolites in the liver.[53-55] Of special interest to phytochemists are plant GSTs that bind compounds such as auxin[56-58] and cytokinins.[59] Although the function of the binding in these cases remains unknown, it seems likely that these GSTs also participate in the intracellular trafficking of metabolites, possibly contributing to the regulation of active and inactive pools of plant hormones in the cell. GSTs may, therefore, play a part in growth and development of plants – an interesting field for future investigation.

What's Next?

The identification and cloning of the pump involved in anthocyanin transport will be an important step in understanding the process of vacuolar sequestration. We anticipate that elucidation of the pump will result from a combination of biochemical and genetic analysis. It is a possibility that different plant species use different types of pumps; that not all species use the same type of enzyme for a given function is evidenced by the GST carrier proteins: *Bz2* is a Tau-type GST and *An9* is a Phi-type enzyme.

Model 3 will need confirmation and refinement. We would like to know the precise function of the GST:anthocyanin binding. The carrier protein hypothesis can be tested by analyzing GST-pump interaction. X-ray crystallography of GST with bound anthocyanin can resolve the binding sites and reveal how the GST prevents substrate oxidation.

Alternative models should be explored. Although it appears unlikely that acylation plays a role in anthocyanin import in maize, it may facilitate import in other plants, such as carrot, where it was first implicated.[23]

We speculate that many other plant GSTs are involved in the subcellular targeting of metabolites. Sequestration of toxic metabolites is of great importance in plants that as a group synthesize many cellular poisons such as colchicine or taxol. Model 3 may, therefore, be fundamental in understanding GST function in general.

ACKNOWLEDGMENTS

Our research is supported by grants from the National Science Foundation. L.A.M. was partially supported by the Swiss Science Foundation. We thank our colleagues Dean Goodman, Rebecca Silady and Savita Shah for participation in this project and for discussions of the models for anthocyanin sequestration.
This is CIW Publication Number 1490.

REFERENCES

1. HARBORNE, J.B., WILLIAMS, C.A., Anthocyanins and other flavonoids, *Natural Products Rep.*, 1995, **12**, 639-657.
2. HARBORNE, J.B., WILLIAMS, C.A., Anthocyanins and other flavonoids, *Natural Products Rep.*, 1998, **15**, 631-652.
3. HOLTON, T.A., CORNISH, E.C., Genetics and biochemistry of anthocyanin biosynthesis, *Plant Cell*, 1995, **7**, 1071-1083.
4. KOES, R.E., QUATTROCCHIO, F., MOL, J.N.M., The flavonoid biosynthetic pathway in plants: Function and evolution, *BioEssays*, 1994, **16**, 123-132.
5. MARKHAM, K., RYAN, K., GOULD, K., RICKARDS, G., Cell wall sited flavonoids in *Lisianthus* flower petals, *Phytochemistry*, 2000, **54**, 681-687.
6. SAUNDERS, J., MCCLURE, J.W., The localization of flavonoids and phenolic enzymes in barley plastids, *Am. J. Bot.*, 1972, **59**, 673.
7. SAUNDERS, J.A., MCCLURE, J.W., WALLACE, J.W., The subcellular localization of enzymes in the flavonoid biosynthetic pathway, *Am. J. Bot.*, 1973, **60**, 29.
8. HARBORNE, J.B., WILLIAMS, C.A., GREENHAM, J., EAGLES, J., Variations in the lipophilic and vacuolar flavonoids of the genus *Vellozia*, *Phytochemistry*, 1994, **35**, 1475-1480.
9. JAY, M., DE LUCA, V., IBRAHIM, R.K., Purification, properties and kinetic mechansim of flavonol 8-*O*-methyltransferase from *Lotus corniculatus* L, *Eur. J. Biochem.*, 1985, **153**, 321-325.
10. DOONER, H.K., ROBBINS, T.P., JORGENSEN, R.A., Genetic and developmental control of anthocyanin biosynthesis, *Annu. Rev. Gen.*, 1991, **25**, 173-199.
11. HELLER, W., FORKMANN, G., *Biosynthesis*, *in The Flavonoids*, J.B. Harborne, Editor, Chapman & Hall: London, 1988, pp. 399-425.

12. HRAZDINA, G., ALSCHER-HERMAN, R., KISH, V.M., Subcellular localization of flavonoid synthesizing enzymes in *Pisum, Phaseolus, Brassica* and *Spinacia* cultivars, *Phytochemistry*, 1980, **19**, 1355-1359.
13. FRITSCH, J.H., GRISEBACH, H., Biosynthesis of cyanidin in cell cultures of *Haplopappus gracilis, Phytochemistry*, 1975, **14,** 2437-2442.
14. HRAZDINA, G., WAGNER, G.J., Metabolic pathways as enzyme complexes: Evidence for the synthesis of phenylpropanoids and flavonoids on membrane associated enzyme complexes, *Arch. Biochem. Biophys.*, 1984, **237**, 88-100.
15. BURBULIS, I.E., WINKEL-SHIRLEY, B., Interactions among enzymes of the *Arabidopsis* flavonoid biosynthetic pathway, *Proc. Natl. Acad. Sci. USA*, 1999, **96**, 12929-12934.
16. NOZUE, M., KUBO, H., NISHIMURA, M., YASUDA, H., Detection and characterization of vacuolar protein (VP24) in anthocyanin-producing cells of sweet potato in suspension culture, *Plant Cell Phys.*, 1995, **36**, 883-889.
17. NOZUE, M., YAMADA, K., NAKAMURA, T., KUBO, H., KONDO, M., NISHIMURA, M., Expression of a vacuolar protein (VP24) in anthocyanin-producing cells of sweet potato in suspension culture, *Plant Physiol.*, 1997, **115**, 1065-1072.
18. TANNER, W., CASPARI, T., Membrane transport carriers, *Annu. Rev. Plant Physiol. Plant Mol. Biol.*, 1996, **47**, 595-626.
19. MARTINOIA, E., GRILL, E., TOMMASINI, R., KREUZ, K., AMRHEIN, N., ATP-dependent glutathione *S*-conjugate 'export' pump in the vacuolar membrane of plants, *Nature*, 1993, **364**, 247-249.
20. REA, P.A., LI, Z.-S., LU, Y.-P., DROZDOWICZ, Y.M., MARTINOIA, E., From vacuolar GS-X pumps to multispecific ABC transporters, *Annu. Rev. Plant Physiol.*, 1998, **49**, 727-760.
21. LU, Y.-P., LI, Z.-S., DORZDOWICZ, Y.M., HÖRTENSTEINER, S., MARTINOIA, E., REA, P.A., AtMRP2, an *Arabidopsis* ATP binding cassette transporter able to transport glutathione *S*-conjugates and chlorophyll catabolites: Functional comparisons with AtMRP1, *Plant Cell*, 1998, **10**, 267-282.
22. KLEIN, M., WEISSENBÖCK, G., DUFAUD, A., GAILLARD, C., KREUZ, K., MARTINOIA, E., Different energization mechanisms drive the vacuolar uptake of a flavonoid glucoside and a herbicide glucoside, *J. Biol. Chem.*, 1996, **271,** 29666-29671.
23. HOPP, W., SEITZ, H.U., The uptake of acylated anthocyanin into isolated vacuoles from a cell suspension culture of *Daucus carota, Planta*, 1987, **170**, 74-85.
24. MARRS, K.A., ALFENITO, M.R., LLOYD, A.M., WALBOT, V., A glutathione *S*-transferase involved in vacuolar transfer encoded by the maize gene *Bronze-2, Nature*, 1995, **375**, 397-400.
25. NUFFER, M.G., Redesignation of pa, *Genet. Maize News Lett.*, 1953, **27**, 68.
26. MARRS, K.A., The functions and regulation of glutathione *S*-transferases in plants, *Annu. Rev. Plant Mol. Biol.*, 1996, **47**, 127-158.
27. LI, Z.-S., ALFENITO, M., REA, P.A., WALBOT, V., DIXON, R.A., Vacuolar uptake of the phytoalexin medicarpin by the glutathione conjugate pump, *Phytochemistry*, 1997, **45**, 689-693.

28. MATERN, U., REICHENBACH, C., HELLER, W., Efficient uptake of flavonoids into parsley (*Petroselinum hortense*) vacuoles requires acylated glycosides, *Planta*, 1986, **167**, 183-189.
29. HARBORNE, J.B., SELF, R., Malonated cyanidin 3-glucosides in *Zea mays* and other grasses, *Phytochemistry*, 1986, **26**, 2417-2418.
30. FUJIWARA, H., TANAKA, Y., YONEKURASAKAKIBARA, K., FUKUCHIMIZUTANI, M., NAKAO, M., FUKUI, Y., YAMAGUCHI, M., ASHIKARI, T., KUSUMI, T., cDNA cloning, gene expression and subcellular localization of anthocyanin 5-aromatic acyltransferase from *Gentiana triflora, Plant J.*, 1998, **16**, 421-431.
31. MUELLER, L.A., SILADY, R.A., GOODMAN, C.D., WALBOT, V., *An9*, a glutathione *S*-transferase from petunia required for vacuolar localization of anthocyanins, is a flavonoid binding protein, *Plant Physiol.*, 2000, **123**, 1561-1570.
32. ALFENITO, M., SOUER, E., GOODMAN, C.D., BUELL, R., MOL, J., KOES, R., WALBOT, V., Functional complementation of anthocyanin sequestration in the vacuole by widely divergent glutathione *S*-transferases, *Plant Cell*, 1998, **10**, 1135-1149.
33. MCLAUGHLIN, M., WALBOT, V., Cloning of a mutable *bz2* allele of maize by transposon tagging and differential hybridization, *Genetics*, 1987, **117**, 771-776.
34. BODEAU, J.P., WALBOT, V., Structure and regulation of the maize *Bronze2* promoter, *Plant Mol. Biol.*, 1996, **32**, 599-609.
35. NASH, J., LUEHRSEN, K.R., WALBOT, V., *Bronze-2* gene of maize: Reconstruction of a wild-type allele and analysis of transcription and splicing, *Plant Cell*, 1990, **2**, 1039-1049.
36. NASH, J., WALBOT, V., *Bronze-2* gene expression and intron splicing patterns in cells and tissues of *Zea mays* L, *Plant Physiol.*, 1992, **100**, 464-471.
37. WALBOT, V., BENITO, M.-I., BODEAU, J., NASH, J., Abscisic acid induces pink pigmentation in maize aleurone tissue in the absence of *Bronze-2, Maydica*, 1994, **39**, 19-28.
38. MANNERVIK, B., DANIELSON, U.H., Glutathione Transferases - Structure and catalytic activity, *CRC Crit. Rev. Biochem.*, 1988, **23**, 283-337.
39. NEUEFEIND, T., HUBER, R., DASENBROCK, H., PRADE, L., BIESELER, B., Crystal-structure of herbicide-detoxifying maize glutathione *S*-transferase-I in complex with lactoylglutathione: evidence for an induced-fit mechanism, *J. Mol. Biol.*, 1997, **274**, 446-453.
40. NISHIHIRA, J., ISHIBASHI, T., SAKAI, M., TSUDA, S., HIKICHI, K., Identification of the hydrophobic ligand-binding region in recombinant glutathione *S*-transferase P and its binding effect on the conformational state of the enzyme, *Arch. Biochem. Biophys.*, 1993, **302**, 128-133.
41. REINEMER, P., PRADE, L., HOF, P., NEUEFEIND, T., HUBER, R., ZETTL, R., SCHELL, J., KOELLN, I., BARTUNIK, H.D., BIESELER, B., Three-dimensional structure of glutathione *S*-transferase from *Arabidopsis thaliana* at 2.2A resolution: Structural characterization of herbicide-conjugating plant glutathione *S*-transferases and a novel active site architecture, *J. Mol. Biol.*, 1996, **255**, 289-309.

42. VANDER JAGT, D.L., HUNSAKER, L.A., GARCIA, K.B., ROYER, R.E., Isolation and characterization of the multiple glutathione *S*-transferases from human liver, *J. Biol. Chem.*, 1985, **260**, 11603-11610.
43. ZHOU, J., GOLDSBROUGH, P.B., An *Arabidopsis* gene with homology to glutathione *S*-transferases is regulated by ethylene, *Plant Mol. Biol.*, 1993, **22**, 517-523.
44. HAYES, J.D., PULFORD, D.J., The glutathione *S*-transferase supergene family: Regulation of GST* and the contribution of the isoenzymes to cancer chemoprotection and drug resistance, *Crit. Rev. Biochem. Mol. Biol.*, 1995, **30**, 445-600.
45. EDWARDS, R., DIXON, D.P., WALBOT, V., Plant glutathione *S*-transferases: enzymes with multiple functions in sickness and health, *Trends Plant Sci.*, 2000, **5**, 193-198.
46. SANDERMANN, H., Plant metabolism of xenobiotics, *Trends Biol. Sci.*, 1992, **17**, 82-84.
47. COLEMAN, J.O.D., RANDALL, R., BLAKE-KALFF, M.M.A., Detoxification of xenobiotics in plant cells by glutathione conjugation and vacuolar compartmentalization: a fluorescent assay using monochlorobimane., *Plant Cell Environ.*, 1997, **20**, 449-460.
48. COLEMAN, J.O.D., Detoxification of xenobiotics by plants: chemical modification and vacuolar compartmentation, *Trends Plant Sci.*, 1997, **2**, 144-151.
49. ISHIKAWA, T., The ATP-dependent glutathione *S*-conjugate export pump, *Trends Biol. Sci.*, 1992, **17**, 463-468.
50. VARIN, L., MARSOLAIS, F., RICHARD, M., ROULEAU, M., Biochemistry and molecular biology of plant sulfotransferases, *FASEB J.*, 1997, **11**, 517-525.
51. LIU, S., PINGHUI, Z., JI, X., JOHNSON, W.W., GILLILAND, G.L., ARMSTRONG, R.N., Contribution of Tyrosine 6 to the catalytic mechanism of isoenzyme 3-3 of glutathione *S*-transferase, *J. Biol. Chem.*, 1992, **267**, 4296-4299.
52. BOARD, P.G., COGGAN, M., WILCE, M.C.J., PARKER, M.W., Evidence for an essential serine residue in the active site of the theta class glutathione transferases, *Biochem. J.*, 1995, **311**, 247-250.
53. LITWACK, G., KETTERER, B., ARIAS, I.M., Ligandin: A hepatic protein which binds steroids, bilirubin, carcinogens and a number of exogenous organic anions, *Nature*, 1971, **234**, 466-467.
54. BHARGAVA, M.M., LISTOWSKY, I., ARIAS, I.M., Ligandin, *J. Biol. Chem.*, 1977, **253**, 4112-4115.
55. HABIG, W.H., PABST, M.J., FLEISCHNER, G., GATMAITAN, Z., ARIAS, I.M., JAKOBY, W.B., The identity of glutathione *S*-transferase B with ligandin, a major binding protein of liver, *Proc. Natl. Acad. Sci. USA*, 1974, **71**, 3879-3882.
56. BILANG, J., MACDONALD, H., KING, P.J., STURM, A., A soluble auxin-binding protein from *Hyoscyamus muticus* is a glutathione *S*-transferase, *Plant Physiol.*, 1993, **102**, 29-34.
57. BILANG, J., STURM, A., Cloning and characterization of a glutathione *S*-transferase that can be photolabeled with 5-azido-indole-3-acetic acid, *Plant Physiol.*, 1995, **109**, 253-260.

58. ZETTL, R., SCHELL, J., PALME, K., Photoaffinity labeling of *Arabidopsis thaliana* plasma membrane vesicles by 5-azido-[7-3H]indole-3-acetic acid: Identification of a glutathione *S*-transferase, *Proc. Natl. Acad. Sci. USA*, 1994, **91**, 689-693.
59. GONNEAU, J., MORNET, R., LALOUE, M., A *Nicotiana plumbaginifolia* protein labeled with an azido cytokinin agonist is a glutathione *S*-transferase, *Physiol. Plant.*, 1998, **103**, 114-124.

INDEX

www.ingramcontent.com/pod-product-compliance
Ingram Content Group UK Ltd.
Pitfield, Milton Keynes, MK11 3LW, UK
UKHW020238290926
15470UKWH00002B/601

9 780080 439778